The Molecular Aspect of Natural Secondary Metabolite Products in Health and Disease

The Molecular Aspect of Natural Secondary Metabolite Products in Health and Disease

Special Issue Editor

Toshio Morikawa

MDPI • Basel • Beijing • Wuhan • Barcelona • Belgrade

MDPI

Special Issue Editor
Toshio Morikawa
Kindai University
Japan

Editorial Office
MDPI
St. Alban-Anlage 66
Basel, Switzerland

This is a reprint of articles from the Special Issue published online in the open access journal *International Journal of Molecular Sciences* (ISSN 1422-0067) from 2017 to 2018 (available at: http://www.mdpi.com/journal/ijms/special_issues/secondary_metabolite_disease)

For citation purposes, cite each article independently as indicated on the article page online and as indicated below:

LastName, A.A.; LastName, B.B.; LastName, C.C. Article Title. *Journal Name* **Year**, *Article Number*, Page Range.

ISBN 978-3-03897-176-4 (Pbk)
ISBN 978-3-03897-177-1 (PDF)

Contents

About the Special Issue Editor

Toshio Morikawa is Professor at the Pharmaceutical Research and Technology Institute, Kindai University, Japan. He was born in Kyoto Prefecture, Japan in 1972 and received his Ph.D. under the supervision of Professor Masayuki Yoshikawa at Kyoto Pharmaceutical University in 2002. In 2001, he started his academic career at Kyoto Pharmaceutical University as an Assistant Professor. He became a Lecturer in 2005, an Associate Professor in 2010, and a Professor in 2015 at the Pharmaceutical Research and Technology Institute, Kindai University. He received The JSP Award for Young Scientists in 2005 and The JSP Award for Scientific Contributions in 2018. His current research program focuses on the search for bioactive constituents from natural resources and the development of new functional foods for the prevention and improvement of lifestyle diseases. He has published over 200 papers in peer reviewed journals and is currently serving on the editorial board of the Journal of Natural Medicines, Traditional and Kampo Medicine, and the Japanese Journal of Food Chemistry and Safety. He has also served twice as a Guest Editor for *IJMS*.

Preface to "The Molecular Aspect of Natural Secondary Metabolite Products in Health and Disease"

Natural secondary metabolite products that are isolated from plants, animals, microorganisms, etc., are classified as polyketides, isoprenoids, steroids, aromatics, alkaloids, etc. Their chemical diversity and variety of biological activities have attracted the attention of chemists, biochemists, biologists, etc. This Special Issue on "The Molecular Aspect of Natural Secondary Metabolite Products in Health and Disease" is intended to present biologically active natural products as candidates and/or leads for pharmaceuticals, dietary supplements, functional foods, cosmetics, food additives, etc. The research fields covered in this Special Issue of IJMS include the chemistry of natural products, phytochemistry, pharmacognosy, food chemistry, bioorganic synthetic chemistry, molecular pharmacology, molecular nutritional sciences, and other research fields related to bioactive natural secondary metabolite products

Toshio Morikawa
Special Issue Editor

International Journal of
Molecular Sciences

MDPI

Article

Ergostane-Type Sterols from King Trumpet Mushroom (*Pleurotus eryngii*) and Their Inhibitory Effects on Aromatase

Takashi Kikuchi [1], Naoki Motoyashiki [1], Takeshi Yamada [1], Kanae Shibatani [2], Kiyofumi Ninomiya [2], Toshio Morikawa [2,*] and Reiko Tanaka [1,*]

[1] Faculty of Pharmaceutical Sciences, Osaka University of Pharmaceutical Sciences, 4-20-1 Nasahara, Takatsuki, Osaka 569-1094, Japan; t.kikuchi@gly.oups.ac.jp (T.K.); mottchan050321@gmail.com (N.M.); yamada@gly.oups.ac.jp (T.Y.)

[2] Pharmaceutical Research and Technology Institute, Kindai University, 3-4-1 Kowakae, Higashi-osaka, Osaka 577-8502, Japan; 1633420001c@kindai.ac.jp (K.S.); ninomiya@phar.kindai.ac.jp (K.N.)

* Correspondence: morikawa@kindai.ac.jp (T.M.); tanakar@gly.oups.ac.jp (R.T.); Tel.: +81-6-4307-4306 (T.M.); +81-72-690-1084 (R.T.); Fax: +81-6-6729-3577 (T.M.); +81-72-690-1084 (R.T.)

Received: 18 October 2017; Accepted: 16 November 2017; Published: 21 November 2017

Abstract: Two new ergostane-type sterols; (22*E*)-5α,6α-epoxyergosta-8,14,22-triene-3β,7β-diol (**1**) and 5α,6α-epoxyergost-8(14)-ene-3β,7α-diol (**2**) were isolated from the fruiting bodies of king trumpet mushroom (*Pleurotus eryngii*), along with eight known compounds (**3–10**). All isolated compounds were evaluated for their inhibitory effects on aromatase. Among them, **4** and **6** exhibited comparable aromatase inhibitory activities to aminoglutethimide.

Keywords: *Pleurotus eryngii*; sterol; ergostane; aromatase inhibitor

1. Introduction

Estrogen is responsible for breast cancer growth. The target genes of an estrogen receptor are in control of cancer cell development in estrogen-dependent breast tumors. Binding of estrogen receptor to estrogen triggers transcription of its target genes [1]. Aromatase is the rate-limiting enzyme in estrogen biosynthesis [1]. This enzyme converts androgens (testosterone and androtestosterone) into estrogens (estradiol and estrone, respectively) [2]. Aromatase inhibitors (AIs) are adjuvant in hormone treatments commonly prescribed for breast cancers that are hormone receptor-positive in the early stage [3]. However, the currently used AIs have several side effects of menopausal symptoms such as hot flashes, vaginal dryness, sexual dysfunction, musculoskeletal symptoms, osteoporosis, bone fracture, fatigue, mood disturbance, nausea, and vomiting [3]. Therefore, natural compounds obtained from safe food resources might be useful in the search for promoter-specific AIs with few side effects [4].

Pleurotus eryngii (Japanese name: eringi, English name: oyster mushroom or king trumpet) is an edible mushroom. *P. eryngii* is native to North Africa, Asia, and Europe [5], and also grown commercially in Japan, China, and the US [6]. Previous studies demonstrated the inhibitory effects on human neutrophil elastase (HNE) [7], antioxidant and antimutagenic activities [8], and inhibitory effects on allergic mediators [9] of *P. eryngii* extracts. *P. eryngii* contains amino acids, vitamins, and dietary fiber [10]. It also includes polysaccharides [11,12], pleurone [7], ergostane-type sterols [13], and eryngiolide A [14]. These chemical constituents exhibit biological activities such as the antioxidant [11] and antitumor activities [12] of a polysaccharide, HNE-inhibitory effects of pleurone [7], and cytotoxicity against human cancer cell lines of eryngiolide A [14]. We recently reported eringiacetal A, which is an ergostane-type sterol with a cage-shaped structure [15], and a 9,11-*seco*-ergostane and five ergostane-type sterols [16] from the fruiting bodies of *P. eryngii*. In a continuing study, we isolated 10 ergostane-type sterols, and elucidated the structures of two new compounds; (22*E*)-5α,6α-epoxyergosta-8,14,22-triene-3β,7β-diol (**1**), and 5α,6α-epoxyergost-8(14)-ene-3β,7α-diol (**2**). In addition, the isolated constituents were evaluated for inhibitory activities on aromatase.

2. Results

2.1. Isolation and Structure Elucidation

(22*E*)-Ergosta-7,22-dien-3β-ol (**3**) [17], (22*E*)-ergosta-5,7,22-trien-3β-ol (**4**) [18], (22*E*)-19-norergosta-5, 7,9,22-tetraen-3β-ol (**5**) [19], ergosterol peroxide (**9**) [18], and 9,11-dehydroergosterol peroxide (**10**) [20] were isolated from sample 1, and Compounds **1**, **2**, (22*E*)-6β-methoxyergosta-7,22-diene-3β,5α-diol (**6**) [21], (22*E*)-3β,5α,9α-trihydroxyergosta-7,22-dien-6-one (**7**) [21], and (22*E*)-3β,5α-dihydroxyergosta-7, 22-dien-6-one (**8**) [22] were obtained from sample 2 (Figure 1). Of these, **1** and **2** were new compounds.

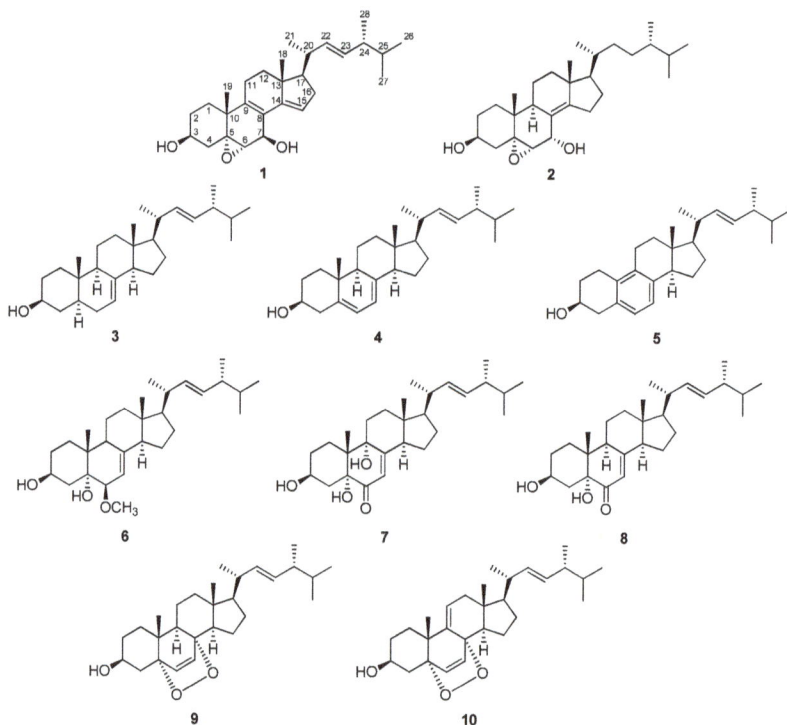

Figure 1. Structures of compounds.

Compound **1** was isolated as an amorphous solid, with a molecular formula of $C_{28}H_{42}O_3$ by HREIMS. The infrared (IR) spectrum indicated the presence of hydroxy groups (ν_{max} 3451 cm^{-1}), and the UV spectrum suggested the presence of a conjugated diene (λ_{max} 242.0 nm). The ^{1}H and ^{13}C NMR spectra (δ_H and δ_C in ppm) in CDCl$_3$ displayed signals for two tertiary methyls (δ_H 0.82 (singlet (s)), 1.30 (s)), four secondary methyls (δ_H 0.83 (doublet (d)), 0.85 (d), 0.93 (d), 1.04 (d)), three oxymethines (δ_H 3.24 (d), 3.96 (triplet of triplets (tt)), 4.85 (broad singlet (br s)); δ_C 59.5 (d), 63.8 (d), 68.4 (d)), an sp^3 oxygenated quaternary carbon (δ_C 63.3 (s)), a tetrasubstituted olefin (δ_C 122.2 (s), 138.8 (s)), a trisubstituted olefin (δ_H 5.55 (br s); δ_C 118.7 (d), 147.7 (s)), and a disubstituted olefin (δ_H 5.20 (doublet of doublets (dd)),

5.28 (dd); δ_C 132.4 (d), 135.1 (d)) (Table 1, Figures S1–S4). In the HMBC spectrum, the correlations were observed as follows; Me-19 (δ_H 0.82 (s))/C-5 (δ_C 63.3 (s)), C-9 (δ_C 138.8 (s)); H-7 (δ_H 4.85 (br s))/C-5 (δ_C 63.3 (s)), C-9 (δ_C 138.8 (s)), C-14 (δ_C 147.7 (s)); H-6 (δ_H 3.24 (d))/C-7 (δ_C 63.8 (d)), C-8 (δ_C 122.2 (s)); Me-18 (δ_H 0.82 (s))/C-14 (δ_C 147.7 (s)); Me-28 (δ_H 0.93 (d))/C-23 and C-25 (Figure 2A and Figure S5). The correlations between H$_2$-1–H$_2$-2–H-3 (δ_H 3.96 (tt))–H$_2$-4; H-6 (δ_H 3.24 (d))–H-7 (δ_H 4.85 (br s)); H-15 (δ_H 5.55 (br s))–H$_2$-16–H-17–H-20–Me-21; H-20–H-22 (δ_H 5.20 (dd))–H-23 (δ_H 5.28 (dd))–H-24–Me-28 (δ_H 0.93 (d)); Me-26 (δ_H 0.85 (d))–H-25–Me-27 (δ_H 0.83 (d)) were observed in the ^{1}H-^{1}H COSY spectrum (Figure 2A and Figure S6). From the above, the planar structure was determined as shown in Figure 2A. The configuration of the hydroxy groups at the C-3 position was determined as β-orientation because of the coupling constant (*J*) (δ_H 3.96 (tt, 11.5, 5.4 Hz)). The NOE correlation of Me-19/H-6β (equatrial) suggested that the epoxy group at C-5,6 was α-oriented, and that of H-7α/H-15 suggested that 7-OH was β-oriented (Figure 2B and Figure S7). The geometry of the double bond at C-22 was determined as *E* from the coupling constants of H-22 (δ_H 5.20 (dd, *J* = 15.2, 7.6 Hz)) and H-23 (δ_H 5.28 (dd, *J* = 15.2, 7.9 Hz)). Comparison of ^{13}C NMR chemical shifts at C-24 (δ_C 42.8) and 28 (δ_C 17.6) with those of 24*R* (δ_C 42.9 (C-24) and 17.7 (C-28)) and 24*S* (δ_C 43.2 (C-24) and 18.1 (C-28)) methylcholestane-type sterols [23,24] established the stereochemistry of C-24 as *R*. Therefore compound **1** was determined as (22*E*)-5α,6α-epoxyergosta-8,14,22-triene-3β,7β-diol (Figure 1, Table S1). Compound **1** was similar to (22*E*)-5α,6α-epoxy-ergosta-8,14,22-triene-7β,7α-diol [25], except for the absence of a 7α-hydroxy group and the presence of a 7β-hydroxy group. There are differences in δ_H value measured with C$_6$D$_6$ such as H-7 (7α-hydroxy-type: δ_H 4.34 (1H, dd, *J* = 11.2, 2.6 Hz) [25] vs. 7β-hydroxy-type (**1**): δ_H 4.74 (br s)), and H-15 (7α-hydroxy-type: δ_H 6.50 (1H, dd, *J* = 3.3, 1.8 Hz) [25] vs. 7β-hydroxy-type (**1**): δ_H 5.33 (br s)).

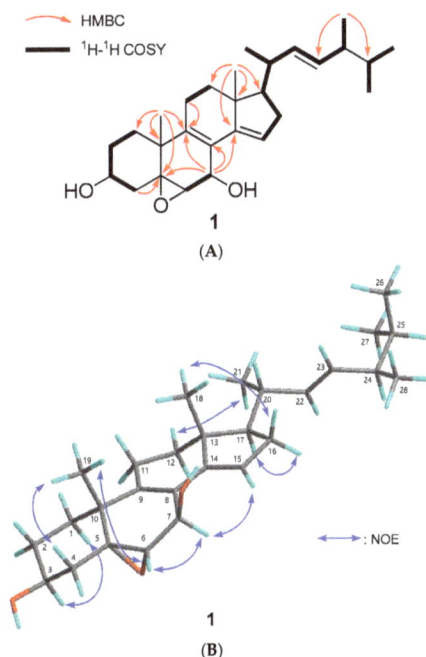

Figure 2. Structure determination of compound **1**. (**A**) Key HMBC and ^{1}H-^{1}H COSY correlations of compound **1**; (**B**) Key NOE correlations of compound **1**. The atoms of C, H, and O were shown in grey, aqua, and red, respectively.

Table 1. ^{1}H and ^{13}C NMR Data for Compounds **1** and **2** in CDCl$_3$ (δ in ppm; J in Hz).

Position		1				2		
	δ_H		δ_C		δ_H		δ_C	
1α	2.01	(1H, multiplet (m))	31.0	t	1.46	(1H, m)	32.2	t
1β	1.86	(1H, m)			1.67	(1H, m)		
2	1.68	(2H, m)	30.9	t	α 1.96	(1H, m)	31.1	t
					β 1.56	(1H, m)		
3	3.96	(1H, tt, J = 11.5, 5.4)	68.4	d	3.92	(1H, tt, J = 11.4, 3.0)	68.7	d
4α	1.50	(1H, m)	39.0	t	1.42	(1H, m)	39.6	t
4β	2.21	(1H, m)			2.13	(1H, dd, J = 13.2, 11.4)		
5			63.3	s			67.8	s
6	3.24	(1H, d, J = 2.4)	59.5	d	3.15	(1H, d, J = 3.5)	61.3	d
7	4.85	(1H, br s)	63.8	d	4.43	(1H, dd, J = 9.6, 3.5)	65.1	d
8			122.2	s			125.1	s
9			138.8	s	2.35	(1H, m)	38.7	d
10			38.3	s			35.8	s
11	2.19	(2H, m)	22.2	t	α 1.49	(1H, m)	19.0	t
					β 1.40	(1H, m)		
12α	1.47	(1H, m)	35.4	t	1.16	(1H, m)	36.7	t
12β	1.99	(1H, m)			1.95	(1H, m)		
13			44.6	s			43.1	s
14			147.7	s			152.7	s
15	5.55	(1H, br s)	118.7	d	α 2.65	(1H, m)	25.0	t
					β 2.30	(1H, m)		
16α	2.27	(1H, m)			1.89	(1H, m)	26.6	t
16β	2.08	(1H, m)	36.8	t	1.41	(1H, m)		
17	1.55	(1H, m)	56.4	d	1.21	(1H, m)	56.6	d
18	0.82	(3H, s)	15.6	quartet (q)	0.85	(3H, s)	17.9	q
19	1.30	(3H, s)	23.6	q	0.87	(3H, s)	16.5	q
20	2.24	(1H, m)	38.8	d	1.46	(1H, m)	34.9	d
21	1.04	(3H, d, J = 6.5)	21.0	q	0.93	(3H, d, J = 6.8)	19.1	q
22	5.20	(1H, dd, J = 15.2, 7.6)	135.1	d	A 1.03	(1H, m)	33.4	t
					B 1.44	(1H, m)		
23	5.28	(1H, dd, J = 15.2, 7.9)	132.4	d	A 0.95	(1H, m)	30.4	t
					B 1.37	(1H, m)		
24	1.88	(1H, m)	42.8	d	1.21	(1H, m)	39.1	d
25	1.48	(1H, m)	33.1	d	1.58	(1H, m)	31.5	d
26	0.85	(3H, d, J = 6.8)	19.9	q	0.85	(3H, d, J = 7.1)	20.5	q
27	0.83	(3H, d, J = 6.8)	19.6	q	0.78	(3H, d, J = 7.0)	17.6	q
28	0.93	(3H, d, J = 6.8)	17.6	q	0.77	(3H, d, J = 6.9)	15.4	q

Compound **2** was isolated as an amorphous solid, with a molecular formula of C$_{28}$H$_{46}$O$_3$. The IR spectrum suggested the presence of hydroxy groups (3387 cm^{-1}). The ^{1}H, ^{13}C NMR and HSQC spectra indicated the presence of two tertiary methyls (δ_H 0.85 (s), 0.87 (s)), four secondary methyls (δ_H 0.77 (d), 0.78 (d), 0.85 (d), 0.93 (d)), two oxymethines (δ_H 3.92 (tt), 4.43 (dd); δ_C 65.1 (d), 68.7 (d)), a trisubstituted epoxy (δ_H 3.15 (d); δ_C 61.3 (d), 67.8 (s)), and a tetrasubstituted olefin (δ_C 125.1 (s), 152.7 (s)) (Table 1, Figures S8–S11). Based on the correlations at Me-18/C-14 (δ_C 152.7 (s)), Me-19/C-5 (δ_C 67.8 (s)), and H-15/C-8 (δ_C 125.1 (s)) and C-14 (δ_C 152.7 (s)) in the HMBC spectrum, and H$_2$-1–H$_2$-2–H-3 (δ_H 3.92 (tt))–H$_2$-4; H-6 (δ_H 3.15 (d))–H-7 (δ_H 4.43 (dd)) in the ^{1}H-^{1}H COSY spectrum (Figure 3A, Figures S12 and S13), oxymethines were at C-3 and C-7 positions, a trisubstituted epoxy group at the C-5, 6 positions, and a tetrasubstituted olefin at the C-8, 14 positions (Figure 3A). The NOE correlation between H-7 and Me-19 demonstrated the configuration of the hydroxy group at the C-7 position as α-orientation (Figure 3B and Figure S14). The NOE correlation between H-4β and Me-19 suggested the orientation of the epoxy group at C-5, 6 was α (Figure 3B and Figure S14). The stereochemistry of C-24 was established as *S* by comparison of the ^{1}H NMR chemical shift at Me-28 (δ_H 0.77) with those of 24*R* (δ_H 0.802) and 24*S* (δ_H 0.781) ergostane-type sterols [26,27]. Therefore, the structure of **2** was established as 5α,6α-epoxyergosta-8(14)-ene-3β,7α-diol (Figure 1, Table S2).

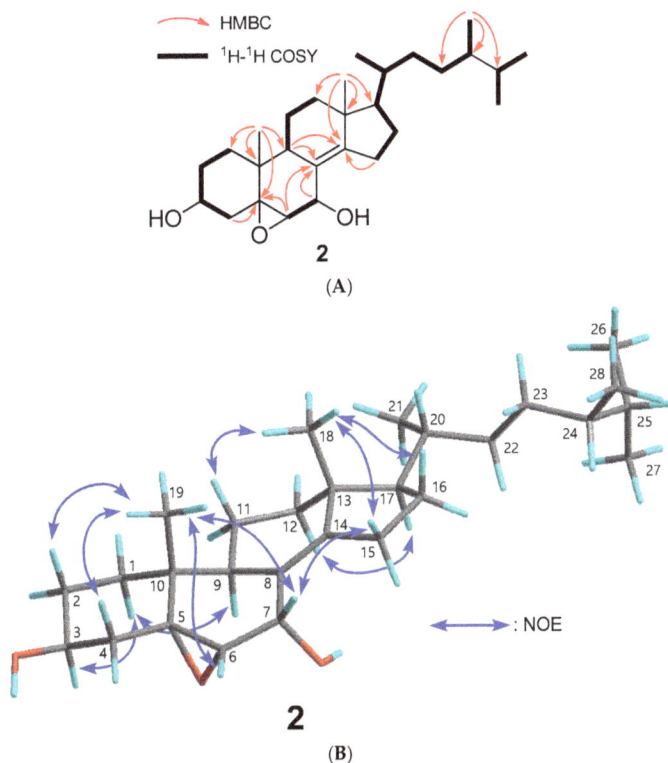

Figure 3. Structure determination of compound **2**. (**A**) Key HMBC and ^{1}H-^{1}H COSY correlations of compound **2**; (**B**) Key NOE correlations of compound **2**. The atoms of C, H, and O were shown in grey, aqua, and red, respectively.

2.2. Evaluation for Aromatase Inhibitory Effects

Compounds **1–10** and aminoglutethimide, a positive control, were evaluated for their aromatase inhibitory activities. Compounds **4** and **6** exhibited comparable inhibitory activities (IC$_{50}$ **4**: 8.1 μM; **6**: 2.8 μM) to aminoglutethimide (IC$_{50}$ 2.0 μM) (Figure 4A). Compounds **1**, **3**, **5**, and **10** showed moderate activities (IC$_{50}$ **1**: 17.3 μM; **3**: 66.1 μM; **5**: 33.8 μM; **10**: 32.6 μM) (Figure 4B). Compounds **2**, **7**, **8**, and **9** weakly inhibited aromatase (Figure 4B). Above results suggested that compounds **4** and **6** can be regarded as potential anti-breast cancer agents targeting aromatase. Based on the results in the figures, the following structure-activity relationship of the compounds can be concluded: (i) The double-bond at C-5, 6 intensifies the aromatase inhibitory activity in ergost-7-ene compounds (**3** (IC$_{50}$ 66.1 μM) vs. **4** (IC$_{50}$ 8.1 μM)); (ii) 9(11)-double-bond enhances the inhibitory activity in 5α,8α-epidioxyergost-6-ene compounds (**9** (IC$_{50}$ > 100 μM) vs. **10** (IC$_{50}$ 32.6 μM)); (iii) 7-ene-6-one compounds did not show this activity (**7** and **8** (IC$_{50}$ > 100 μM)).

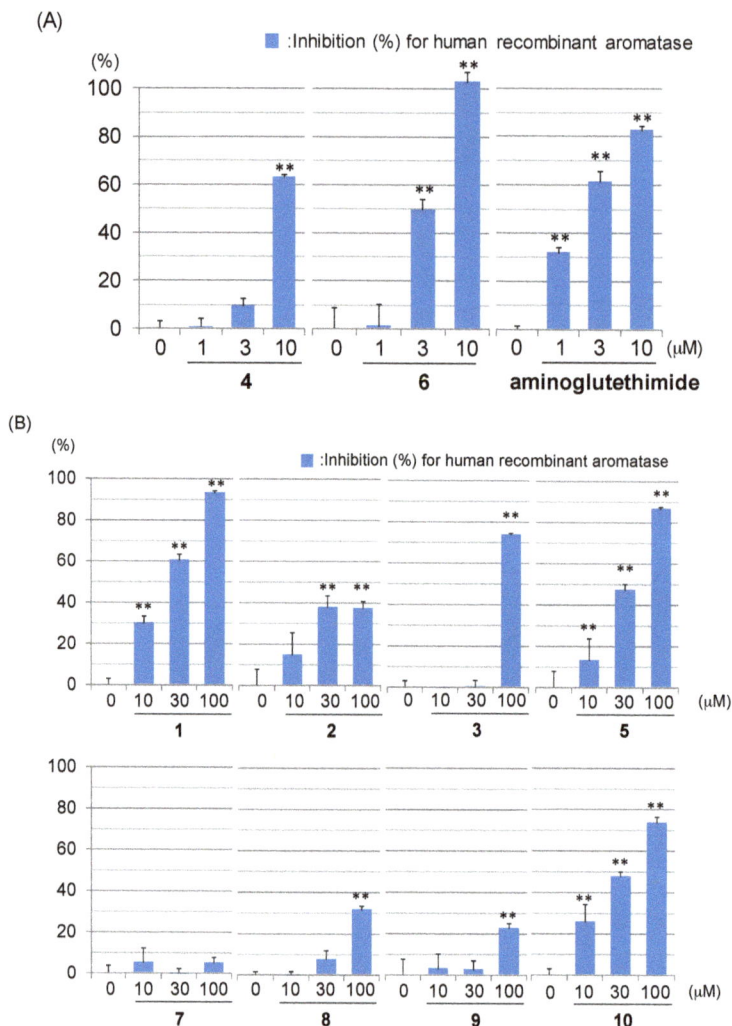

Figure 4. Inhibitory effects of sterols (**1–10**) from *P. eryngii* against human recombinant aromatase. (**A**) Inhibitory effects of sterols (**4**, **6**) and aminoglutethimide at 1, 3, and 10 μM. (**B**) Inhibitory effects of sterols (**1–3**, **5**, **7–10**) at 10, 30, and 100 μM. Each value represents the mean ± the standard error (S.E.) of three determinations. Significant differences from the vehicle control (0 μM) group shown as ** $p < 0.01$.

3. Experimental Section

3.1. General Methods

Dibenzylfluorescein (DBF) and Human CYP19 + P450 Reductase SUPERSOMES (human recombinant aromatase) were obtained from BD Biosciences (Heidelberg, Germany). The physical data were obtained by the following instruments: a Yanagimoto micro-melting point apparatus for melting points (uncorrected); a JASCO DIP-1000 digital polarimeter for Optical rotations; a Perkin-Elmer 1720X FTIR spectrophotometer for IR spectra; an Agilent-NMR-vnmrs600 for the ^{1}H and ^{13}C NMR spectra (^{1}H: 600 MHz; ^{13}C: 150 MHz) in CDCl$_3$ with tetramethylsilane as the internal standard; a Hitachi

M-4000H double-focusing mass spectrometer for EIMS (70 eV). Column chromatography was carried out by Silica gel (70–230 mesh, Merck, Darmstadt, Germany) and silica gel 60 (230–400 mesh, Nacalai Tesque, Inc., Kyoto, Japan). HPLC was performed by the following systems; system I: *Cosmosil 5SL-II column* (25 cm × 20 mm i.d.) (Nacalai Tesque, Inc.), hexane/EtOAc (5:1), 8.0 mL/min, 35 °C; system II: *Shimpack PREP-ODS* (25 cm × 20 mm i.d.) (Shimadzu corp., Kyoto, Japan), MeOH, 8.0 mL/min, 35 °C; system III: *Cosmosil 5C₁₈-MS-II column* (25 cm × 20 mm i.d.) (Nacalai Tesque, Inc.), MeOH/H$_2$O (95:5), flow rate, 4.0 mL/min, 35 °C; system IV: *Cosmosil 5C₁₈-MS-II column*, MeOH/H$_2$O (9:1), 4.0 mL/min, 35 °C.

3.2. Materials

The fruiting bodies of *P. eryngii* were purchased from HOKUTO Corp. They were cultivated in Kagawa, Japan (Sample 1 in 2011, and Sample 2 in 2014). A voucher material has been deposited in the Herbarium of the Laboratory of Medicinal Chemistry, Osaka University of Pharmaceutical Sciences.

3.3. Extraction and Isolation

3.3.1. Sample 1

Sample 1 (fruiting bodies of *P. eryngii* (21 kg, fresh weight)) was extracted with MeOH under reflux (1 week, 4 times). The MeOH extract (170 g) was then divided into EtOAc and H$_2$O fractions by liquid-liquid partition. The EtOAc fraction (60 g) was separated into 20 fractions (Fr. *S1-A* to *S1-T*) with SiO$_2$ column chromatography (CC) (SiO$_2$ (3.5 kg); CHCl$_3$/EtOAc (1:0 to 0:1), and EtOAc/MeOH (5:1, and 0:1)).

Fr. *S1-H* (836.5 mg), CHCl$_3$/EtOAc (10:1)-eluted fraction, was separated with SiO$_2$ CC to yield 8 fractions, *S1-H1* to *S1-H8*. Preparative HPLC (system I) of *S1-H3* (185.7 mg), hexane/EtOAc (5:1)-eluted fraction, provided 7 fractions, *S1-H3-1* to *SF3-7*. *S1-H3-4* was identified as **4** (31.8 mg; retention time (t_R) 19.2 min). Preparative HPLC (system II) of *S1-H3-5* (5.4 mg, t_R 36.5 min) provided **3** (1.9 mg; t_R 37.5 min). Preparative HPLC (system IV) of *S1-H6* (14.3 mg), hexane/EtOAc (3:1)-eluted fraction, provided **10** (1.3 mg, t_R 95.4 min) and **9** (1.7 mg, t_R 120.2 min).

Fr. *S1-I* (1072.3 mg), CHCl$_3$/EtOAc (10:1)-eluted fraction, was separated with SiO$_2$ CC to give 8 fractions, *S1-I1* to *S1-I10*. Preparative HPLC (system I) of *S1-I5* (45.2 mg), hexane/EtOAc (5:1)-eluted fraction, provided **5** (2.8 mg, t_R 42.7 min).

3.3.2. Sample 2

Sample 2 (fruiting bodies of *P. eryngii* (120 kg, fresh weight)) was extracted with MeOH under reflux (3 days, 4 times). The MeOH extract (2625 g) was divided into EtOAc and H$_2$O fractions by liquid-liquid partition. The EtOAc fraction (240 g) was separated into 37 fractions (Fr. *S2-A* to *S2-Z*, and *S2-a* to *S2-k*) with SiO$_2$ column chromatography (CC) (SiO$_2$ (2.8 kg); CHCl$_3$/EtOAc (1:0 to 0:1), and MeOH).

Fr. *S2-V* (3964.9 mg), CHCl$_3$/EtOAc (1:1)-eluted fraction, was separated by SiO$_2$ CC to give 8 fractions, *S2-V1* to *S2-V21*. Preparative HPLC (system III) of Fr. *S2-V4* (110.9 mg), hexane/EtOAc (1:1)-eluted fraction, provided **8** (1.5 mg; t_R 36.9 min) and **6** (6.3 mg; t_R 49.5 min). Preparative HPLC (system IV) of Fr. *S2-V6* (451.6 mg), hexane/EtOAc (1:1)-eluted fraction, provided **7** (1.1 mg; t_R 59.8 min). Preparative HPLC (system IV) of Fr. *S2-V7* (270.5 mg), hexane/EtOAc (1:1)-eluted fraction, provided **2** (3.0 mg; t_R 76.8 min). Preparative HPLC (system III) of Fr. *S2-V10* (270.5 mg), hexane/EtOAc (1:1)-eluted fraction, provided **1** (1.9 mg; t_R 36.3 min).

3.3.3. (22E)-5α,6α-Epoxyergosta-8,14,22-triene-3β,7β-diol (**1**)

[α]^{20}$_D$ −23.6 (*c* = 0.13, EtOH); IR ν_{max}^{KBr} cm^{-1}: 3451, 2960, 1697, 1557, 1456; UV λ_{max}^{EtOH} nm (logε): 206.0 (3.75), 242.0 (3.62); EIMS *m/z*: 426 [M]$^+$ (71), 315 (27), 300 (57), 172 (83), 69 (100); HREIMS *m/z*: 426.3127 [M]$^+$ (calcd for 426.3134: C$_{28}$H$_{42}$O$_3$); ^{1}H NMR (400 MHz, C$_6$D$_6$) δ_H ppm: 0.80 (s, H-19),

0.91 (d, 6.8 Hz), 0.93 (d, 6.4 Hz), 1.01 (d, 6.8 Hz), 1.07 (d, 6.4 Hz), 1.12 (s, H-18), 3.11 (d, 2.4 Hz, H-6), 3.83 (tt, 11.6, 4.8 Hz, H-3), 4.74 (br s, H-7), 5.24 (dd, 15.6, 8.4 Hz, H-22), 5.32 (overlapped, H-23), 5.33 (br s, H-15).

3.3.4. 5α,6α-Epoxyergost-8(14)-ene-3β,7α-diol (2)

$[\alpha]^{20}_D$ −112.2 (*c* = 0.13, EtOH); IR ν_{max}^{KBr} cm^{-1}: 3387, 2959, 2936, 2871, 1466, 1377; EIMS *m/z*: 430 [M]$^+$ (5), 412 (100), 394 (57), 379 (58), 267 (23), 213 (21); HREIMS *m/z*: 430.3450 [M]$^+$ (calcd for 430.3447: $C_{28}H_{46}O_3$).

3.4. Inhibitory Effects against Human Recombinant Aromatase

Inhibitory assay against human recombinant aromatase was performed as described previously [28,29].

3.5. Statistics

Values are described as the mean ± standard error of the mean (S.E.M.). Statistical analysis was performed by one-way analysis of variance, followed by Dunnett's test. Probability (*p*) values less than 0.05 were regarded as significant.

4. Conclusions

In this study, we isolated two new sterols (**1** and **2**) and elucidated their structures. They have 5α,6α-epoxy-7-hydroxy ergostane structure. In aromatase inhibitory assay, compounds **4** and **6** possessed comparable inhibitory effects (IC$_{50}$ **4**: 8.1 μM; **6**: 2.8 μM) against human recombinant aromatase to aminoglutethimide (IC$_{50}$ 2.0 μM). These results suggested that compounds **4** and **6** have potential as anti-breast cancer agents.

Supplementary Materials: Supplementary materials can be found at www.mdpi.com/1422-0067/18/11/2479/s1.

Acknowledgments: We thank Katsuhiko Minoura and Mihoyo Fujitake (Osaka University of Pharmaceutical Sciences) for the NMR and MS measurements.

Author Contributions: Takashi Kikuchi, Takeshi Yamada, Kiyofumi Ninomiya, Toshio Morikawa and Reiko Tanaka designed the experiments; Takashi Kikuchi, Naoki Motoyashiki, Takeshi Yamada and Reiko Tanaka isolated compounds, and elucidated their structures. Kanae Shibatani, Kiyofumi Ninomiya and Toshio Morikawa evaluated aromatase inhibitory effects of compounds. Takashi Kikuchi, Toshio Morikawa and Reiko Tanaka wrote the paper.

Conflicts of Interest: The authors declare no conflict of interest.

Abbreviations

NMR	Nuclear magnetic resonance
HREIMS	High resolution electron ionization mass spectrometry
CDCl$_3$	Duterated chloroform
HMBC	Heteronuclear multiple bond coherence
COSY	Correlation spectroscopy
NOE	Nuclear overhauser effect
HSQC	Hetero nuclear single quantum coherence

References

1. Hong, Y.; Li, H.; Yuan, Y.-C.; Chen, S. Molecular characterization of aromatase. *Ann. N. Y. Acad. Sci.* **2009**, *1155*, 112–120. [CrossRef] [PubMed]
2. Johnston, S.R.D.; Dowsett, M. Aromatase inhibitors for breast cancer: Lessons from the laboratory. *Nat. Rev. Cancer* **2003**, *3*, 821–831. [CrossRef] [PubMed]
3. Bae, K.; Yoo, H.-S.; Lamoury, G.; Boyle, F.; Rosenthal, D.S.; Oh, B. Acupuncture for aromatase inhibitor-induced arthralgia: A systematic review. *Integr. Cancer Ther.* **2015**, *14*, 496–502. [CrossRef] [PubMed]

4. Balunas, M.J.; Kinghorn, A.D. Natural compounds with aromatase inhibitory activity: An update. *Planta Med.* **2010**, *76*, 1087–1093. [CrossRef] [PubMed]
5. Shimizu, K.; Yamanaka, M.; Gyokusen, M.; Kaneko, S.; Tsutsui, M.; Sato, J.; Sato, I.; Sato, M.; Kondo, R. Estrogen-like activity and prevention effect of bone loss in calcium deficient ovariectomized rats by the extract of *Pleurotus eryngii*. *Phytother. Res.* **2006**, *20*, 659–664. [CrossRef] [PubMed]
6. Rodriguez Estrada, A.E.; Royse, D.J. Yield, size and bacterial blotch resistance of *Pleurotus eryngii* grown on cottonseed hulls/oak sawdust supplemented with manganese, copper and whole ground soybean. *Bioresour. Technol.* **2007**, *98*, 1898–1906. [CrossRef] [PubMed]
7. Lee, I.-S.; Ryoo, I.-J.; Kwon, K.-Y.; Ahn, J.S.; Yoo, I.-D. Pleurone, a novel human neutrophil elastase inhibitor from the fruiting bodies of the mushroom *Pleurotus eryngii* var. ferulae. *J. Antibiot.* **2011**, *64*, 587–589. [CrossRef] [PubMed]
8. Kang, M.Y.; Rico, C.W.; Lee, S.C. In vitro antioxidative and antimutagenic activities of oak mushroom (*Lentinus edodes*) and king oyster mushroom (*Pleurotus eryngii*) byproducts. *Food Sci. Biotechnol.* **2012**, *21*, 167–173. [CrossRef]
9. Han, E.H.; Hwang, Y.P.; Kim, H.G.; Choi, J.H.; Im, J.H.; Yang, J.H.; Lee, H.-U.; Chun, S.-S.; Chung, Y.C.; Jeong, H.G. Inhibitory effect of *Pleurotus eryngii* extracts on the activities of allergic mediators in antigen-stimulated mast cells. *Food Chem. Toxicol.* **2011**, *49*, 1416–1425. [CrossRef] [PubMed]
10. Kawai, J.; Ouchi, K.; Inatomi, S.; Andoh, T. *Pleurotus eryngii* ameliorates lipopolysaccharide-induced lung inflammation in mice. *Evid.-Based Complement. Altern. Med.* **2014**, *2014*, 532389. [CrossRef] [PubMed]
11. Zhang, A.; Li, X.; Xing, C.; Yang, J.; Sun, P. Antioxidant activity of polysaccharide extracted from *Pleurotus eryngii* using response surface methodology. *Int. J. Biol. Macromol.* **2014**, *65*, 28–32. [CrossRef] [PubMed]
12. Yang, Z.; Xu, J.; Fu, Q.; Fu, X.; Shu, T.; Bi, Y.; Song, B. Antitumor activity of a polysaccharide from *Pleurotus eryngii* on mice bearing renal cancer. *Carbohydr. Polym.* **2013**, *95*, 615–620. [CrossRef] [PubMed]
13. Yaoita, Y.; Yoshihara, Y.; Kakuda, R.; Machida, K.; Kikuchi, M. New sterols from two edible mushrooms, *Pleurotus eryngii* and *Panellus serotinus*. *Chem. Pharm. Bull.* **2002**, *50*, 551–553. [CrossRef] [PubMed]
14. Wang, S.-J.; Li, Y.-X.; Bao, L.; Han, J.-J.; Yang, X.-L.; Li, H.-R.; Wang, Y.-Q.; Li, S.-J.; Liu, H.-W. Eryngiolide A, a cytotoxic macrocyclic diterpenoid with an unusual cyclododecane core skeleton produced by the edible mushroom *Pleurotus eryngii*. *Org. Lett.* **2012**, *14*, 3672–3675. [CrossRef] [PubMed]
15. Kikuchi, T.; Masumoto, Y.; In, Y.; Tomoo, K.; Yamada, T.; Tanaka, R. Eringiacetal A, 5,6-*seco*-(5*S*,6*R*,7*R*,9*S*)-5, 6:5,7:6,9-triepoxyergosta-8(14),22-diene-3β,7β-diol, an unusual ergostane sterol from the fruiting bodies of *Pleurotus eryngii*. *Eur. J. Org. Chem.* **2015**, *2015*, 4645–4649. [CrossRef]
16. Kikuchi, T.; Maekawa, Y.; Tomio, A.; Masumoto, Y.; Yamamoto, T.; In, Y.; Yamada, T.; Tanaka, R. Six new ergostane-type steroids from king trumpet mushroom (*Pleurotus eryngii*) and their inhibitory effects on nitric oxide production. *Steroids* **2016**, *115*, 9–17. [CrossRef] [PubMed]
17. Keller, A.C.; Maillard, M.P.; Hostettmann, K. Antimicrobial steroids from the fungus *Fomitopsis pinicola*. *Phytochemistry* **1996**, *41*, 1041–1046. [CrossRef]
18. Seo Hyo, W.; Hung Tran, M.; Na, M.; Jung Hyun, J.; Kim Jin, C.; Choi Jae, S.; Kim Jung, H.; Lee, H.-K.; Lee, I.; Bae, K.; et al. Steroids and triterpenes from the fruit bodies of *Ganoderma lucidum* and their anti-complement activity. *Arch. Pharm. Res.* **2009**, *32*, 1573–1579. [CrossRef] [PubMed]
19. Barrero, A.F.; Oltra, J.E.; Poyatos, J.A.; Jimenez, D.; Oliver, E. Phycomysterols and Other Sterols from the Fungus *Phycomyces blakesleeanus*. *J. Nat. Prod.* **1998**, *61*, 1491–1496. [CrossRef] [PubMed]
20. Du, Z.-Z.; Shen, Y.-M. A rare new cleistanthane diterpene from the pericarp of *Trewia nudiflora*. *Helv. Chim. Acta* **2006**, *89*, 2841–2845. [CrossRef]
21. Kawagishi, H.; Katsumi, R.; Sazawa, T.; Mizuno, T.; Hagiwara, T.; Nakamura, T. Cytotoxic steroids from the mushroom *Agaricus blazei*. *Phytochemistry* **1988**, *27*, 2777–2779. [CrossRef]
22. Ishizuka, T.; Yaoita, Y.; Kikuchi, M. Sterol constituents from the fruit bodies of *Grifola frondosa* (Fr.) S.F. Gray. *Chem. Pharm. Bull.* **1997**, *45*, 1756–1760. [CrossRef]
23. Yan, X.-H.; Liu, H.-L.; Huang, H.; Li, X.-B.; Guo, Y.-W. Steroids with Aromatic A-Rings from the Hainan Soft Coral Dendronephthya studeri Ridley. *J. Nat. Prod.* **2011**, *74*, 175–180. [CrossRef] [PubMed]
24. Li, W.; Zhou, W.; Song, S.B.; Shim, S.H.; Kim, Y.H. Sterol Fatty Acid Esters from the Mushroom *Hericium erinaceum* and Their PPAR Transactivational Effects. *J. Nat. Prod.* **2014**, *77*, 2611–2618. [CrossRef] [PubMed]

25.   Ohnuma, N.; Amemiya, K.; Kakuda, R.; Yaoita, Y.; Machida, K.; Kikuchi, M. Sterol constituents from two edible mushrooms, Lentinula edodes and Tricholoma matsutake. *Chem. Pharm. Bull.* **2000**, *48*, 749–751. [CrossRef] [PubMed]

26.   Rubinstein, I.; Goad, L.J.; Clague, A.D.H.; Mulheirn, L.J. The 220 MHz NMR spectra of phytosterols. *Phytochemistry* **1976**, *15*, 195–200. [CrossRef]

27.   Kobayashi, M.; Krishna, M.M.; Haribabu, B.; Anjaneyulu, V. Marine sterols.   XXV. Isolation of 23-demethylgorgost-7-ene-3β,5α,6β-triol and (24S)-ergostane-3β,5α,6β,7β,15β-pentol from soft corals of the Andaman and Nicobar coasts. *Chem. Pharm. Bull.* **1993**, *41*, 87–89. [CrossRef]

28.   Ninomiya, K.; Shibatani, K.; Sueyoshi, M.; Chaipech, S.; Pongpiriyadacha, Y.; Hayakawa, T.; Muraoka, O.; Morikawa, T. Aromatase Inhibitory Activity of Geranylated Coumarins, Mammeasins C and D, Isolated from the Flowers of Mammea siamensis. *Chem. Pharm. Bull.* **2016**, *64*, 880–885. [CrossRef] [PubMed]

29.   Tanabe, G.; Tsutsui, N.; Shibatani, K.; Marumoto, S.; Ishikawa, F.; Ninomiya, K.; Muraoka, O.; Morikawa, T. Total syntheses of the aromatase inhibitors, mammeasins C and D, from Thai medicinal plant *Mammea siamensis*. *Tetrahedron* **2017**, *73*, 4481–4486. [CrossRef]

International Journal of
Molecular Sciences

MDPI

Article

The Evaluation of Pro-Cognitive and Antiamnestic Properties of Berberine and Magnoflorine Isolated from Barberry Species by Centrifugal Partition Chromatography (CPC), in Relation to QSAR Modelling

Wirginia Kukula-Koch [1,*], Marta Kruk-Słomka [2], Katarzyna Stępnik [3], Radosław Szalak [4] and Grażyna Biała [2]

1   Chair and Department of Pharmacognosy with Medicinal Plant Unit, Medical University in Lublin, 20-093 Lublin, Poland
2   Department of Pharmacology and Pharmacodynamics, Medical University of Lublin, 20-093 Lublin, Poland; marta.kruk@umlub.pl (M.K.-S.); grazyna.biala@umlub.pl (G.B.)
3   Faculty of Chemistry, Chair of Physical Chemistry, Department of Planar Chromatography, Maria Curie-Skłodowska University, 20-031 Lublin, Poland; katarzyna.stepnik@poczta.umcs.lublin.pl
4   Department of Animal Anatomy and Histology, Faculty of Veterinary Medicine, University of Life Science, 20-950 Lublin, Poland; radek.szalak@up.lublin.pl
*   Correspondence: virginia.kukula@gmail.com; Tel.: +48-814-487-080

Received: 4 September 2017; Accepted: 20 November 2017; Published: 24 November 2017

Abstract: Civilization diseases associated with memory disorders are important health problems occurring due to a prolonged life span. The manuscript shows the results of an in vivo study targeting the emergence of two drug candidates with anti-amnestic properties. The preceding quantitative structure–activity relationship (QSAR) studies provided information on the ability of berberine and magnoflorine to cross the blood–brain barrier (BBB). In the light of these findings, both compounds were purified from crude plant extracts of barberries: berberine—from *Berberis siberica* using a method published earlier, and magnoflorine—from *Berberis cretica* by centrifugal partition chromatography (solvent system: ethyl acetate:butanol:water-0.6:1.5:3 $v/v/v$). Both the compounds were evaluated for their memory enhancing and scopolamine inhibitory properties in an in vivo passive avoidance (PA) test on mice towards short-term and long-term memory. Cognition enhancing properties were observed at the following doses: 5 mg/kg (i.p.) for berberine and 20 mg/kg (i.p.) for magnoflorine. In addition, both the tested isoquinolines with the co-administered scopolamine were found to block long-term but not short-term memory impairment. No influence on the locomotor activity was observed for the tested doses. The results confirmed a marked central activity of magnoflorine and showed the necessity to lower the dosage of berberine. Optimized purification conditions have been elaborated for magnoflorine.

Keywords: *Berberis*; alkaloids; passive avoidance test; counter-current chromatography; memory and learning; mice

1. Introduction

Dementia disorders constitute a growing problem affecting inhabitants of developed countries. The advancement of knowledge has contributed to the prolongation of life, which undoubtedly favors various disorders of the central nervous system (CNS) such as Alzheimer's disease (AD). More than 18 million patients suffer from AD and, according to the World Health Organization (WHO), this number is predicted to grow to as much as ca. 70 million patients by 2050 [1]. An improvement on the efficacy of existing and an introduction of new therapeutic strategies in the treatment of CNS disorders associated with dementia slow the progression of the disease.

Some natural products are being used in the basic treatment of AD [2,3]. Thus, this situation encourages the authors to search for new drug candidates from similar groups of natural products that could possess improved pharmacokinetic and toxicological profiles.

The currently available drugs often bear an increased toxicity or considerable side effects in proportion to the prescribed dosage. In addition, the development of tolerance for their activity has been observed, which triggers the necessity to increase the drug dose and in turn induce the occurrence of side effects.

In a former screening study [1], a marked anticholinesterase activity of two alkaloids: berberine and magnoflorine, in an in vitro thin-layer chromatography (TLC)-based bioautographic study of *Argemone mexicana* extracts was revealed. To confirm these preliminary results, a reference to the in vivo studies on both compounds was needed.

Published in vivo studies of berberine performed after traumatic brain injury in mice showed its neuroprotective properties toward brain as quick as 10 min after administration [4]. This action resulted from the reduction in permeability of leukocytes to the injury site. In addition, rats treated with the extract of *Coptis chinensis*, which contains a high quantity of berberine, showed its antiamnesic action in the scopolamine-induced amnesia model [5]. Another study confirmed its ability to improve spatial memory in rats in a Morris water maze. Hydrochloride salt of berberine at a dose of 50 mg/kg (p.o.) reduced the symptoms of AD, and decreased the synthesis of Interleukin-I β (IL-β) and nitric oxide synthase (iNOS) [6].

In addition, berberine has been documented to increase the viability of cells in the hippocampus and to possess the ability to regenerate neurons located peripherally in the nervous system by an accelerated remyelination of nerve cells, when administered at a dose of 20 mg/kg (i.p.) daily for a week [7].

Scientific reports on the biological activity of magnoflorine are scarce, which is why this compound was of interest to the authors and, in this study, was compared with the activity of a model isoquinoline alkaloid—berberine. Schiff [8] showed a pronounced activity of magnoflorine toward the muscarinic and serotoninergic receptors. This finding encouraged the authors to evaluate its CNS activity further.

To achieve all the goals of the study, in the first step, it was important to check the ability of berberine and magnoflorine to cross the blood–brain barrier (BBB). For this purpose, detailed quantitative structure–activity relationship (QSAR) studies were performed to construct new models of the pharmacological behavior assessment, such as the pharmacokinetic descriptors of the brain, namely, logBB, logPS, logPS*$_{fu,brain}$ fractions unbound in plasma and brain as well as some physicochemical and linear solvation energy relationship (LSER) parameters to further discuss the application of previously obtained in vitro results and about the ability of studied compounds to exert any central effects.

Then, a semi-preparative purification of these molecules by hydrostatic counter-current chromatography was planned and optimized to sustain the need for a high quantity of natural products, which would cover the scheduled dosage in the in vivo study. This technique, however, is characterized by high capacity, low solvent consumption and a possible scale-up, resulting in a one-step purification of high loads of crude samples.

The extracts from *Berberis cretica* and *Berberis siberica*, two deciduous shrubs occuring in South East Europe/Asia Minor and Central Asia, respectively, known to produce isoquinoline alkaloids (including berberine and magnoflorine), were selected for the purification of these two alkaloids due to a higher content of the desired molecules in relation to the previously studied *Argemone mexicana* [1,9].

For the assessment of antiamnestic and pro-cognitive properties of alkaloids, a passive avoidance (PA) test was performed on several groups of mice. In addition, the influence of alkaloids on the locomotor activity of mice for all studied doses was evaluated, in order to obtain the final results on both pharmacological activity and on the ranges of therapeutic windows of both compounds, which are expressed as a disturbance into the mice motility.

2. Results

2.1. Quantitative Structure–Activity Relationship (QSAR) Studies

To reduce the differences between the actual and the predicted values, the QSAR models were based on the multiple linear regression (MLR) methodology with backward elimination of variables. The best relationship between the BBB descriptors (logBB, logPS) and various physicochemical parameters have been established as a result of many attempts. In each case, the following statistical parameters were determined: determination coefficient, significance level, Fisher criterion, and mean squared error. Then, during the analysis of the above parameters, the best models have been selected (high correlation coefficient, low mean squared error, low *p*-value, etc.).

Based on the parameters from Table 1, some entirely new QSAR models for protoberberine and aporphine alkaloids have been constructed.

Optimization of the structures of the studied alkaloids was conducted before the calculation of both LSER and physicochemical parameters. To provide a structural similarity between the tested compounds, biological activity of the substances should be determined against the chemical group of which they are a part. Therefore, the following compounds have been considered among protoberberine alkaloids: berberine, protoberberine, sanguinarine, coptisine, stylopine, chelidonine, and jatrorrhizine. All of these alkaloids have a 5,6-dihydrodibenzo[*a,g*]quinolizinium ($C_{17}H_{14}N^+$) skeleton as the basic skeleton of the quaternary protoberberine alkaloids. The group of aporphine alkaloids included: magnoflorine, tetrandrine, glaucine, thalrugosine, protopine, boldine, corydine, and norcorydine. All the above-mentioned compounds follow the Lipiński rule of five [10]; thus, BBB parameters, LSER descriptors, and some physicochemical parameters have been calculated using the ACD/Percepta software (version 2012, Advanced Chemistry Development, Inc., Toronto, ON, Canada).

For protoberberine alkaloids, the following models were obtained:

$$logBB = -0.433 + 0.502A - 0.543B + 0.732S - 0.011E - 0.392V$$
$$n = 7, R^2 = 95.95\%, S = 0.113 \tag{1}$$

Considering the chosen structural parameters, that is, lipophic descriptor ($logP_{o/w}$), molar mass (M) and topological polar surface area (TPSA), the rate of penetration into the brain (logPS) has been estimated as follows:

$$logPS = -1.981 + 0.6184\, logP_{o/w} - 0.003552M + 0.00221TPSA$$
$$n = 7, R^2 = 99.90\%, S = 0.414 \tag{2}$$

Likewise, the following models have been constructed for aporphine alkaloids:

$$logBB = -2.34 - 1.94A - 3.98B + 1.84S + 3.65E - 1.41V$$
$$n = 8, R^2 = 74.98\%, S = 0.252 \tag{3}$$

$$log\ PS = -2.118 + 0.5703\, logP_{o/w} - 0.00128M - 0.0164\ TPSA$$
$$n = 8,\ R^2 = 96.85\%,\ S = 0.244 \tag{4}$$

Table 1. The collation of the blood–brain barrier (BBB) pharmacokinetic parameters, the Abraham linear solvation energy relationship (LSER) descriptors and the chosen physicochemical data.

Compound	logBB	logPS	logPS*$_{fu,brain}$	Fraction Unbound in Plasma	Fraction Unbound in Brain	Molar Mass (M)	Topological Polar Surface Area (TPSA)	Logarithm of n-Octanol-Water Partition Coefficient (logP$_{o/w}$)	Abraham LSER Descriptors				
									A	B	S	E	V
Berberine	−0.35	−3.9	−3.95	0.44	0.98	336.366	40.80	−1.33	0	1.09	2.25	2.25	3957
Magnoflorine	0.23	−4.4	−4.4	0.58	0.98	342.414	58.92	−1.38	0.55	0.94	1.62	1.62	2.5903

The logPS and logBB values calculated from the newly constructed Models (1)–(4) for berberine and magnoflorine have been compared with the predicted proper values predicted in silico (Figure 1). The values obtained from the newly formed models have been found to be very close to those predicted in silico. This can indicate that the newly formed models are relevant for description of BBB permeability of the studied compounds. The best fit was obtained for Model (1) based on the Abraham parameters.

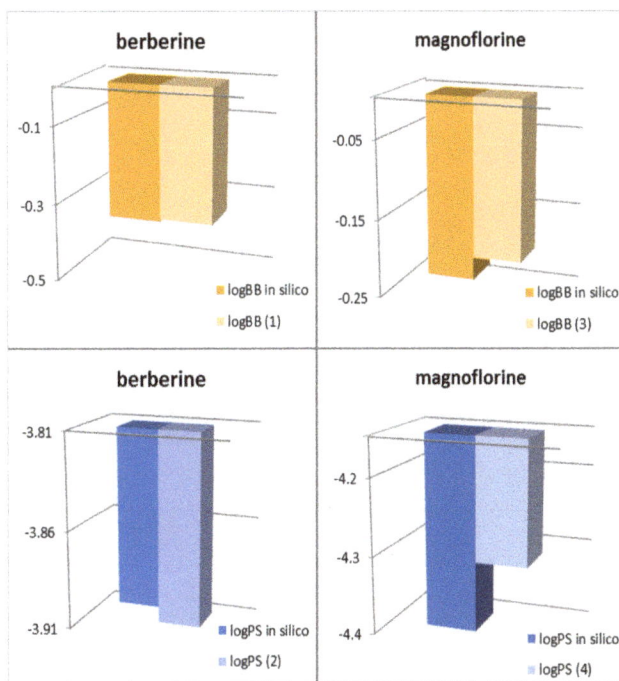

Figure 1. Comparison between the logBB (logPS) calculated in silico and the values obtained from the newly constructed models.

Applicability Domain

The applicability domain has to be essentially defined to evaluate the reliability of the QSAR models. For this purpose, the Williams plot was plotted, illustrating the relationships between standardized residuals and leverages. In our study, standardized residuals are the differences between logBB (logPS) values in silico calculated and predicted from the newly constructed models. All the compounds that fall into that category may be treated as reliable. For each tested compound, the leverage h_i value has been calculated:

$$h_i = \left(X^T X\right)^{-1} x_i{}^T (i = 1, ..., n) \tag{5}$$

where x_i is the row vector of the descriptors, T is the matrix/vector transposed, X and i are the variable matrices deduced from the training set variable values.

Moreover, the critical leverage values h^* were estimated for each group of the tested compounds using the following general equation:

$$h^* = 3(k+1)/N \tag{6}$$

where k is the number of model parameters and N is the number of training compounds.

The h^* values were calculated for logBB models (for protoberberines = 2.57; aporphines = 2.25). Analogously, h^* values for logPS models are as follows: 1.71 (protoberberines) and 1.5 (aporphines). Because all the obtained leverage values for each tested compound (h_i values are less than the critical h^* values, and the predicted models can be treated as reliable).

Based on the obtained relations, it was deduced that both berberine and magnoflorine are in the applicability domain of the models.

To check which of the tested structural parameters is the most significant for permeation into the brain, analysis of variance (ANOVA) was performed (Table 2) using the Minitab 17 Statistical Software (Minitab Inc., State College, PA, USA). In the group of aporphine alkaloids, the obtained p-values for molar mass and TPSA parameters were found to be greater than the significance level; therefore, the influence of these descriptors on permeation into the brain is statistically insignificant. However, p-value for the logP$_{o/w}$ parameters is less than the significance level; therefore, lipophilicity of aporphine alkaloids can play a key role in BBB permeability. Similarly, for protoberberine alkaloids, all of the tested parameters are statistically significant for BBB permeability. The above correlations confirm (according to the Hansch model) that lipophilicity as well as steric and electronic properties of the molecule affect BBB permeability of protoberberine alkaloids.

Table 2. Statistical p- and F-values obtained for the tested alkaloids.

Parameter	Aporphines		Protoberberines	
	F-Value	p-Value	F-Value	p-Value
logP$_{o/w}$	94.87	0.001	1010.13	0.000
M	1.50	0.288	1405.09	0.000
TPSA	1.86	0.245	46.28	0.0006

To confirm these assumptions, principal components analysis (PCA) was conducted (Minitab 17 software). The results of PCA were obtained using in silico calculated logBB, logPS, logPS*f$_{u,brain}$, logP$_{o/w}$, molar mass, and TPSA values. The first two principal components explain more than 87% of the variance in the data in both the aporphines and protoberberines groups. In Figure 2, loading plots of the above-mentioned parameters are presented. As we can see from the loading plot corresponding to the first two principal components, there is a correlation between the BBB parameters and logP$_{o/w}$ values for aporphines and between the BBB parameters and all the tested physicochemical descriptors for protoberberines. Therefore, the previous correlations between the BBB descriptors and the chosen molecular parameters were proven. Thereby, the initial in silico BBB studies as well as the newly constructed model and the PCA analysis indicated the ability of both berberine and magnoflorine to cross the BBB.

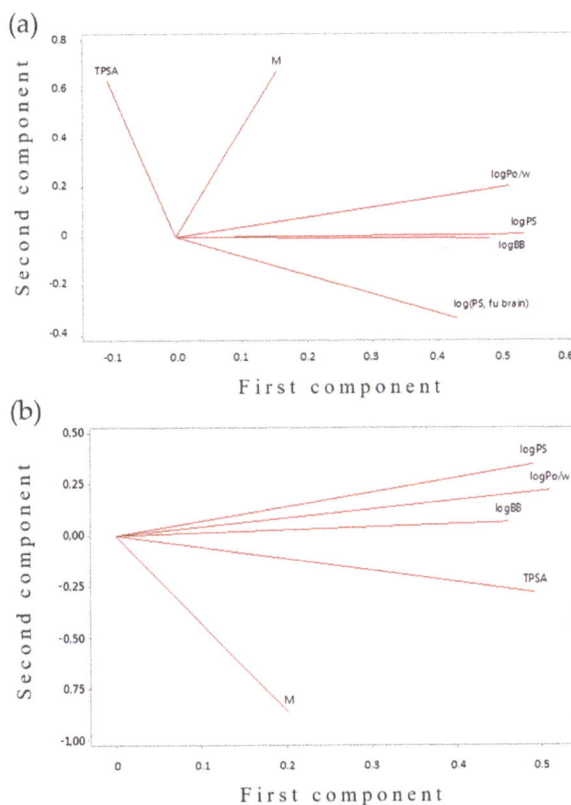

Figure 2. Loading plots of the chosen parameters for: (**a**) aporphine; (**b**) protoberberine alkaloids.

2.2. HPLC and HPLC-MS Analysis of the Extract Composition

A high-resolution mass spectrometer coupled with a liquid chromatograph was used in order to recognize the major secondary metabolites present in a variety of extracts obtained from the selected representatives of Papaveraceae and Berberidaceae in the search for magnoflorine and berberine. Two species were selected as potential sources of these compounds for their isolation for in vivo studies: Cretan barberry (*Berberis cretica*), whose roots contained a high quantity of magnoflorine, and Siberian barberry (*Berberis siberica*) for the purification of berberine in its overground parts, as a successful centrifugal partition chromatography (CPC) separation protocol have been elaborated earlier for berberine by the authors [11]. The overground parts of Siberian barberry were characterized by a high variety of metabolites and a marked content of phenols (a variety of simple phenolic acids and flavonoids such as luteolin and apigenin) present next to alkaloids (e.g., berberine, palmatine, pronuciferine, pakistanamine, and magnoflorine), whereas Cretan barberry contained high quantities of isoquinoline alkaloids, such as an aporphine alkaloid-magnoflorine (which was found to be the major alkaloid in the extract), and protoberberines–berberine, jatrorrhizine, and palmatine, with a low contribution of phenolics (Figure S1) [11].

2.3. Purification of Magnoflorine and Berberine from Plant Extracts by CPC Chromatography

According to the rules of counter-current chromatography, the partition coefficient values (expressed as the peak area of a given peak in the upper phase chromatogram, divided by the peak area of the same peak in the lower phase chromatogram) should range around 0.5–2 to provide a successful separation [12]. The authors' idea was to construct a biphasic solvent system with satisfactory enough K-values to achieve a high-purity magnoflorine after a single injection of a total extract into the column. Because of a high polarity of the alkaloid, determined in the initial test tubes' trials with Arizona system solvents, a water-containing biphasic system had to be considered. Considering the above requirements, the following solvent system was selected for the purification of magnoflorine from Cretan barberry's root: ethyl acetate:butanol:water (0.6:1.5:3 $v/v/v$). It delivers the best separation conditions among all tested systems and provides a short settling time, which did not disturb the countercurrent separation at high rotation speeds.

The calculated K-values for the major components of the extract in the most favourable solvent system were the following: 0.6 (for magnoflorine), 8.9, 3.4, 2.8, 2.3, and 1.8.

Due to a higher affinity of magnoflorine to the aqueous (lower) phase, the purification was performed in the descending mode of operation, sustaining relatively high rotation speeds and a medium-speed flow rate. The remaining compounds (among them berberine, palmatine, and jatrorhizine) were eluted from the column together with a stationary phase. After the separation, the composition of every second fraction was analyzed on an HPLC chromatograph to join the samples with a similar content. In case of doubts, further injections were performed.

As a consequence, the optimized purification protocol resulted in the collection of magnoflorine from the eluate quickly—in the 35th minute, a purity of 91% directly from the crude extract was obtained (see Figure 3).

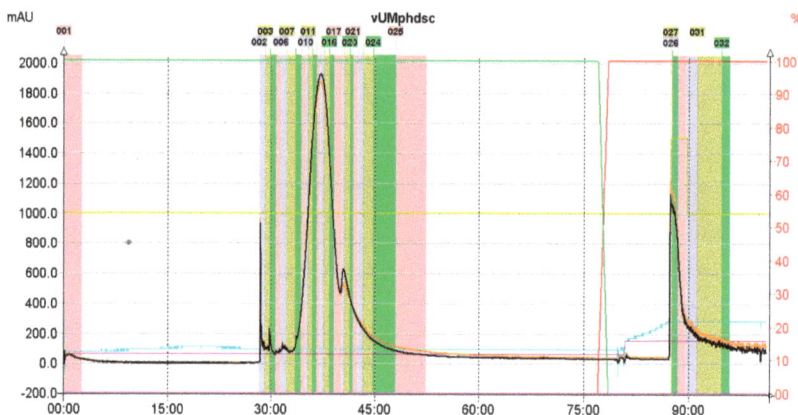

Figure 3. CPC chromatogram obtained from the separation of methanolic extract from Cretan barberry with a prominent magnoflorine peak in the 35th minute (above).

To increase the purity of magnoflorine, the resultant fractions containing the alkaloid were mixed, evaporated on a rotary evaporator at 45 °C and transferred into a column filled halfway with Sephadex LH-20, conditioned in a mixture of methanol:water (30:70 v/v). The residue containing magnoflorine was then washed with the same mixture of solvents. The collected fractions delivered enriched magnoflorine-containing fractions with a purity of 97.4% (see Figure S2).

2.4. Locomotor-Activity

2.4.1. The Influence of Berberine on the Locomotor Activity of Mice

One-way ANOVA analyses revealed that an acute i.p. administration of berberine (2.5–50 mg/kg) had a statistically significant effect on locomotion of mice ($F(5.47) = 15.70$; $p < 0.0001$). Indeed, the Tukey's test confirmed that injection of berberine significantly decreased locomotor activity of mice

($p < 0.05$ for the doses of 10 and 20 mg/kg; $p < 0.01$ for the dose of 50 mg/kg) as compared with the control vehicle-injected group (Figure 4).

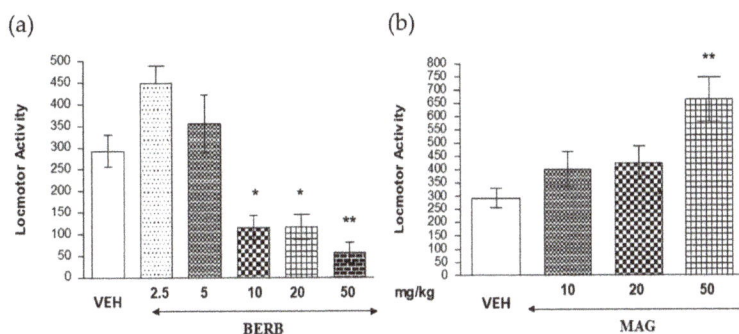

Figure 4. Effects of berberine (BERB) (**a**) and magnoflorine (MAG) (**b**) on the locomotor activity. The data are shown as the means ± standard error of the mean (SEM), photocell beam breaks of mice measured immediately after injection for 60 min; $n = 9$; * $p < 0.05$; ** $p < 0.01$ vs. vehicle (VEH)-treated control group; Tukey's test.

2.4.2. The Influence of Magnoflorine on the Locomotor Activity of Mice

One-way ANOVA revealed that an acute i.p. administration of magnoflorine (5–20 mg/kg) had a statistically significant effect on locomotion of mice ($F(3.31) = 5.821$; $p < 0.0032$). Indeed, Tukey's test confirmed that injection of magnoflorine at a dose of 50 mg/kg significantly increased the locomotor activity of mice ($p < 0.01$) as compared with that of the control vehicle-injected group (Figure 4).

2.5. Memory-Related Responses

2.5.1. The Influence of Berberine on the Short- and Long-Term Memory Acquisition in the Passive Avoidance (PA) Test in Mice

One-way ANOVA revealed that the administration of acute i.p. doses of berberine (2.5, 5, 10 and 20 mg/kg) had a statistically significant effect on latency index (LI) values for short-term memory acquisition ($F(4.41) = 2.770$; $p = 0.0413$) and long-term memory acquisition ($F(4.33) = 4.472$; $p = 0.0062$).

Indeed, treatment with berberine at the dose of 5 mg/kg significantly increased LI values in the PA test in mice compared with those in the vehicle-treated control group ($p < 0.05$, a post-hoc Tukey's test—for short-term memory acquisition (Figure 5a) and $p < 0.01$, a post-hoc Tukey's test, for long-term memory acquisition (Figure 5b), indicating that berberine, at the used dosage, improves the short-term as well as long-term acquisition of memory and learning.

2.5.2. The Influence of Magnoflorine on the Short- and Long-Term Memory Acquisition in the PA Test in Mice

The statistical analysis (one-way ANOVA test) showed that an acute i.p. administration of magnoflorine (5, 10, and 20 mg/kg) exhibited a statistically significant effect on the measured LI values calculated for the acquisition of both short-term ($F(3.30) = 3.491$; $p = 0.0292$) and long-term memory ($F(3.30) = 3.083$; $p = 0.0441$).

Indeed, treatment with magnoflorine at the dose of 20 mg/kg significantly increased LI values in the PA test in mice compared with those in the vehicle-treated control group ($p < 0.05$, a post-hoc Tukey's test—for short-term memory acquisition (Figure 6a) and for long-term memory acquisition (Figure 6b), indicating that magnoflorine at this used dose, improves the short-term as well as long-term acquisition of memory and learning.

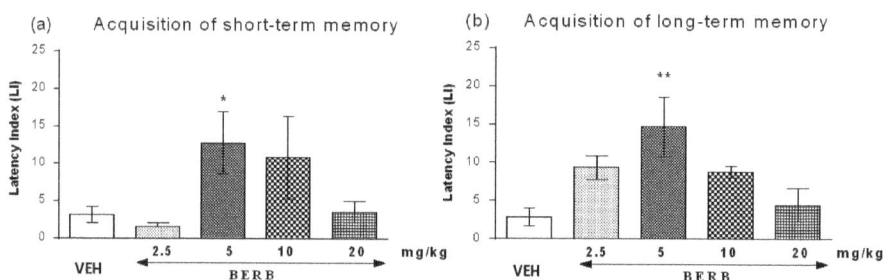

Figure 5. Effects of an acute berberine (BERB) on the latency index (LI) during the short-term or long-term acquisition trials using the PA test in mice. BERB (2.5, 5, 10 and 20 mg/kg; i.p.) or vehicle (VEH) were administered: 30 min before the first trial and animals were re-tested 2 h (for short-term memory (**a**)) or 24 h (for long-term memory (**b**)) later; n = 10–11; the means $\pm$ SEM; * $p < 0.05$; ** $p < 0.01$ vs. VEH-treated control group; Tukey's test.

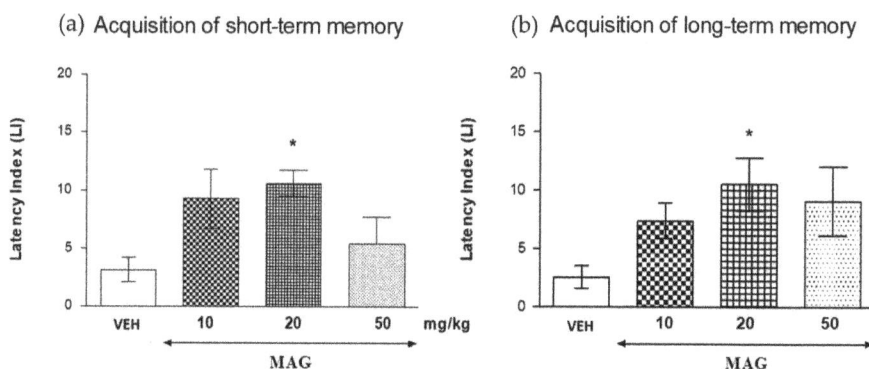

Figure 6. Effects of an acute magnoflorine (MAG) injection on the latency index (LI) during the short-term or long-term acquisition trial using the PA test in mice. MAG (10, 20 and 50 mg/kg; i.p.) or vehicle (VEH) were administered: 30 min before the first trial and animals were re-tested 2 h (for short-term memory (**a**)) or 24 h (for long-term memory (**b**)) later; n = 11–12; the means $\pm$ SEM; * $p < 0.05$ vs. VEH-treated control group; Tukey's test.

Next, based on the results obtained from the above pilot experiments, the non-effective (in PA test) dose of berberine (2.5 mg/kg), magnoflorine (10 mg/kg), as well as the effective (in PA test) dose of berberine (5 mg/kg) and magnoflorine (20 mg/kg) were then chosen for the next behavioral experiment, evaluating the influence of these plant compounds on the memory impairment, provoked by an acute injection of scopolamine (1 mg/kg), using the PA test in mice. The used doses of berberine or magnoflorine had no influence on the locomotion of the mice.

2.5.3. Influence of an Acute Administration of Noneffective and Effective Dose of Berberine on the Memory Impairment Induced by an Acute Administration of Scopolamine in the PA Test in Mice

For short-term memory acquisition, one-way ANOVA analyses revealed that there is a statistically significant effect on LI values ($F(5.49) = 5.635$; $p = 0.0004$). Post-hoc Tukey's test confirmed that berberine at a dose of 5 mg/kg significantly increased LI values in mice in the PA test in comparison with the vehicle/vehicle-treated mice, indicating that berberine at this dose improves the acquisition of long-term memory ($p < 0.05$). In addition, scopolamine at the dose of 1 mg/kg significantly decreased LI values in the PA test in comparison with the vehicle/vehicle treated mice, confirming an amnestic effect of this drug ($p < 0.05$) (Figure 7a). Berberine at any dose used did not influence scopolamine-induced disturbances in the acquisition of short-term memory in mice.

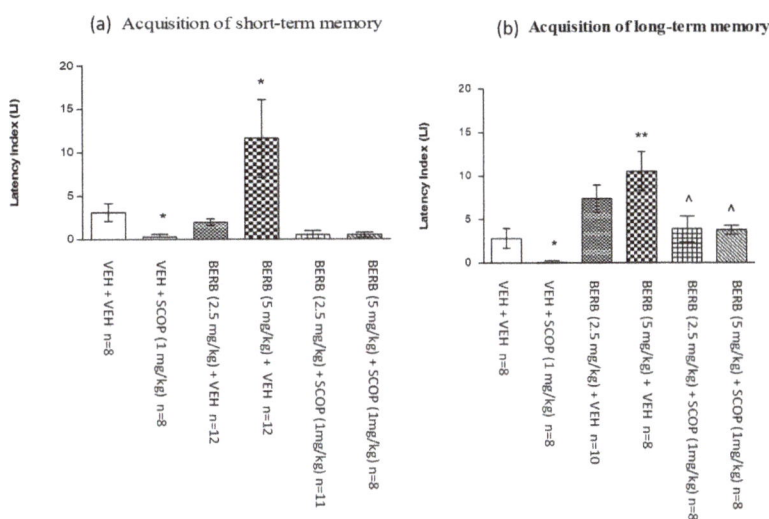

Figure 7. The influence of an acute administration of berberine (BERB) on the memory impairment induced by an acute injection of scopolamine (SCOP), expressed as latency index (LI) during the short-term (**a**) or long-term (**b**) memory acquisition using the PA test in mice. A non-effective dose of BERB (2.5 mg/kg), effective dose of BERB (5 mg/kg) or vehicle were administered 15 min before vehicle or SCOP (1 mg/kg) injection. All drugs were administered 15 min before the first trial and animals were retested 2 h (for short-term memory) or 24 h (for long-term memory) later; $n = 8$–12; the means ± SEM; * $p < 0.05$; ** $p < 0.01$ vs. VEH/VEH-treated group; ˆ $p < 0.05$ vs. VEH/SCOP (1 mg/kg)-treated group; Tukey's test.

For long-term memory acquisition, one-way ANOVA analyses revealed that there is a statistically significant effect on LI values ($F(5.45) = 7.237$; $p < 0.0001$). The post-hoc Tukey's test confirmed that berberine at the dose of 5 mg/kg significantly increased LI values in mice in the PA test in comparison to the vehicle/vehicle mice, indicating that berberine at this dose used improves acquisition of short-term memory ($p < 0.01$). Additionally, scopolamine at the dose of 1 mg/kg significantly decreased LI values in the PA test in comparison with the vehicle/vehicle treated mice, confirming an amnestic effect of this drug ($p < 0.05$). Furthermore, this amnestic effect provoked by an acute injection of scopolamine (1 mg/kg) was attenuated by non-effective (2.5 mg/kg) ($p < 0.05$) as well as by effective (5 mg/kg) ($p < 0.05$) dose of berberine in comparison with the vehicle/scopolamine (1 mg/kg)-treated mice (Figure 7b).

2.5.4. Influence of the Non-Effective and Effective dose of Magnoflorine on the Memory Impairment Induced by an Acute Administration of Scopolamine in the PA Test in Mice

For short-term memory acquisition, one-way ANOVA analyses revealed that there is a statistically significant effect on LI values ($F(5.46) = 14.35$; $p < 0.0001$). Post-hoc Tukey's test confirmed that magnoflorine at a dose of 20 mg/kg significantly increased LI values in mice in the PA test in comparison with the vehicle/vehicle mice, indicating that magnoflorine at this dose improves acquisition of short-term memory ($p < 0.001$). In addition, scopolamine at a dose of 1 mg/kg significantly decreased LI values in the PA test in comparison with the vehicle/vehicle treated mice, confirming an amnestic effect of this drug ($p < 0.05$) (Figure 8a). Similarily to berberine, magnoflorine did not influence scopolamine-induced disturbances in the acquisition of short-term memory in mice at any applied dose.

(a) Acquisition of short-term memory   (b) Acquisition of long-term memory

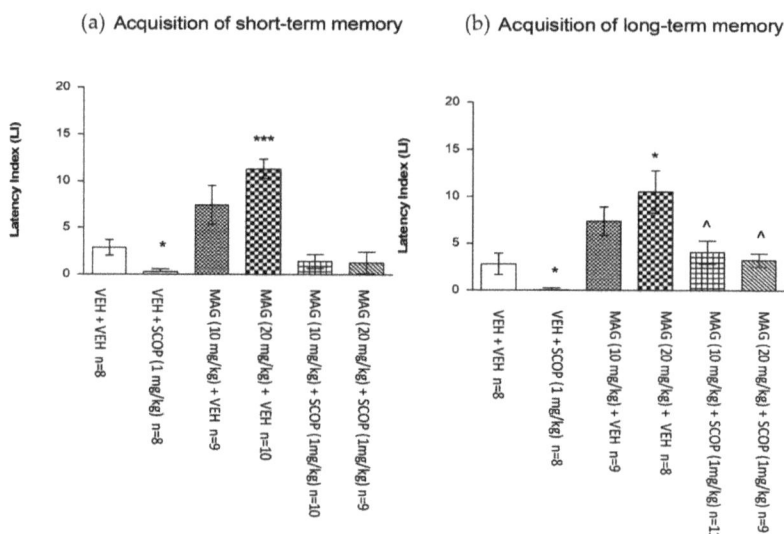

Figure 8. Influence of an acute administration of magnoflorine (MAG) on the memory impairment induced by an acute injection of scopolamine (SCOP), expressed as latency index (LI) during the short-term (**a**) or long-term (**b**) memory acquisition using the PA test in mice. A non-effective dose of MAG (10 mg/kg), effective dose of MAG (20 mg/kg) or vehicle were administered 15 min before vehicle or SCOP (1 mg/kg) injection. All drugs were administered 15 min before the first trial and animals were re-tested 2 h (for short-term memory) or 24 h (for long-term memory) later; $n = 8$–11; the means ± SEM; * $p < 0.05$; *** $p < 0.001$ vs. VEH/VEH-treated group; ^ $p < 0.05$ vs. VEH/SCOP (1 mg/kg)-treated group; Tukey's test.

For long-term memory acquisition, one-way ANOVA analyses revealed that there is a statistically significant effect on LI values ($F(5.46) = 7.557$; $p < 0.0001$). Post-hoc Tukey's test confirmed that magnoflorine at a dose of 20 mg/kg significantly increased LI values in mice in the PA test in comparison with the vehicle/vehicle mice, indicating that magnoflorine at this dose improves acquisition of long-term memory ($p < 0.05$). In addition, scopolamine at a dose of 1 mg/kg significantly decreased LI values in the PA test in comparison with the vehicle/vehicle treated mice, confirming an amnestic effect of this drug ($p < 0.05$). Furthermore, this amnestic effect provoked by an acute injection of scopolamine (1 mg/kg) was attenuated by a non-effective (10 mg/kg) ($p < 0.05$) as well as by an effective (20 mg/kg) ($p < 0.05$) dose of magnoflorine in comparison with the vehicle/scopolamine (1 mg/kg)-treated mice (Figure 8b).

3. Discussion

3.1. QSAR Studies

The QSAR used in studying the dependence between the structures and a wide range of compounds' properties, plays an important role in different fields of sciences including drug design, toxicology, etc. In a QSAR approach, a biological activity results from various compound–receptor interactions accompanying the transport of an active compound through biological membranes [13]. These interactions are assumed to be governed by the chemical structure characterized by lipophilic, steric and electronic parameters of a compound according to the Hansch approach. The QSAR also uses LSER as well as the empirical model of Abraham. This model describes different physicochemical processes in living organisms including blood–brain barrier permeability, which was evaluated for the sake of this study [14,15].

The Hansch approach [16,17] being the property–property relationship model [18–23], correlates the biological activity of a compound with its physicochemical properties by linear or nonlinear regression analysis.

The general model proposed by Hansch involves biological response of a substance, which can be expressed through physicochemical properties [18]:

$$\frac{d(BR)}{dt} = ae^{-\frac{(\log P - \log P_0)^2}{b}}(C)(k_1) \tag{7}$$

where:

BR—the biological response in a constant time interval;

C—the molar concentration of compound producing a standard response in a constant time interval;

k_1—the coefficient determined by the method of least squares;

a, b—the constants;

$\log P_{o/w}$—the logarithm of n-octanol/water partition coefficient of a derivative $d\log(1/C)/d\log P = 0$.

Biological activities of compounds are characterized by some basic intermolecular interaction forces: mainly steric, hydrophobic, and electronic effects. Therefore, the LSER model tends to describe different biological and physicochemical processes. The general LSER equation originally used by [24–26] is expressed as follows:

$$SP = c + vV + sS + bB + aA + eE \tag{8}$$

where *SP* is the dependent solute property in a given system. The independent variables are solute descriptors: *V* is the solute McGowan volume in units of $cm^3 \ mol^{-1}/100$, *S* is the polarizability/dipolarity, *B* is the overall hydrogen–bond basicity, *A* is the overall hydrogen–bond acidity, and *E* is an excess molar refraction. The coefficients: *v, s, b, a, e* reflect the differences in the two phases between which the compound is transferred.

In the presented paper, the QSAR studies were applied to determine the BBB penetration of both alkaloids, which was expressed as the logarithm of the ratio between the brain and blood concentrations of the tested substances [13]:

$$\log \ BB = \log(conc. \ in \ brain/conc. \ in \ blood) \tag{9}$$

To investigate BBB permeability more thoroughly, the most important pharmacokinetic parameters of brain, namely, the BBB permeability–surface area product (PS), usually given as the logPS, the brain–plasma equilibration rate (logPS*$f_{u,brain}$) and the fractions unbound in plasma and brain have been calculated (with ACD/Percepta software). The obtained BBB permeability parameters as well as LSER ones were collated with the most important physicochemical descriptors (Table 1).

In addition, other isoquinoline alkaloids tested, which belonged to the group of aporphines and protoberberines, were found to exhibit the ability to penetrate the BBB. These properties may depend on physicochemical characteristics of a molecule i.e., lipophilicity, ionization state, molecular volume, molecular mass, and so on. These characteristics are important determinants of distribution and potential accumulation of a substance within brain tissues. Physicochemical characteristics can impact cellular and subcellular distribution e.g., entry into the CNS via the BBB [27].

The performed calculations clearly showed that both selected alkaloids—berberine and magnoflorine—are prone to penetrate the BBB and exhibit some pharmacological properties. These findings let the authors continue the studies on the purification and testing of the selected alkaloids.

However, it must be stated that the performance of all newly-constructed QSAR models depends on the availability and quality of experimental in vivo data, which are, in fact, rarely available. In vivo datasets are dependent on various factors like solubility and stability of a compound, detection technique of BBB penetration, dose and compound formulation [28]. To apply QSAR methodology, chemicals located within the applicability domain of the model need to have an appropriate number of measured experimental values of certain biological activity (in our case: BBB permeability). Therefore, in our study, the limitation of the models that we proposed should be pointed out. Among them, unavailability of sufficiently large datasets of both logBB and logPS values are worth mentioning. In addition, it is necessary to remember that logBB values reflect also processes other than BBB permeability including plasma protein binding, tissue binding, local metabolism and others [29]. As regards logPS values, being considered the most distinctive in vivo measure of BBB permeability, it should be emphasized that logPS does not provide information on the substance concentration in the brain [28]. Due to the complex nature of the brain penetration process, it is required to treat in silico obtained values as a kind of modeling assumption. Nevertheless, our in vivo findings that both compounds provide anti-amnesic and pro-cognitive properties suggest that they both cross the BBB as our QSAR model predicted.

3.2. The Purification of Alkaloids by CPC Chromatography

During the presented research, both magnoflorine and berberine were successfully purified using counter-current chromatography in sufficient quantities for behavioral studies and a purity exceeding 90%.

A proper separation was influenced by a number of parameters. The most important of them included a good selection of solvent system, but also proper column rotation speed and flow rate settings, which provided an equilibrium between the stationary and mobile phase volumes on the column (good retention of the stationary phase) [30]. To a large extent, the latter parameter depends on how soon after mixing the selected solvent system reaches equilibrium again and creates two separate, immiscible layers [31].

For Cretan barberry, the results of this study clearly indicated that magnoflorine was strongly distributed in the mobile aqueous phase, contrary to other alkaloids present in the extract. This is why it could be easily obtained purified, directly from the crude extract. In addition, the applied parameters enabled high sample loads (1 g of an injected extract). Berberine collected from Siberian barberry aerial parts was washed out in a pH-zone refining mode counter-current chomatography, according to a previous publication of the authors [11]. This mode of operation characterized by the addition of a base to organic phase and an acid to water phase uses the ability of alkaloids to exist in two forms—a nonpolar base and a polar salt, which helps in their purification from other metabolites.

3.3. In Vivo Studies

As mentioned in the Introduction section, the tested compounds had been previously found to be acetylcholinesterase inhibitors in in vitro tests [1]. The aim of the study was to assess the influence of berberine and magnoflorine on the acquisition of memory in PA test. In addition, their potency against memory disturbances provoked by scopolamine was tested.

First, the initial locomotor activity tests were performed using a variety of berberine and magnoflorine dosages to well determine the highest non-active and the lowest active dosages. For this purpose, 2.5, 5, 10, 20 and 50 mg/kg (i.p.) of berberine and 5, 10, 20, 50 mg/kg (i.p.) of magnoflorine were administered and the animals reactions were measured in the actometer cages. Based on the results of locomotor activity tests, the doses that did not affect the locomotor activity of animals were selected, to be sure, that no disturbance would affect the results of memory tests. The applied PA test was finally conducted on two selected lowest doses: 2.5 and 5 mg/kg (i.p.) of berberine, and 10 and 20 mg/kg (i.p.) of magnoflorine.

Other studies on berberine showed its memory enhancing activity when administered to rats or mice [13]; however, the majority of them were performed using higher doses of berberine, for example, 10, or even 20 mg/kg (i.p.) [7,31]. This matter concerns the authors as this dosage seems to be too high; in this study, the 10 or 20 mg/kg berberine aggravated the locomotor activity of the tested mice [32].

PA test performed for berberine clearly demonstrated that already a single injection of effective (5 mg/kg) and non-effective (2.5 mg/kg) dose showed the memory enhancing effects; however, the response was stronger for the long-term memory than for the short-term memory after one injection.

With the influence of scopolamine in all the tested groups, the animals were found to be excessively excited, which was reflected in an increased locomotor activity that does not influence cognitive performance. This behavior is induced by the ability of scopolamine to block muscarinic receptors and, based on this ability, evoke disturbed strategy elaboration and place recognition in animals. Those results are consistent with some previous experiments of the authors [33].

The dose of 5 mg/kg of berberine was able to overcome the action of the co-administered scopolamine in the scopolamine model of memory impairment. However, higher doses of this alkaloid did not increase the LI values, which may be connected with the disturbances with previously studied locomotor activity.

Berberine, which had been proven to exhibit memory enhancing properties, was treated as a model compound for the initial tests of the other compound—magnoflorine. This aporphine alkaloid was found less toxic and did not interfere with the locomotor activity of mice at a lower non-effective dose of 10 mg/kg or a higher effective dose of 20 mg/kg, which may indicate that the alkaloid is well tolerated at these selected doses.

Magnoflorine at a dose of 20 mg/kg, improved the cognitive processes of short and long term memory in mice in the PA test significantly, as it increased the passage time of animals to the dark room during the applied test. The animals associated the dark room with a negative electric stimulus, which explains their prolonged stay in a bright room and unwillingness to enter the dark one. Only a single dose of this alkaloid (20 mg/kg, i.p.) improved the cognitive function, assessed on the same day— 2 h after the training session but also 24 h after the test, which confirmed the effect of magnoflorine on both short- and long-term memory processes and prevented the animals from entering the dark compartment. A dosage of 10 mg/kg exhibited a similar pro-cognitive tendency; however, the obtained results were not statistically significant. The effect of this dose might be observed in the long-term administration studies, which will be performed by the authors in the near future. In addition, this compound was found to reverse the scopolamine-induced long-term memory impairment, both at a dose of 10 and 20 mg/kg, compared with the control group, which demonstrates the anti-amnestic properties of magnoflorine.

Considering the effect of magnoflorine on acetylcholinesterase assessed earlier [1] and the current behavioral studies, it may be suggested that the effect of cholinergic transmission and related cognitive processes is strongly influenced by this compound. Although magnoflorine may act as an inhibitor of scopolamine-induced memory effects, its probable mechanism of action may be based on the inhibition of acetylcholinesterase. Even though these mechanisms of action are probable, the authors cannot exclude other activities of this compound.

This is the first time the procognitive activity of this compound was evaluated. The obtained results confirm the proprioceptive effects of magnoflorine and qualify it for further pharmacological studies, as, possibly in the future, it could be perceived as a drug candidate directed to the therapeutical strategies of various diseases progressing with memory deficits such as dementia, or AD.

An interesting phenomenon was observed for the tested compounds. Their influence on long-term memory was stronger than on the short-term memory. This might have been due to the actual activity of their metabolites, which appear within a longer time. According to Wang and co-investigators [34], berberine forms a variety of metabolites in vivo during its demethylation, demethylenation, reduction, hydroxylation and conjugation in vivo. In fact, the metabolites present in the forms of glucuronides or sulfates are more abundant in blood samples and more active than the original compound. A similar occurrence might be observed for magnoflorine.

4. Methods

4.1. Chemicals

All gradient grade solvents were obtained from Avantor Performance Materials (Center Valley, PA, USA). Chromatographic and spectroscopic grade solvents used in the quality assessment procedures: acetonitrile, water, acetic acid and formic acid were produced by J.T. Baker (Center Valley, PA, USA). Scopolamine was purchased from Sigma Aldrich (St. Louis, MO, USA).

4.2. Plant Material

The root of *Berberis cretica* and the overground parts of *Berberis siberica* were used for the purification of alkaloids for the bioactivity studies. The former was collected and identified by Dr. Eleftherios Kalpoutzakis from the University of Athens in the central part of Crete Island in the fall of 2007. The latter species was obtained and identified by Dr. Ottonbataar Urjin in September of 2015 in the Bayan Province in Mongolia. The gathered plant material was roughly cut, dried at a temperature of 40 °C, later ground in a mortar and directed to extraction. Voucher specimens of both species are stored in the Chair and Department of Pharmacognosy of the Medical University of Lublin, Poland.

4.3. Extraction of Plant Material

The extraction of plant material was performed by accelerated solvent extraction in 100 mL stainless steel vessels on 150 g of powdered roots of both species, according to the previously published methods [1,28,35], as described in the Supplementary file.

4.4. Identification of Major Constituents of the Extracts by HPLC and HPLC-MS

The composition of both extracts and fractions obtained in the separation process was constantly monitored to provide data on the purification efficiency of the applied techniques. First, the major secondary metabolites—alkaloids and phenolics—present in the extracts were recognized in a high-resolution mass spectrometry analysis, according to the previous studies of the authors [11,35]. Later, based on the obtained HPLC-MS data, the purity control was performed by HPLC for every second fraction from CPC purification. Similar fractions were joined. In addition, HPLC chromatograms delivered suitable information for the calculations of partition coefficient values (*K*) in the optimization of the counter-current separation method. Detailed chromatographic and spectrometric conditions are presented in the Supplementary file. They are based on the previous studies of authors [35].

4.5. Purification of Alkaloids by Means of Hydrostatic Counter-Current Chromatography

Centrifugal partition chromatography (CPC) was applied in the separation process of two alkaloids—berberine and magnoflorine. The former was purified from *Berberis siberica* overground parts according to the previously published method [11], whereas the latter was from the powdered root of *Berberis cretica* in an optimized separation protocol described in the scientific literature for the first time.

4.5.1. Purification of Magnoflorine

The CPC separation was performed using an Armen SCPC-250-L instrument (Brittany, France), equipped in a 250 mL rotor, a quaternary pump, a UV detector (DAD 600 Flesh06S), and an automatic fraction collector (LS-5600). Prior to the analysis, a mixture of solvents was freshly prepared in a large separating funnel. The upper and lower phases were collected to separate amber glass bottles. The following solvent system was selected for the fractionation of Cretan barberry's root extract: ethyl acetate:butanol:water (0.6:1.5:3 *v*/*v*/*v*), which was pumped into the rotating column in the descending mode at the rotation speed of 1300 rpm. The solvent flow was set at 6 mL/min until the end of the process—for 150 min. After 90 min, the stationary phase was pumped in the same operation mode to elute the components with a higher affinity to the stationary phase. In addition, 1 g of extract dissolved in 5 mL of a 50:50 mixture of upper and lower phases was introduced immediately after the column was filled with stationary phase. Subsequent injections were made one by one, without stopping the column's rotations.

4.5.2. Separation of Berberine

Berberine was obtained for the study in a pH-refining mode of CPC chromatography according to the previously described protocol [11], using a biphasic solvent system composed of methyl-*tert*-buthyl ether (MtBE) and water (1:1 *v*/*v*) with the addition of modifiers—10 mM of HCl and 10 mM of trimethylamine (TEA) into the aqueous and organic phases, respectively.

The separation was carried out in an ascending mode in the flow rate of 5 mL/min, rotation speed of 1050 rpm within 220 min.

4.6. Animals

A group of naïve male Swiss mice were used in the scheduled experiments (Farm of Laboratory Animals, Warszawa, Poland). Each of them weighted ca. 20–30 g. Standard laboratory conditions were applied to maintain the animals (12-h light/dark cycle, controlled room temperature at $21 \pm 1\,°C$). They were provided free access to laboratory chow (Agropol, Motycz, Poland) and tap water. Before the study, they were acquainted with their home cages for a period of one week. Each experimental group contained 8–12 animals. The behavioral experiments were conducted regularly between 8:00 and 15:00, according to the National Institute of Health Guidelines for the Care and Use of Laboratory, and also based on the directions of Animals and to the European Community Council Directive for the Care and Use of laboratory animals of 22 September 2010 (2010/63/EU). The performed tests were first approved by the Ethics Committee at Medical University of Lublin, Poland.

4.7. Drugs

The compounds tested were:

Berberine (2.5, 5, 10, 20, 50 mg/kg) from methanolic extract of *Berberis siberica* herb, and magnoflorine (5, 10, 50 mg/kg)—obtained from methanolic extract of *Berberis cretica* root.

Scopolamine hydrochloride (1 mg/kg) was produced by Sigma-Aldrich (St. Louis, MO, USA).

Magnoflorine, berberine and scopolamine used for the study were dissolved in 0.9% NaCl with an addition of 0.2% of DMSO (vehicle). The tested compound was administered i.p. at a constant volume of 10 mL/kg. Each day prior to the experimentation, the drug solutions were prepared. In parallel, a control group of mice was given i.p. injections of saline with an addition of 0.2% DMSO (later referred to as vehicle), at the very same volume as tested drugs.

4.8. Experimental Procedures

Experimental doses of drugs used and procedures were selected based on the literature data [6,7] and our preliminary studies. For each stage of memory as well for both short-term and long-term memory, independent and different groups of mice were used.

4.8.1. Locomotor Activity

The locomotor activity test was recorded for each tested animal in a separate actometer cage (Multiserv, Lublin, Poland; 32 cm in diameter, two light beams) in a low-noise room. In the apparatus, two separate photocell beams, which were located perpendicularily to each other, measured the movements of animals [33]. Horizontal locomotor activity was measured immediately after a single injection of tested drugs, berberine (2.5–50 mg/kg, i.p.), or magnoflorine (5–20 mg/kg, i.p.), scopolamine (1 mg/kg, i.p.) or vehicle for the control group, immediately after the last injection for 60 min.

4.8.2. Memory Related Responses

The applied PA test measured the memory-related responses in the apparatus, which contained two compartments—a lightened one (10 × 13 × 15 cm) and a darkened one (25 × 20 × 15 cm). Fluorescent light (8 W) was illuminating the light chamber. The dark compartment contained an electric grid floor. The entrance of an animal triggered an electric foot shock for 2 s (0.2 mA).

First, a pre-test was arranged to acknowledge the animals with the apparatus. Each mouse was individually introduced to the PA apparatus and allowed to explore the bright compartment for 30 s. Later, a guillotine was raised to let the animals escape to the dark room. When in the dark room, the guillotine door was closed and immediately an electric foot shock at the duration of 2 s (0.2 mA) was applied on the grid floor. The time measured until the opening of the guillotine door to the entrance of an animal to the dark compartment was measured and calculated as latency time (TL1). If the time was longer than 300 s, the experiment was stopped and the animal was transferred to the dark compartment and delivered an electric shock. The time of 300 s was put down in this case.

Then, another trial was conducted and called retention. After a 30 s time for adaptation, the guillotine door were opened in the PA apparatus and the time needed for the animal to enter the dark compartment was measured as TL2. No foot-shock was given in this trial. In addition, 300 s were noted for the animals, which did not move from the light compartment to the dark one.

The PA test enabled the measurement of both short-term and long-term memory, based on the time passed from the pre-test to the actual retention test. Short-term memory was assessed, when the trial was performed 2 h after the training test. Latency time measured after 24 h delivered information on the long-memory process assessment. In both cases, the drugs were administered before the pre-test, and they were expected to interfere with the physiological acquisition of information [33].

The first step of experiment was designed to estimate the influence of tested drugs, berberine and magnoflorine on the acquisition of short- as well as long-term memory in mice, using the PA test.

Berberine (2.5–20 mg/kg, i.p.), magnoflorine (5–20 mg/kg, i.p.), or vehicle, for the control group, were administered 30 min before the first trial and re-tested after 2 h (short-term memory) or after 24 h (long-term memory).

Finally, we evaluated the impact of the combining administration of tested drugs, berberine and magnoflorine on the memory impairment induced by an acute administration of scopolamine.

Berberine (2.5–20 mg/kg, i.p.), magnoflorine (5–20, i.p.) or vehicle, for the control group, were administered 15 min before injection of scopolamine (1 mg/kg, i.p.) or vehicle. Fifteen minutes after the last injection, the mice were tested in the PA test (first trial) and re-tested 2 h (short-term memory) or 24 h (long-term memory) later.

4.9. Statistical Analysis

The statistical analysis was performed using one-way analysis of variance (ANOVA). Post-hoc comparison of means was carried out with the Tukey's test for multiple comparisons when appropriate. The data were considered statistically significant at a confidence limit of $p < 0.05$. ANOVA analysis with a Tukey's post-hoc test were performed using GraphPad Prism version 5.00 for Windows (GraphPad Software, San Diego, CA, USA). Moreover, to check the usefulness of newly constructed QSAR models, the Principal Components Analysis (PCA) has been done (Minitab 17).

For the locomotor activity, the number of photocell beam breaks was measured for 60 min and put down [33].

For the memory-related behaviors, the results obtained in the PA test were expressed as latency index (LI). It is expressed as the difference between the time measured in the second test (retention test) and the previous test (pre-test, training test) for each animal: $LI = TL2 - TL1/TL1$.

TL1—the time needed to enter the dark compartment in the training session (pre-test); TL2—the time needed to re-enter the dark part of the apparatus during the retention test [36].

5. Conclusions

Berberine and magnoflorine were successfully isolated from two selected barberry species by centrifugal partition chromatography at high purity. They were found to exhibit antiamnestic and cognition enhancing properties in the PA test on mice at the doses of 5 and 20 mg/kg, respectively, towards long-term and short-term memory. No impact on the locomotor activity of mice in the studied dosage was noted, which shows that the selected doses were within the therapeutic windows of both alkaloids. The possible mechanism of action of these two compounds may be related to the inhibition of acetylcholinesterase, as it has been proven in the earlier in vitro studies of the authors. In addition, both compounds block long-term but not short-term memory impairment affected by scopolamine.

Supplementary Materials: Supplementary materials can be found at www.mdpi.com/1422-0067/18/12/2511/s1.

Acknowledgments: The manuscript was partially supported by the National Centre for Research and Development—project number: 4/POLTUR-1/2016.

Author Contributions: The study was designed by Wirginia Kukula-Koch, Katarzyna Stępnik, Marta Kruk-Słomka and Grażyna Biała. The analytical part was performed by Wirginia Kukula-Koch (counter-current separation and HPLC-MS analysis), Marta Kruk-Słomka (in vivo experiments), Katarzyna Stępnik (QSAR studies), and Radosław Szalak (in vivo experiments). Data were analyzed by all authors. The manuscript was written by Wirginia Kukula-Koch, Marta Kruk-Słomka and Katarzyna Stępnik in consultation with all co-authors. All authors accepted the final version of the manuscript.

Abbreviations

AChE	Acetylcholinesterase
AD	Alzheimer's Disease
ASE	Accelerated Solvent Extraction
BBB	Blood–Brain Barrier
CC	Column Chromatography
CNS	Central Nervous System
CPC	Centrifugal Partition Chromatography
ESI	Electrospray Ionization
HPLC	High Performance Liquid Chromatography
i.p.	Intraperitoneally
iNOS	Nitric Oxide Synthase
K	Partition Coefficient
LI	Latency Index
LSER	Linear Solvation Energy Relationship
MS	Mass Spectrometry
PA	Passive Avoidance Test
QSAR	Quantitative Structure–Activity Relationship
TEA	Trimethylamine
TL	Latency Time
TPSA	Topological Polar Surface Area

References

1. Kukula-Koch, W.; Mroczek, T. Application of hydrostatic CCC-TLC-HPLC-ESI-TOF-MS for the bioguided fractionation of anticholinesterase alkaloids from *Argemone mexicana* L. roots. *Anal. Bioanal. Chem.* **2005**, *407*, 2581–2589. [CrossRef] [PubMed]

2. Argyropoulou, A.; Aligiannis, N.; Trougakos, I.P.; Skaltsounis, A.L. Natural compounds with anti-ageing activity. *Nat. Prod. Rep.* **2013**, *30*, 1412–1437. [CrossRef] [PubMed]

3. Libro, R.; Giacoppo, S.; Rajan, T.S.L.; Bramanti, P.L.; Mazzon, E. Natural phytochemicals in the treatment and prevention of dementia: An overview. *Molecules* **2016**, *21*, 518. [CrossRef] [PubMed]

4. Chen, C.C.; Hung, T.H.; Lee, C.Y.; Wang, L.F.; Wu, C.H.; Ke, C.H.; Chen, S.F. Berberine protects against neuronal damage via supression of glia-mediated inflammation in traumatic brain injury. *PLoS ONE* **2014**, *9*, e115694. [CrossRef] [PubMed]

5. Hsieh, M.T.; Peng, W.H.; Wu, C.R.; Wang, W.H. The ameliorating effect of the cognitive-enhancing chinese herbs on scopolamine-induced amnesia in rats. *Phytother. Res.* **2000**, *14*, 375–377. [CrossRef]

6. Zhu, F.; Qian, C. Berberine chloride can ameliorate the spatial memory impairment and increase the expression of interleukin-1β and inducible nitric oxide synthase in the rat model of Alzheimer's disease. *BMC Neurosci.* **2006**, *7*, 78–87. [CrossRef] [PubMed]

7. Han, A.M.; Heo, H.; Kwon, Y.K. Berberine promotes axonal regeneration in injured nerves of the peripheral nervous system. *J. Med. Food* **2012**, *15*, 413–417. [CrossRef] [PubMed]

8. Schiff, P.L. The Thalictrum Alkaloids: Chemistry and Pharmacology. In *Alkaloids: Chemical and Biological Perspectives*; Pelletier, S.W., Ed.; Pergamon: Oxford, UK, 1996; Volume 11, pp. 1–237, ISBN 0080427979.

9. Angelis, A.; Hamzaoui, M.; Aligiannis, N.; Nikou, T.; Michailidis, D.; Gerolimatos, P.; Termentzi, A.; Hubert, J.; Halabalaki, M.; Renault, J.H.; et al. An integrated process for the recovery of high added-value compounds from olive oil using solid support free liquid-liquid extraction and chromatography techniques. *J. Chromatogr. A* **2017**, *1491*, 126–136. [CrossRef] [PubMed]

10. Lipinski, C.A.; Lombardo, F.; Dominy, B.W.; Feeney, P.J. Experimental and computational approaches to estimate solubility and permeability in drug discovery and development settings. *Adv. Drug Deliv. Rev.* **2001**, *46*, 3–26. [CrossRef]

11. Kukula-Koch, W.; Koch, W.; Angelis, A.; Halabalaki, M.; Aligiannis, N. Application of pH-zone refining hydrostatic countercurrent chromatography (hCCC) for the recovery of antioxidant phenolics and the isolation of alkaloids from Siberian barberry herb. *Food Chem.* **2016**, *203*, 394–401. [CrossRef] [PubMed]

12. Berthod, A.; Hassoun, M.; Ruiz-Angel, M.J. Alkane effect in the Arizona liquid systems used in countercurrent chromatography. *Anal. Bioanal. Chem.* **2005**, *383*, 327–340. [CrossRef] [PubMed]

13. Berthod, A.; García-Alvarez-Coque, M.C. *Micellar Liquid Chromatography*; Marcel Dekker: New York, NY, USA, 2000.

14. Platts, J.A.; Abraham, M.H.; Zhao, Y.H.; Hersey, A.; Ijaz, L.; Butina, D. Correlation and prediction of a large blood-brain distribution data set—An LFER study. *Eur. J. Med. Chem.* **2001**, *36*, 719–730. [CrossRef]

15. Abraham, M.H. The factors that influence permeation across the blood-brain barrier. *Eur. J. Med. Chem.* **2004**, *39*, 235–240. [CrossRef] [PubMed]

16. Hansch, C.; Maloney, P.P.; Fujita, T.; Muir, R.M. Correlation of biological activity of phenoxyacetic acids with hammett substituent constants and partition coefficients. *Nature* **1962**, *194*, 178–180. [CrossRef]

17. Hansen, O.R. Hammett series with biological activity. *Acta Chem. Scand.* **1962**, *16*, 1593–1600. [CrossRef]

18. Hansch, C. Quantitative approach to biochemical structure-activity relationships. *Acc. Chem. Res.* **1969**, *2*, 232–239. [CrossRef]

19. Hansch, C. Quantitative Structure-Activity Relationships. In *Drug Design*; Ariens, E.J., Ed.; Academic Press: New York, NY, USA, 1971; Volume 1, ISBN 0120603063.

20. Hansch, C. Quantitative structure-activity relationships and the unnamed science. *Acc. Chem. Res.* **1993**, *26*, 147–153. [CrossRef]

21. Hansch, C.; Fujita, T. p-σ-π analysis. A Method for the correlation of biological activity and chemical structure. *J. Am. Chem. Soc.* **1964**, *86*, 1616–1626. [CrossRef]

22. Hansch, C.; Quinlan, J.E.; Gary, L.L. Linear free-energy relationship between partition coefficients and the aqueous solubility of organic liquids. *J. Org. Chem.* **1968**, *33*, 347–350. [CrossRef]

23. Kubinyi, H. *QSAR: Hansch Analysis and Related Approach, Methods and Principles in Medicinal Chemistry*; VCH Publishers: New York, NY, USA, 1961; Volume 1, pp. 1–240, ISBN 978–3527300358.

24. Abraham, M.H.; Chadha, H.S.; Mitchell, R.C. Hydrogen bonding. 33. Factors that influence the distribution of solutes between blood and brain. *J. Pharm. Sci.* **1994**, *83*, 1257–1268. [CrossRef] [PubMed]

25. Abraham, M.H.; Chadha, H.S.; Martins, F.; Mitchell, R.C.; Bradbury, M.W.; Gratton, J.A. Hydrogen bonding part 46: A review of the correlation and prediction of transport properties by an LFER method: Physicochemical properties, brain penetration and skin permeability. *Pest Manag. Sci.* **1999**, *55*, 78–88.

26. Abraham, M.H.; Ibrahim, A.; Zissimos, A.N. Determination of sets of solute descriptors from chromatographic measurements. *J. Chromatogr. A* **2004**, *1037*, 29–47. [CrossRef] [PubMed]

27. Chico, L.K.; Van Eldik, L.J.; Watterson, D.M. Targeting protein kinases in central nervous system disorders. *Nat. Rev. Drug Discov.* **2009**, *8*, 892–909. [CrossRef] [PubMed]

28. Vastag, M.; Keserű, G. Current in vitro and in silico models of blood-brain barrier penetration: A practical view. *Curr. Opin. Drug Discov. Dev.* **2009**, *12*, 115–124.

29. Bickel, U. How to measure drug transport across the blood-brain barrier. *Neurotherapeutics* **2005**, *2*, 15–26. [CrossRef] [PubMed]

30. Garrard, I. Simple approach to the development of a CCC Solvent Selection Protocol Suitable for Automation. *J. Liq. Chromatogr. Relat. Technol.* **2005**, *28*, 1923–1935. [CrossRef]

31. Marston, A.; Hostettmann, K. Separation and quantification of flavonoids. In *Chemistry, Biochemistry and Application*; Oyvind, M.A., Markham, K.R., Eds.; CRC Press Taylor and Francis Group: Boca Raton, FL, USA, 2006; pp. 2–20, ISBN 9780849320217.

32. Zhang, Z.; Li, X.; Li, F.; An, L. Berberine alleviates postoperative cognitive dysfunction by suppressing neuroinflammation in aged mice. *Int. Immunopharmacol.* **2016**, *38*, 426–433. [CrossRef] [PubMed]

33. Kruk-Słomka, M.; Budzyńska, B.; Biała, G. Involvement of cholinergic receptors in the different stages of memory measured in the modified elevated plus maze test in mice. *Pharmacol. Rep.* **2012**, *64*, 1066–1080. [CrossRef]

34. Wang, K.; Feng, X.; Chai, L.; Cao, S.; Qiu, F. The metabolism of berberine and its contribution to the pharmacological effects (Review). *Drug Metab. Rev.* **2017**, *49*, 139–157. [CrossRef] [PubMed]

35. Kukula-Koch, W.; Aligiannis, N.; Halabalaki, M.; Skaltsounis, A.L.; Glowniak, K.; Kalpoutzakis, E. Influence of extraction procedures on phenolic content and antioxidant activity of Cretan barberry herb. *Food Chem.* **2013**, *138*, 406–413. [CrossRef] [PubMed]

36. Bejar, C.; Wang, R.H.; Weinstock, M. Effect of rivastigmine on scopolamine-induced memory impairment in rats. *Eur. J. Pharmacol.* **1999**, *383*, 231–240. [CrossRef]

International Journal of
Molecular Sciences

MDPI

Review

Antidiabetic Potential of Monoterpenes: A Case of Small Molecules Punching above Their Weight

Solomon Habtemariam

Pharmacognosy Research Laboratories & Herbal Analysis Services, University of Greenwich, Central Avenue, Chatham-Maritime, Kent ME4 4TB, UK; s.habtemariam@herbalanalysis.co.uk;
Tel.: +44-208-331-8302 or +44-208-331-8424

Received: 14 November 2017; Accepted: 18 December 2017; Published: 21 December 2017

Abstract: Monoterpenes belong to the terpenoids class of natural products and are bio-synthesized through the mevalonic acid pathway. Their small molecular weight coupled with high non-polar nature make them the most abundant components of essential oils which are often considered to have some general antioxidant and antimicrobial effects at fairly high concentrations. These compounds are however reported to have antidiabetic effects in recent years. Thanks to the ingenious biosynthetic machinery of nature, they also display a fair degree of structural complexity/diversity for further consideration in structure-activity studies. In the present communication, the merit of monoterpenes as antidiabetic agents is scrutinized by assessing recent in vitro and in vivo studies reported in the scientific literature. Both the aglycones and glycosides of these compounds of rather small structural size appear to display antidiabetic along with antiobesity and lipid lowering effects. The diversity of these effects vis-à-vis their structures and mechanisms of actions are discussed. Some key pharmacological targets include the insulin signaling pathways and/or the associated PI3K-AKT (protein kinase B), peroxisome proliferator activated receptor-γ (PPARγ), glucose transporter-4 (GLUT4) and adenosine monophosphate-activated protein kinase (AMPK) pathways; proinflammatory cytokines and the NF-κB pathway; glycogenolysis and gluconeogenesis in the liver; glucagon-like-1 receptor (GLP-1R); among others.

Keywords: monoterpenes; diabetes; obesity; multiple mechanisms

1. Introduction

Several estimates on the current level of diabetes and its projected cases in the next few decades have been made in recent years. According to the World Health Organization's (WHO), there were 422 million cases in 2014 with 8.5% prevalence that rose from 4.7% in 1980 [1]. Similarly, the major risk factor of diabetes has been recognized as obesity which prevalence in 2014 was 600 million; while the number of adults (18 years and older) reported as overweight were more than 1.9 billion [2]. As we are living in an era where more of the world population are suffering from consequences of excess calories than undernutrition, the cost of diabetes and obesity will remain the major social burden to societies in the next few generations. To date, diabetes is already a direct cause of death for millions of people annually but its impact is even more severe in its association with other diseases such as cardiovascular complications, organ failure (e.g., kidney) and disabilities, such as blindness and limb amputations [1].

There is as yet no drug of cure for diabetes and the current therapeutic approaches are directed on the management of the disease, primarily on glycemic control. The disease itself, also called diabetes mellitus, is manifested when there is persistent hyperglycemia in the blood arising from either insufficient (or not at all) amount of insulin released from the pancreatic β-cells and/or resistance to insulin is developed by vital organs. The most uncomplicated diabetes case is type-1 diabetes (T1D) were pancreatic β-cells are destroyed through an autoimmune-mediated reaction. On the other hand, type-2 diabetes (T2D) is characterized by insulin resistance with major risk factors being age and obesity. The disease is complex and may involve impaired insulin secretion and β cell death along with various metabolic dysregulations.

Since its discovery in the 1920's, insulin has been taking the center stage in the therapy of both T1D and T2D [3,4]. Other classes of antidiabetic drugs include the α-glucosidase inhibitors (e.g., acarbose, miglitol and voglibose) that target carbohydrate digestion in the gut thereby limiting the availability of glucose taken up by the blood; the biguanides (the metformin-like compounds) that suppress glucose release/production in the liver; the thiazolidinediones such as glitazones which increase the sensitivity of insulin target organs; the insulin secretagogue sulfonylureas and the meglitinides; the glucose-dependent insulinotropic polypeptides (GLP-1) analogues; the DDP-4 inhibitors, among others [5]. These drugs have numerous side effects including gastrointestinal complications and loss of efficacy after pronged usage [6,7]. The high cost and patient compliance have also been other issues in diabetes therapy considering the required long period of treatment and repeated dosage. Hence, the urgency of identifying novel antidiabetic drugs with safer, more efficacious and cheaper profile cannot be overemphasized. Given diabetes being a complex disease, the rational for the search of multifunctional compounds that target multiple mechanisms of the diffused pathology has been advocated [8]. Furthermore, the antidiabetic potential of many natural products especially polyphenols that combine antioxidant, antiinflammatory, enzyme inhibition and various other general and specific insulin signaling modulatory effects have been outlined in recent years [9–11]. In this review, attention is given to structurally the smallest molecular weight group of compounds of the terpenoids classes of natural products, monoterpenes. Attempt is also made to scrutinize their antidiabetic activity profile built from their in vitro and in vivo data. As antidiabetic agents, monoterpenes now, somehow unexpectedly, appears to be recognized as small molecules punching a lot more than their weight.

2. Chemistry

Like all other terpenoids, the monoterpenes are constructed from repeating units of basic 5-carbon building blocks called isoprenes [12,13]. Their biosynthesis start from the simplest primary metabolite acetyl-CoA that goes through a series of biosynthetic pathways involving the mevalonic acid. Accordingly, terpenoids are often called products of the mevalonic acid pathway of secondary metabolites [14–16]. The conversion of the primary metabolites into the secondary, terpenoids, natural products involve the synthesis of the key 5-carbon reactive intermediates, isopentenyl diphosphate (IPP) and dimethylallyl diphosphate (DMAPP). The condensation of these two isoprene units gives rise to a 10-carbon skeleton geranyl diphosphate (GPP): the immediate precursor of all monoterpenes (Figure 1). Sequential addition of further isoprene units to GPP leads to sesqueterpenes (15 carbon) and diterpenes (20 carbon) or their dimers, triterpenes (30 carbon) and tetraterpenes (40 carbon) respectively. The two 5-carbon terpenoid skeletons (IPP and DMAPP) could themselves give rise to a handful of 5 carbon derivatives that exist in alcohol, acid or hydrocarbon forms but they mostly incorporate into other secondary metabolites as ether or ester derivatives. Hence, the smallest but most structurally diverse group of terpenoids are represented by the monoterpenes.

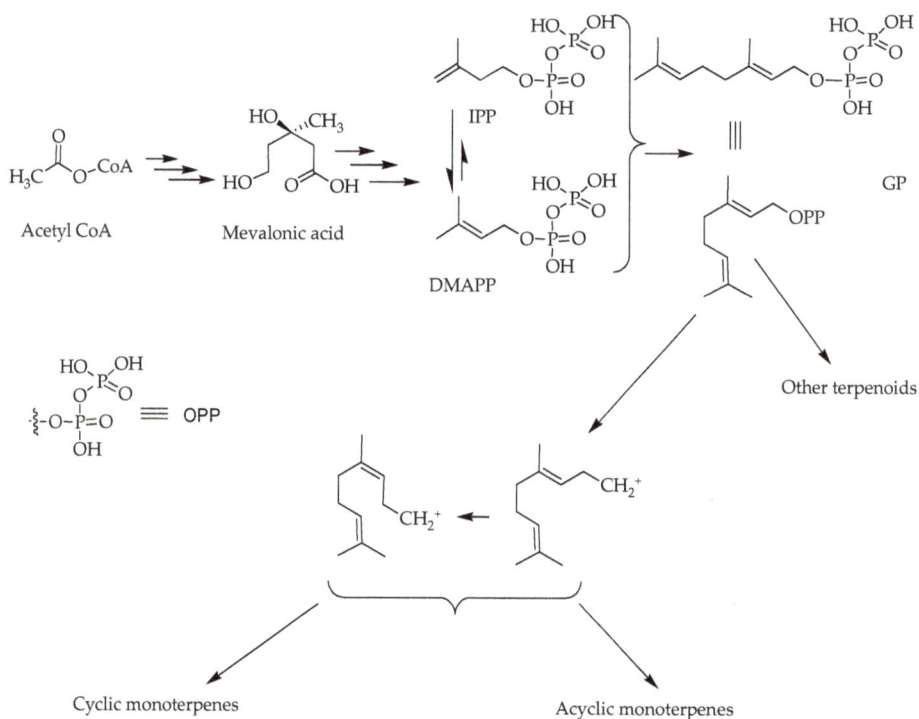

Figure 1. An overview of biosynthesis pathway of monoterpenes; isopentenyl diphosphate (IPP) and dimethylallyl diphosphate (DMAPP); glycogen phosphorylase (GP); OPP: diphosphate leaving group.

Despite their small atomic mass (10 carbon), the monoterpenes have remarkable structural diversity owing to the diphosphate leaving group of the GPP. The resulting unstable cations (Figure 1) undergoes a series of reactions including addition reaction typically leading to incorporation of a hydroxyl group, double bond rearrangements and unsaturation that all eventually lead to stable structures of the monoterpenoid class (Figure 2). In fact, the hallmark of diversity in monoterpenoids that arose from just one precursor GPP (or its isomer neryl diphosphate) through a serious of enzyme catalyzed reactions including cyclization, hydroxylation, dehydrogenation, oxidation and/or reduction, isomerization, and conjugation have been the subject of over a century-old research for natural product chemists.

For most monoterpenoids, their small molecular weight coupled with their nonpolar nature means that they are easily extracted by either non-polar solvents (like hexane), supercritical CO_2 or steam distillation. There are however monoterpenes mostly existing in glycosylated form and hence are very polar. The iridoids (Figure 2) are such group of compounds with *cis*-fused cyclopentano pyran ring system in their structure [16]. The monoterpenes can also be incorporated into other structures such as polyphenol (e.g., phenolic acids and flavonoids). In this review, biological activities that solely attributes to the monoterpenes are discussed.

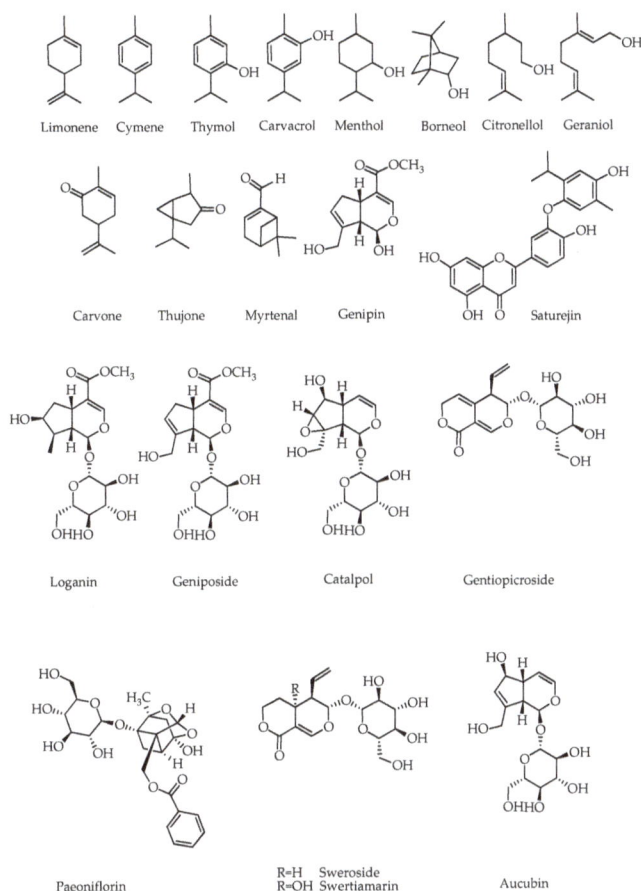

Limonene   Cymene   Thymol   Carvacrol   Menthol   Borneol   Citronellol   Geraniol

Carvone   Thujone   Myrtenal   Genipin   Saturejin

Loganin   Geniposide   Catalpol   Gentiopicroside

Paeoniflorin   R=H Sweroside   Aucubin
R=OH Swertiamarin

Figure 2. Structures of monoterpenes discussed in this review as antidiabetic agents.

3. General Function in Nature

Terpenoids play important roles in the organisms (bacteria, fungi, plants, and animals) that produce them. The large volume of volatile compounds that plants utilize to attract pollinator insects or deter herbivores belong to the monoterpene class of compounds. As such, the fragrances of herbs, spices, flowers and fruits are attributed to the essential oils which are dominated by monoterpenes. The various roles played by monoterpenes as signaling molecules in plant metabolism, plant-plant and plant-animal interactions have been extensively reviewed [12,17–19]. The monoterpenes are also among the best exploited group of compounds by mankind. The most prominent use once again relates to them being the major component of essential oils that make them valuable in the perfumery industry. Within the food industry, the flavoring and preservative potential of essential oils have been well understood along with their general antimicrobial and antioxidant effects. They are also exploited as pharmaceutical agents [19–21] and in agrochemical industries primarily as insecticides [22].

The general mechanism of action of monoterpenes such as their antimicrobial and cough therapy have been mainly associated to their volatile nature allowing them to freely move through space inclining biological membranes and interact with various biomolecules. Some of these actions have long been perceived as non-specific that lack drug-like selectivity to particular receptors. For example, the hydrophobicity of these compounds and essential oils in general have been shown to account for their disruptive effect on bacterial cellular structures (e.g., cell membrane) and hence leading to cell death [23]. There are also reports on non-specific and non-competitive mechanism of action for the muscle relaxant effects of some monoterpenes. For example, Boskabady and Jandaghi [24] have shown that the relaxant effect of carvacrol on tracheal smooth muscles of guinea pigs could not be accounted by effects via β2-adrenergic stimulatory, histamine H1, and muscarinic blocking. On the other hand, there are reports on specific effects of monoterpenes on receptors, such as in the cooling effect of menthol through action on thermoreceptors, even though the lack of specific mechanism for nasal decongestant action by monoterpenes is still the subject of intense discussion [25,26]. A non-competitive manor of receptor inhibition by monoterpenes have also been reported and exemplary findings include the direct interaction of linalool with the N-methyl-D-aspartate receptors [27] and nicotinic acetylcholine receptor inhibition by borneol [28]. Not surprisingly, the antidiabetic effects of monoterpenes were only coming to light in the last decade but now appear to gain lots of momentum. In the following sections, some exemplary effects collected from in vitro and in vivo studies are presented to highlight their promise in diabetes as well as insulin resistance and obesity.

4. Antidiabetic Potential of Monoterpenes

4.1. In Vitro Protective Effects

In vitro experiments (Table 1) on the antidiabetic effects of monoterpenes include cell culture studies using pancreatic β cells, muscles, adipocytes and liver cells among others; as well as enzyme/protein-based assays [29–48]. Of these, the study by Tan et al. [45] included direct comparison of activity of a number of commercially available monoterpenes: geraniol, nerol, citral, (R)-(−)-linalool, (R)-(+)-limonene, (S)-(−)-perillyl alcohol, (R)-(+)-β-citronellol, (S)-(−)-β-citronellol, α-terpineol, L-menthol, γ-terpinene and terpinolene. Their study employed in vitro antioxidant; α-amylase and α-glucosidase enzyme inhibition; and glucose uptake and lipid metabolism in 3T3-L1 adipocytes. Even though these compounds have been shown to have some radical scavenging properties (DPPH and ABTS radicals) at higher concentrations such as 10 and 100 mM, the observed activity is not of therapeutic relevance. In fact, considering their structure (Figure 1), that lack a phenolic structural moiety (with few exceptions), they are not expected to display potent antioxidant effect and/or direct radical scavenging. Over the years, we have shown the potent direct radical scavenging effects of polyphenolic compounds particularly those with catechol functional groups and those optimized with the flavonoids skeleton [11,49–62]. Hence, apart from compounds like carvacrol and thymol with phenolic skeleton, most monoterpenes (Figure 2) do not seem to have direct radical scavenging mechanism that account for their antidiabetic effects. With respect to inhibition of key carbohydrate enzymes (α-amylase and α-glucosidase) inhibition, these compounds were reported to act at 10 mM range [45] and is once again of no therapeutic relevance. In this connection, we have shown that many promising natural products such as flavonoids and polyphenols are acting at micromolar range [9,50,52,53,62–65]. On the other hand, the most prevalent effect of these monoterpenoid compounds were on glucose uptake and lipid metabolism in 3T3-L1 adipocytes cell culture [45]. (S)-(−)-β-citronellol, terpinolene and (R)-(−)-linalool did not affect the glucose uptake in 3T3-L1 adipocytes; while geraniol, citral, (R)-(+)-limonene, (R)-(+)-β-citronellol, nrerol, (S)-(−)-perillyl alcohol, γ-terpinene and α-terpineol showed some degree of activity (up to 21% inhibition) when tested at 1 µM. Readers should bear in mind that this concentration is so small and hence promising but a dose-dependent effect profile was not reported for these compounds. Similarly, the free glycerol released into the medium in the cell culture was measured, and at 1 µM concentrations, the lipolysis effect of (R)-(+)-limonene, (S)-(−)-perillyl alcohol, (R)-(+)-β-citronellol and geraniol were shown [45]. Once again, this is a screening result and does not show the concentration range and total profile of activity for these compounds. On the other hand, (R)-(+)-limonene treatment did not affect the mRNA expression of PPAR-γ in 3T3-L1 adipocytes; it increased the mRNA expression of GLUT1 by 1.2-fold whereas the mRNA expression of GLUT4 remained unchanged. These were all at a fixed concentration of one dose (1 µM).

Table 1. In vitro antidiabetic effect of monoterpenes.

Compound	Model	Outcome	References
Carvacrol	H_2O_2-induced cellular injury on isolated pancreas islets—Following 20, 40 and 80 mg/kg/day in vivo treatment	Cytoprotective	[29]
Cymene	Advanced glycation end *products* (AGEs)	100 μM—Inhibit AGE formation; inhibit glycation specific decline in BSA α-helix content and β-sheet.	[30]
Genipin	C2C12 myotubes	10 μM—Stimulate glucose uptake; promote GLUT4 translocation; increase insulin receptor IRS-1, AKT, and GSK3β phosphorylation; increase ATP levels, close K(ATP) channels; increase intracellular calcium level; effect blocked by wortmannin and EGTA *.	[31]
Geniposide	Rat INS-1 pancreatic β cells	Prevent cell damage induced by high (25 mM) glucose through the AMPK pathway	[32,33]
Geniposide	Pancreatic β-cells—cultured primary cells of rats origin	10 μM—Potentiate insulin secretion via activating the glucagon-like-1 receptor (GLP-1R) as well as the adenylyl cyclase (AC)/cAMP signaling pathway; inhibit voltage-dependent potassium channels; activate Ca^{2+} channels.	[34]
Geniposide	Primary cortical neurons; PC12 cells	Enhance PPARγ phosphorylation; accelerate the release of phosphorylated FoxO1 (forkhead box O1) from nuclear fraction to the cytosol; activate the activity of insulin-degrading enzyme promoter in PC12 cells	[35]
Geniposide	INS-1 pancreatic β cells	10 μM—Increase phosphorylation of PDK1 and Akt473; inhibit the phosphorylation of downstream target GSK3β; increase expression of GLUT2; effect abolished by inhibitor of PI3K (LY294002).	[36]
Geniposide	INS-1 pancreatic β cells	Up to 10 μM—Enhance glucose-stimulated insulin secretion in response to low or moderately high glucose concentrations; promote glucose uptake and intracellular ATP levels; modulate pyruvate carboxylase expression.	[37]
Geniposide	Pancreatic INS-1 cells	Attenuate palmitate-induced β-cell apoptosis and caspase-3 expression; improve the impaired GLP-1R signaling by enhancing the phosphorylation of Akt and Foxo1; increase the expression of PDX-1; effect inhibited by exendin (9–39), an antagonist for GLP-1 receptor.	[38]
Geniposide	Pancreatic INS-1 cells	10 μmol/L—Enhance acute insulin secretion in response to both the low (5.5 mmol/L) and moderately high levels (11 mmol/L) of glucose; Effect inhibited by GLP-1R antagonist exendin (9–39) or knock-down of GLP-1R with shRNA interference in INS-1 cells.	[39]
Geniposide	HepG2 fatty liver model- free fatty acid treatment	Suppress the intracellular lipid accumulation; increase the intracellular expression of a fatty acid oxidation-related gene (PPARα).	[40]
Gentiopicroside	HL1C hepatoma cells	50 and 100 μM—Suppress Pck1 expression; induce phosphorylation of components in the insulin signaling cascade (Akt and Erk1/2 phosphorylation).	[41]
Paeoniflorin	3T3-L1 adipocytes treated with tumour necrosis factor (TNF)-α	50 μg/mL—Increase insulin-stimulated glucose; promote serine phosphorylation of IRS-1 and insulin-stimulated phosphorylation of AKT; inhibit the expressions and secretions of IL-6 and MCP-1; attenuate TNF-α-mediated suppression of the expressions of PPARγ and PPARγ target gene; effect reversed by antagonist of PPARγ activity.	[42]

Table 1. *Cont.*

Compound	Model	Outcome	References
Paeoniflorin	3T3-L1 adipocytes and RAW 264.7 macrophages	12.5–100 µg/mL—Inhibit TNF-α and FFA production; inhibit TNF-α-stimulated adipocyte lipolysis; suppress phosphorylation of TNF-α-activated ERK1/2; attenuate (partially) palmitate-induced macrophage TNF-α production.	[43]
Paeoniflorin derivatives (methoxyl and glucoside analogues)	Human HepG2 cells and HUVECs	10 µM—Increase glucose uptake; reverse glucose-induced inhibition of glycogen synthesis in HepG2; increase AMPK and GSK-3β phosphorylation; phosphorylate AMPK and increase phosphorylation of GSK-3β while suppressing lipogenic expression (acetyl-CoA carboxylase and fatty acid synthase); induced eNOS phosphorylation in HUVECs.	[44]
(R)-(+)-limonene	3T3-L1 cell culture; α-amylase and α-glucosidase enzymes	Increase GLUT1 expression at mRNA level; Weak enzyme inhibition (mM range).	[45]
Saturejin (3'-(2,5-dihydroxy-p-cymene) 5,7,4'-trihydroxy flavone) from *Satureja khuzistamica* Jamzad	Antioxidant activity; α- and β-glucosidase inhibitory	10 µg/mL—Significant in vitro radical (DPPH) scavenging and enzyme inhibitory effects.	[46]
Sweroside	HL1C hepatoma cells	Suppress Pck1 expression and induce phosphorylation of components in the insulin signaling cascade (Akt and Erk1/2 phosphorylation).	[41]
Swertiamarin	Steatosis in HepG2 cells induced by 1 mM oleic acid	25 µg/mL—Maintain membrane integrity; prevent apoptosis; increase the expressions of major insulin signaling proteins (insulin receptor, PI3K and pAkt) with concomitant reduction in p307 IRS-1; activate AMPK; modulate PPAR-α; decrease the levels of the gluconeogenic enzyme, PEPCK.	[47]
Thujone	Palmitate-induced insulin resistance in skeletal muscle (Soleus muscles)	Ameliorate palmitate oxidation and enhance insulin-stimulated glucose transport; restore (partially) GLUT4 translocation and AS160 phosphorylation; increase AMPK phosphorylation.	[48]

* EGTA represent ethylene glycol bis(2-aminoethyl ether)tetraacetic acid.

The general antidiabetic effect of monoterpenes through specific biological targets are illustrated in Table 1. Even though essential oils are known for general antioxidant and enzyme inhibitory activities including α-glucosidases, such effect may not be therapeutically relevant as it occur at relatively high concentrations. There is a great deal of diversity among the monoterpenes however as some such as thymol are phenolic in nature while the vast majority are highly non-polar compounds (Figure 2). As discussed above, direct reactive oxygen species and radical scavenging (ROS) is mainly a function of polyphenolic compounds, which (with few exceptions such as thymol and carvacrol) are poorly represented in monoterpenes. Hence, such biological effect is not addressed here as the major mechanism of action of these compounds for their antidiabetic effects. It is interesting to note however that when monoterpenes are incorporated into other structural groups such as flavonoids as exemplified by saturejin, a far better direct antioxidant and enzyme inhibitory effects were obtained [46].

A number of other in vitro experiments have been carried out to assess the effect of monoterpenes on cultured pancreatic β cells. This has shown remarkable results as concentrations in micromolar range have been demonstrated to potentiate insulin secretion (Table 1). Various experiments on adipocytes, hepatocytes and muscle cells also confirmed that these compounds can effectively increase glucose uptake through upregulation of the glucose transporter (GLUT4) translocation. As discussed in the following sections, some key pharmacological targets at molecular level including the insulin signaling pathways have been implicated. Genipin and geniposide as well as iridoids and their glycosides have also been shown to display potential antidiabetic effects in vitro. In the cultured muscle myotube cells, an increase in the phosphorylation of insulin receptor substrate-1 (IRS-1) and other signal transduction pathways leading to increased intracellular calcium concentrations have been reported (Table 1). The AMPK pathway has also been shown to be involved in the protection of pancreatic β cells by geniposide [32,33]. Furthermore, activation of the glucagon-like-1 receptor has been implicated in the insulin secretion promotion by geniposide [34]. The role of PPARγ and phosphatidyl inositol 3-kinase (PI3K) are also shown to be involved in the effect of these compounds in neuronal cells. As discussed in the following texts, the PI3K involvement, which has been confirmed by using specific inhibitor (wortmannin), is a further interesting insight into the possible mechanism of action of monoterpenes [31]. Hence, at concentrations as low as 10 μM, these compounds can protect pancreatic β cells and other cells such as neurons; promote insulin secretion and facilitate glucose uptake. Moreover, amelioration of the pro-inflammatory cytokines (e.g., TNF) in adipocytes and lipid accumulation in liver cells have been shown to be induced by geniposide. The involvement of the GLP-1R signaling pathway in the antidiabetic effect of geniposide has also been confirmed by specific receptor antagonist, exendin [38,39]. The production of TNF and free fatty acids (FFAs) in adipocytes and macrophages in vitro could also be ameliorated by monoterpenes such as paeoniflorin [42,43], swertiamarin [47] and thujone [48].

4.2. In Vivo Antidiabetic Effects

The remarkable feature of monoterpenes is that they have shown promising antidiabetic effect in vivo despite their rather simplistic structural appearance [35,66–98]. The list of monoterpenes with good effects on in vivo antidiabetic models including in the streptozotocin (STZ)-induced diabetic, high fat diet (HFD) fed and spontaneously obese mouse models are shown in Table 2. Some of these compounds are presented as simple hydrocarbons such as limonene and cymene; ketone or hydroxy derivatives; aromatic or non-aromatic skeletons; or glycosides of iridoids (Figure 2). In all cases, a promising antidiabetic effect at doses as low as 5 mg/kg have been observed. The variability in dose regimen and duration of study in these studies, however, do not allow direct comparison of potency between the various structural groups of the monoterpenes. The following observation were however evident as a general antidiabetic agents.

Table 2. In vivo antidiabetic effects of monoterpenes.

Compound	Model	Outcome	References
Aucubin	STZ-induced diabetic rats—5 mg/kg, i.p. twice daily for the first 5 days followed by single injections daily for 10 days.	Lower blood glucose; reverse lipid peroxidation and the decreased in activities of antioxidant enzymes in liver and kidneys; increase immunoreactive beta cells.	[66]
Borneol	STZ-induced diabetic rats—25 or 50 mg/kg, p.o. for 30 days.	Lower blood glucose and HbA1c; increase blood insulin; restore body weight loss; increase liver glycogen level; reverse the diabetes-induced increase in the levels of TC, TGs LDL-C, VLDL-C; restore urea and ALT and AST levels; increase antioxidant status (SOD, catalase, GSH) in the liver and kidney; reduce MDA level.	[67]
Carvacrol	HFD-induced C57BL/6J diabetic mice—20 mg/kg p.o. for 35 days.	Suppress elevated TC, TG, phospholipids and FFAs, VLDL-C, LDL-C in plasma and tissues; Suppress liver tissue inflammatory cytokines (TNF-α and IL-6); increase high density lipoproteins-cholesterol (HDL-C)	[68]
Carvacrol	HFD-induced type 2 diabetic C57BL/6J mice—20 mg/kg, p.o. for 35 days.	Ameliorate the increased glucose-6-phosphatase and fructose-1,6-bisphosphatase, decreased glucokinase and glucose-6-phosphate dehydrogenase activities; normalize hepatic markers (ASP, ALA, ALP, and γ-glutamyl transpeptidase).	[69]
Carvacrol	STZ-induced diabetic rats—25, 50, and 100 mg/kg, p.o. for 7 weeks or 20, 30 and 40 i.p.	Improve diabetes-associated cognitive deficit; suppress oxidative stress (increased MDA level and decreased SOD as well as reduced GSH) and inflammatory and apoptosis markers (NF-κB p65 unit, TNF-α, IL-1β, and caspase-3).	[70,71]
Carvacrol	STZ-induced diabetes in rats—25 and 50 mg/kg, p.o. for 7 days.	Suppress serum glucose, total cholesterol, ALA, AST and lactate dehydrogenase; no effect on serum insulin levels, food-water intake values and body weight changes.	[72]
Carvone	STZ-induced diabetic rats—50 mg/kg, p.o. for 30 days.	Reduce plasma glucose, HbA1c; improve the levels of hemoglobin and insulin. Revers activities of carbohydrate metabolic enzymes, enzymatic antioxidants and hepatic marker enzymes.	[73]
Catalpol	STZ-induced diabetic rats—10 mg/kg, i.p. for 14 days.	Improve impaired renal functions; ameliorate pathological changes in kidneys; abolish the diabetes induced elevation of Grb10 expression in the kidneys; increase IGF-1 mRNA levels and IGF-1R phosphorylation in kidneys.	[74]
Catalpol	HFD-fed mice receiving 100 mg/kg, p.o. for 4 weeks	No effect on body weight; improve fasting glucose and insulin levels, glucose tolerance and insulin tolerance; reduce macrophage infiltration into adipose tissue; reduce mRNA expressions of M1 pro-inflammatory cytokines while increasing M2 anti-inflammatory gene expressions in adipose tissue; suppress the JNK and NF-κB signaling pathways in adipose tissue.	[75]
Catalpol	STZ-diabetic rats—0.1 mg/kg, i.p.	Enhance glucose uptake in the isolated soleus muscle of diabetic rats; increase glycogen synthesis.	[76]
Catalpol	STZ-induced diabetic rats—10, 50 and 100 mg/kg, p.o. for 6 weeks.	Improve neuronal injury and cognitive dysfunction; increase the nerve growth factor concentration and decrease the blood glucose.	[77]
Citronellol	STZ-induced diabetic rats—25, 50, and 100 mg/kg, p.o. for 30 days.	Improve the levels of insulin, hemoglobin and hepatic glycogen with significant decrease in glucose and HbA1c levels. Restore altered activities of carbohydrate metabolic enzymes and level of hepatic and kidney markers; improve morphology of hepatic cells and insulin-positive β-cells.	[78]
Cymene	STZ-induced diabetic rats—20 mg/kg, p.o. for 60 days	Improve HbA1c and nephropathic parameters (like albumin excretion rate, serum creatinine and creatinine clearance rate).	[30]
Genipin	Aging rats—25 mg/kg, i.p. for 12 days	Ameliorate systemic and hepatic insulin resistance, alleviate hyperinsulinemia, hyperglyceridemia and hepatic steatosis, relieve hepatic oxidative stress and mitochondrial dysfunction; improve insulin sensitivity by promoting insulin-stimulated glucose consumption and glycogen synthesis; inhibit cellular ROS overproduction and alleviate the reduction of levels of MMP and ATP.	[79]

Table 2. *Cont.*

Compound	Model	Outcome	References
Geniposide	Insulin-deficient—APP/PS1 transgenic mouse model of Alzheimer's disease. 5, 10, and 20 mg/kg, intragastric for 4 weeks.	Decrease the phosphorylation of tau protein.	[80]
Geniposide	STZ-induced diabetic rats—injection (50 μM, 10 μL) to the lateral ventricle	Prevent spatial learning deficit; reduce tau phosphorylation.	[81]
Geniposide	Transgenic mouse model with Streptozotocin—5, 10 and 20 mg/kg, intragastric for 4 weeks	Decreased level of β-amyloid peptides (Aβ1-40 and Aβ1-42); up-regulate the protein levels of β-site APP cleaving enzyme (BACE1) and insulin-degrading enzyme (IDE); decrease the protein levels of ADAM10; enhance the effects of insulin by reducing Aβ1-42 levels in primary cultured cortical neurons.	[35]
Geniposide	STZ-induced diabetic rats—800 mg/kg/day, p.o. for 46 days.	Improve insulin and blood glucose; decrease Aβ1-42 level; improve the expression of insulin-degrading enzyme.	[82]
Geniposide	Spontaneously obese Type 2 diabetic TSOD mice	Suppress body weight, visceral fat and intrahepatic lipid accumulation; alleviate abnormal lipid metabolism; alleviate abnormal glucose tolerance and hyperinsulinemia.	[40]
Geniposide	High fat diet—25, 50 or 100 mg/kg, p.o. for six weeks.	Improve liver histology through reducing the elevated liver index (liver weight/body weight), serum alanine aminotransferase and aspartate aminotransferase; decrease total cholesterol, triglycerides and FFAs in serum and liver; increased serum insulin levels but reduced serum TNF-α level; suppressed expression of CYP2E1 and increased PPARα expression	[83]
Geniposide	HFD and STZ-induced diabetic mice —200 and 400 mg/kg for 2 weeks	Decrease blood glucose, insulin and TG levels; decrease the expression of glycogen phosphorylase and glucose-6-phosphatase at mRNA level and immunoreactive protein levels, as well as enzyme activity.	[84]
Geraniol	STZ-induced diabetic rats—100, 200 and 400 mg/kg, p.o. for 45 days	Improve the levels of insulin, hemoglobin and decrease plasma glucose, HbA1c; improve hepatic glycogen content; preserve the normal histological appearance of hepatic cells and pancreatic β-cells.	[85]
D-Limonene	STZ-induced diabetic rats—50, 100 and 200 mg/kg, p.o. for 45 days.	Reverse the following diabetic effect: increased blood glucose and glycosylated hemoglobin levels, increased activity of gluconeogenic enzymes (glucose 6-phosphatase and fructose 1,6-bisphosphatase) and decreased activity of glycolytic enzyme, glucokinase and liver glycogen.	[86]
D-Limonene	STZ-induced diabetic rats—50 mg/kg, p.o. for 28 days.	Decrease DNA damage, glutathione reductase enzyme activities and MDA levels; increase GSH levels and CAT, SOD and GSH-Px enzyme activities and altered lipid and liver enzyme parameters in diabetic rats.	[87]
Logonin	STZ-induced diabetic mice, 20 mg/kg, p.o. for 12 weeks.	Reduce kidney/body weight ratio, 24 h urine protein levels, serum levels of urea nitrogen and creatinine; improve histology of pancreas and kidney; alleviate structural alterations in endothelial cells, mesangial cells and podocytes in renal cortex; reduce AGE levels in serum and kidney; downregulate mRNA and protein expression of receptors for AGEs in kidney; reduce the levels of MDA; increase the levels of SOD in serum and kidney.	[88]
Menthol	STZ-nicotinamide induced diabetes in rats—25, 50, and 100 mg/kg, p.o. for 45 days	Reduce blood glucose and glycosylated hemoglobin levels; increase the total hemoglobin, plasma insulin and liver glycogen levels; protect hepatic and pancreatic islets; modulating glucose metabolizing enzymes, suppression of pancreatic β-cells apoptosis and altered hepatic, pancreatic morphology	[89]
Myrtenal	STZ-induced diabetic rats—80 mg/kg, p.o. for 28 days	Decrease plasma glucose; increase plasma insulin levels; up-regulate IRS2, Akt and GLUT2 in liver; increase IRS2, Akt and GLUT4 protein expression in skeletal muscle.	[90]

Table 2. *Cont.*

Compound	Model	Outcome	References
Myrtenal	STZ-induced diabetic rats—20, 40, and 80 mg/kg, p.o. for 28 days	Reduce plasma glucose, haemoglobin A1c (HbA1c); increase the levels of insulin and hemoglobin; reverse body weight loss; normalize hexokinase, glucose-6-phosphatase, fructose-1,6-bisphosphatase, glucose-6-phosphate dehydrogenase, and hepatic enzymes AST, ALT, and ALP levels; improve hepatic and muscle glycogen content; restore islet cells and liver histology.	[91]
Myrtenal	STZ-induced diabetic rats—80 mg/kg, p.o. for 28 days	Improve plasma glucose, pancreatic insulin and lipid profiles (TC, TG, FFAs, phospholipids, LDL, VLDL, atherogenic index); improve histopathological feature of the liver.	[92]
Paeoniflorin	High-sucrose, HFD rat receiving low dose STZ—15 and 30 mg/kg, p.o. for 4 weeks.	Reduce brain inflammatory cytokines (IL-1β and TNF-α), decrease suppressor of cytokine signaling 2 expressions and promote (IRS-1 activity and phosphorylation levels of protein kinase B (Akt) and glycogen synthase kinase-3β (GSK-3β).	[93]
Paeoniflorin	HFD—induced obese mice—Diet containing 0.05% (*w/w*) of paeoniflorin	Lower body weight, hyperlipidemia, and insulin resistance; block inflammation; inhibiting lipid ectopic deposition; lower lipid synthesis pathway (de novo pathway, ^{3}HMG-CoAR), promote fatty acid oxidation (peroxisome proliferator-activated receptor-alpha (PPARα), carnitine palmitoyltransferase-1); increase cholesterol output (PPARγ-liver X receptor-α-ATP-binding cassette transporter-1); block inflammatory genes activation and reduce gluconeogenic genes expression (phosphoenolpyruvate carboxykinase and G6Pase).	[82]
Swertiamarin	STZ-induced diabetic rats—50 mg/kg, i.p. for 6 weeks	Reduce serum triglycerides, cholesterol and low-density lipoprotein levels; decrease serum fasting glucose; increase insulin sensitivity index.	[94]
Thymol	HFD-induced type 2 diabetes in C57BL/6J mice—10, 20 and 40 mg/kg, intragastric for 5 weeks.	Antihyperglycaemic; lower plasma TG, TC, FFAs, LDL and increase HDL cholesterol; lower hepatic lipid contents TG, total cholesterol, FFAs and phospholipids.	[95]
Thymol	HFD- induced type 2 diabetes in C57BL/6J mice—40 mg/kg, intragastric for 5 weeks.	Inhibit diabetic nephropathy; inhibit the activation of transforming growth factor-β1 and vascular endothelial growth factor; increase level of antioxidants and suppress lipid peroxidation markers in erythrocytes and kidney tissues; downregulate the expression level of sterol regulatory element binding protein-1c and reduce lipid accumulation in the kidney.	[96]
Thymol	HFD-induced obese C57BL/6 J mice—20, 40 mg/kg daily	Reverse body weight gain; ameliorate peripheral insulin resistance; improve cognitive impairments in the Morris Water Maze test; decrease HFD-induced Aβ deposition and tau hyperphosphorylation in the hippocampus; down-regulate the level of P-Ser307 IRS-1 and enhance the expression of P-Ser473 AKT and P-Ser9 GSK3β; up-regulate nuclear respiratory factor /heme oxygenase-1pathway.	[97]
Thymol	HFD-fed rats—14 mg/kg, p.o. for 4 weeks.	Decrease body weight gain, visceral fat -pad weights, lipids, ALT, AST, LDH, blood urea nitrogen, glucose, insulin, and leptin levels; decrease serum lipid peroxidation and increase antioxidant levels.	[98]

As shown in Table 2, hyperglycemia is the primary criteria assessed in all studies and the indicated monoterpenes have shown promising effects in all studied models. In addition to lowering the blood glucose level, the common antidiabetic effect assessment involves the measurement of glycated hemoglobin, primarily HbA1c. This is also indicated for many compounds studied (Table 2). In the liver, one of the main antidiabetic drugs target, increasing the level of glycogen level has been reported for borneol [67], citronellol [78] and myrtenal [90,91]. Modulation of key hepatic enzymes of glucose metabolism such as glucose-6-phosphatase and fructose-1,6-bisphosphatase; decreased glucokinase and glucose-6-phosphate dehydrogenase activities, have been reported for carvacrol [69], carvone [73], citronellol [78], geniposide [84], myrtenal [91] and paeoniflorin [82,93]. The common markers of hepatic function aspartate aminotransferase (ASP), alanine aminotransferase (ALA), alkaline phosphatase (ALP), and gamma-glutamyl transpeptidase have also been routinely measured in diabetes and their raised status in diabetes have been suppressed by carvacrol [69] and carvone [73]. Other organ functions primarily the kidney (cymene and logonin) and the brain (geniposide, thymol and D-limonene) have also been reported (Table 2).

It is important to note the effect of the studied compounds on gross weight of the animals especially given administration of STZ-induced diabetes is associated with weight loss. For example, administration of borneol [67], carvacrol [68] and swertiamarin [94] have shown to reverse the body weight loss in diabetes. On the other hand, the body weight gain in HFD-obese model could be ameliorated by monoterpenes. Even though, catalpol had no effect at the dose of 100 mg/kg (although it improved fasting glucose and insulin levels) [75], geniposide [83], paeoniflorin [82,93] and thymol [97,98] were among the monoterpenes that showed good potential antiobesity effect. Geniposide also displayed antiobesity effect in spontaneously obese Type 2 diabetic Tsumura Suzuki Obese Diabetes (TSOD) mice model [40].

Given that STZ administration primarily targets pancreatic β cells leading to insulin depletion and hence diabetic condition, enhancing insulin secretion is one the well-defined mechanism of antidiabetic agents. While carvacrol did not have significant effect on the serum insulin level at the tested dose [72], carvone [73], citronellol [78], geniposide [82], geraniol [85], myrtenal [90] and swertiamarin [94] have all been shown to increase insulin level in the STZ-induced diabetes. As demonstrated for menthol [89] and myrtenal [91], improvement in the pancreatic and hepatic cellular and structures, as evidenced from histological studies, have been reported. On the other hand, in the HFD-obese model, hyperglycemia is associated with a rise in insulin level which appeared to be targeted/normalized by monoterpenes. Good examples are geniposide [84] and thymol [98].

As discussed above, antidiabetic effect is often measured by assessing the level of glycation of proteins including hemoglobin which has been shown to be suppressed by monoterpenes. Considering the link between glycation and oxidative stress, it is worth looking at the antioxidant effect of these compounds in vivo. Even though most monoterpenes are not known for their potent direct ROS scavenging effect in vitro (Table 1), their in vivo antidiabetic effect is associated with an improved antioxidant status (Table 2). For example, borneol [67] has been shown to enhance antioxidant status in diabetic animals including augmented level/activity of superoxide dismutase (SOD), catalase, reduced glutathione (GSH) in the liver and kidney, while concomitantly suppressing the level of oxidation marker, malondialdehyde (MDA). Similar results were obtained for aucubin [66], carvacrol [70,71], genipin [79], D-limonene [87], logonin [88], and thymol [96].

The antihyperlipidemic and antiobesity effect of monoterpenes appears to be well demonstrated in vivo. The diabetes-induced increase in the plasma or tissue levels of TC, TGs LDL-C, VLDL-C are often measured to assess the potential lipid lowering effect of drugs while raising the level of HDL is regarded as valuable parameter of lipid modulation. Borneol [67], carvacrol [72], genipin [79], geniposide [40], D-limonene [87], swertiamarin [94] and thymol [95,96] are classical examples of monoterpenes shown to display such effects (Table 2). The accumulation of fat in tissues as evidenced for geniposide and paeoniflorin have also been demonstrated along with the various effects in the HFD-induced diabetes (Table 2). Hand in hand with all these effects are the amelioration of the diabetes inflammation by various monoterpenes including effects on proinflammatory cytokines such as TNF along with the NF-κB pathway (Table 2). The various other effects of monoterpenes in animal models of diabetes including central effects and neuroprotection are listed in Table 2.

5. Bioavailability

Several studies on the absorption, distribution, metabolism, and excretion of monoterpenes have been published in recent years. Liu et al. [99] have studied the oral bioavailability of paeoniflorin in a single-pass "four-site" rat intestinal perfusion model and cultured Caco-2 cells. As expected, the absorption of this compound in cultured cells was slower than its aglycone (paeoniflorigenin). They have found that poor permeation, P-gp-mediated efflux, and hydrolysis via a glucosidase contributed to the poor bioavailability of paeoniflorin. When the crude plant extract preparation of *Paeonia lactiflora* roots (Chishao) that contain paeoniflorin as a major component (85.5%) was administered in human subjects by intravenous and multiple infusions, an elimination half-lives of 1.2–1.3 h was noted for paeoniflorin. After exposure to major organs, glomerular-filtration-based renal excretion has been found to be the major elimination pathway for this compound [100]. The bioavailability of paeoniflorin has been generally considered low in rabbit (7.24%) and rat (3.24%) after oral administration [101].

Oral administration of the monoterpene aglycones such as thymol and carvacrol has also been shown to result in their slow absorption into the blood stream [102]. Both unchanged and their glucuronide and sulfate conjugates have also been detected. For example, Dong et al. [103] have shown that carvacrol, as a substrate to uridine diphosphate-glucuronosyl transferase, can inhibit the enzyme. In another study of a clinical trial involving 12 healthy volunteers, no thymol could be detected in plasma or urine [104]. However, its metabolites thymol sulfate and thymol glucuronide were found in urine and the reported mean terminal elimination half-life was 10.2 h. Moreover, thymol sulfate was detectable up to 41 h after administration and urinary excretion normally followed over 24 h [104].

In a potential anticancer therapy study, administration of lemonene in women with newly diagnosed operable breast cancer indicated preferential concentration of the compound in the breast tissue, reaching high tissue concentration (mean of 41.3 µg/g tissue) while its major active circulating metabolite was reported to be perillic acid [105]. The accumulation of limonene on adipose tissues has also been described in other studies [106]. Oral administration of swertiamarin in rats has been shown to be associated with rapid and wide distribution in tissues with the highest amount obtained in the liver and kidney, perhaps indicating the possible metabolism and/or elimination sites [107]. As demonstrated for geniposide, administration of these compounds in the form of crude plant extract may be better for their bioavailability than the purified compounds [108].

There is no doubt that more data is needed to establish a clear bioavailability and/or pharmacokinetic profile of monoterpenes. The fact that these compounds display antidiabetic effect when administered orally and at even modest doses suggest that they can be absorbed/bioavailable to exert their action in the various organ systems. There distribution profile in tissues reported so far has also been promising although further studies on optimization and formulation studies are critical. The extreme nonpolar nature of the monoterpene aglycones imply that they can easily travel across cell membranes but their poor water solubility is a challenge, and given this fact, the observed antidiabetic effect should be considered remarkable. On the other hand, the glycosides of these compounds are also shown to display antidiabetic effects and may constitute another route for optimizing their antidiabetic effects.

6. General Summary and Discussion

The various in vitro and in vivo data now suggest that monoterpenes, despite their structural simplicity, have a promise to be seriously considered as antidiabetic lead compounds either by their own or as part of structural moieties in complex structures. Hence, the diverse mechanism of action reported in these studies deserve further scrutiny. A growing body of evidence show that the PPARγ play a vital role in the regulation of carbohydrate, lipid, and protein metabolism both in healthy and disease states (e.g., diabetes, obesity, metabolic syndrome, cardiovascular diseases, etc.). In the case of diabetes, the PPARγ with its predominant expression in adipose tissues (but also in macrophages and intestine) has profound effects in the regulation of adipogenesis, insulin sensitivity, and inflammation [109–111]. The role of PPARγ in key other insulin-target organs/tissues such as the liver and muscles has also been well understood. Even though the thiazolidinediones (TZDs) groups of antidiabetic drugs have been reported to have numerous undesirable side effects [112,113], targeting the PPARγ by receptor agonists remains an attractive pharmacological target for the treatment and prevention of metabolic disorders including diabetes. In this regard, the role of natural products as a source of potential drugs that lack the undesirable effects of thiazolidinediones have been advocated [114,115]. Hence, the modulatory effect of monoterpenes on the PPARγ appear to be a well-defined mechanism of antidiabetic action. Furthermore, monoterpenes such as carvacrol, thymol and others that showed antidiabetic effects are common ingredients of foods and flavors and hence not expected to display the undesirable effects of thiazolidinediones.

The role of the AMPK in regulating glucose metabolism and as a validated target for diabetes has gained lots of momentum in recent years. Drugs that activate the AMPK have been shown to enhance the phosphorylation of insulin receptor substrate 1 (IRS-1) Ser789 phosphorylation leading to the cascade of activation of phosphoinositide 3 kinase/protein kinase B (PI3K/PKB) signaling [116]. Such effects have been shown to be both insulin dependent (potentiation) and independent (in unstimulated state). The classical antidiabetic drug, metformin, also appear to act through activation of the AMPK pathway in various target organs [117]. Interestingly, the activation of AMPK in adipocytes has been shown to suppress lipogenesis while promoting energy dissipation suggesting potential antiobesity effect [118]. Even though some studies still show the paradoxical nature of the AMPK activation in glucose and lipid metabolism, mostly related to the level of stimulation (chronic verses acute) (e.g., [119,120]), the role of AMPK activators in various disease condition have been well documented [121]. In a recent review [122], the antidiabetic potential of natural products such as anthocyanins through effect on the AMPK and insulin-signaling pathways have been highlighted. Interestingly, monoterpenes have been shown to share this crucial mechanism of antidiabetic effect (Table 1).

The GLUT4 is an ATP-independent glucose transport system across cell membranes and predominantly expressed in adipose tissues and skeletal muscles. While under expression of the GLUT4 can predispose animals to diabetes and insulin resistance, its overexpression has been shown to overcome such physiological/pathological disorder [123,124]. The reduced level of GLUT4 has also been noted in muscles of diabetic patients. The coupling between insulin receptor activation in target organs with the known insulin action in GLUT4-mediated glucose uptake has been the subject of intense research in recent years. The insulin receptor substrate proteins phosphorylation has been shown to lead to the activation of PI3K which intern leads to the generation of phosphatidylinositol 3,4,5-trisphosphate (PIP3). The PIP3 in turn is linked to recruit protein kinase B (PKB), also known as Akt, that play a pivotal role in the insulin-dependent GLUT4 mobilization. Review articles on this cascade of insulin transduction pathways have been published [125,126]. As discussed above, metformin that modulates the AMPK pathway affects the translocation of GLUT4 in insulin target cells [127]. Hence, agents that facilitate the expression or mobilize GLUT4 could have antidiabetic effect. In this regard, the various effect of monoterpenes in signaling pathways appear to be consistent with antidiabetic effect through upregulation of GLUT4 expression and/or translocation (Table 1). The modulation of the AMPK pathway by natural products that enhance glucose uptake while concomitantly inhibiting gluconeogenesis and stimulation of glycogen synthesis have previously been reported (e.g., [128–132]. Hence, possible multiple mechanisms involving insulin signaling and other pathways involving the AMPK, GLUT4 and PPARγ are at play in the antidiabetic effect of monoterpenes.

The liver plays a central role in glucose homeostasis by regulating glucose production and storage in the form of glycogen. In terms of drug therapy, agents that suppress the glycolysis pathways (glycogenolysis inhibitors) or hepatic glucose production (gluconeogenesis) have antidiabetic properties [133–136]. These classical examples of antidiabetic (T2D) drugs therapy approaches target key enzymes involved in the cascades of reaction pathways in hepatic cells. The most important validated targets constitute the first important step of glycogenolysis that requires the enzymatic action of glycogen phosphorylase (GP) to convert glycogen to glucose-1-phosphate. On the other hand, glucose-6-phosphatase (G6Pase) at the final step of gluconeogenesis serves as an important drug target. In this regard, the effects of many monoterpenes as modulators of these pathways have been reported (Table 2). Hence, the antidiabetic mechanism of monoterpenes may include gluconeogenesis and glycolysis as potential targets.

The role of inflammation as a link between diabetes and obesity has been documented in recent years [137]. The low grade persistent/chronic state of inflammation in obesity is generally considered to increase insulin resistance in target organs [138]. The closer link between inflammation under obesity condition and insulin resistance in diabetes has become clearer with the identification of proinflammatory cytokines such as tumour necrosis factor (TNF)-α and interleukin-6 (IL-6) as major culprits [139,140]. For example, the above-mentioned PI3K-AKT (or PKB) pathway of the insulin-signaling cascade can be inhibited by these cytokines via activation of the NF-κB. This crosstalk between the inflammation pathway and obesity/diabetes as mechanisms of action for natural products such as anthocyanins has also been described recently [122]. As various monoterpenes appear to suppress the inflammation associated with diabetes (Tables 1 and 2), their potential therapeutic effect could be in part attributed to such pharmacological effects. Hand in hand with all these multiple mechanism, the lipid lowering and/or antiobesity effect of monoterpenes appear to be demonstrated (Table 2). In this connection, the direct effect of monoterpenes on lipogenic expression via targeting key fatty acid synthesis enzymes (acetyl-CoA carboxylase and fatty acid synthase) have been demonstrated in vitro [42].

Besides the diverse possible mechanisms of actions that accounts to the pharmacological effects of monoterpenes, the reported specific effect through modulation of the GLP-1 receptor is very interesting. The best studied compound in this regard is the iridoid glycoside, geniposide [38]. By using the receptor antagonist, exenidin, the effect of geniposide through this mechanism has been confirmed [38,39].

In summary, in vitro and in vivo studies on the antidiabetic effects of monoterpenes so far have largely been demonstrated to show their promises as potential antidiabetic along with potential antiobesity and lipid lowering agents. One crucial evidence still missing is that on human/clinical studies. These compounds as structurally simple as they seem to be have good set of effects with a further opportunity to optimize their activity through structure-activity studies. Such an approach is also largely unexplored so far as these compounds appear to be relatively newcomers in antidiabetic study. The incorporation of these compounds with other structural groups such flavonoids [46] is an interesting development. As such, flavonoids have been demonstrated to have potent antioxidant effect that is of great interest in ameliorating both the diabetes-induced glycation and oxidative injuries. Flavonoids and other phenolic compounds are also known for their α-glucosidase inhibition and other mechanisms involved in diabetes pathology. The formulation of monoterpenes is also another avenue of development in their potential antidiabetic therapeutic effects. Their extreme non-polar nature, though often compensated by glycosylation, would need further studies in modifying their bioavailability and general pharmacokinetic profiles. In this direction, the potential of in silico studies in lead prediction and optimization should also be fully employed. Exemplary indexing model on filtering and mapping discriminative physicochemical properties for identifying antidiabetic natural products have also been published [141]. With all these developments awaiting, the current level of evidences suggest that monoterpenes indeed have a promise as antidiabetic lead compounds through multiple mechanisms as evidenced from both in vitro and in vivo studies.

Conflicts of Interest: The author declares no conflict of interest.

Abbreviations

Aβ	Amyloid Beta
AMPK	Adenosine Monophosphate-Activated Protein Kinase
AKT	Protein Kinase B (PKB)
AGE	Advanced Glycation End Products
ALP	Alkaline Phosphatase
ALT	Alanine Transaminase
AST	Aspartate Aminotransferase

BACE1	Beta-Secretase 1
CAT	Catalase
CYP2E1	Cytochrome P450 2E1
DMAPP	Dimethylallyl Diphosphate
DPPH	2,2-Diphenyl-1-Picrylhydrazyl
EGTA	Ethylene Glycol-bis(β-Aminoethyl Ether)-*N,N,N',N'*-Tetraacetic Acid
Erk	Extracellular Signal–Regulated Kinases
HFD	High Fat Diet
HDL	High Density Lipoproteins
HMG-CoAR	3-Hydroxy-3-Methyl Glutaryl Coenzyme A Reductase
FFAs	Free fatty acids
FoxO1	Forkhead Box O1
GLP-1R	Glucagon-Like-1 Receptor
GLUT	Glucose Transporter
GPx	Glutathione Peroxidase
Grb10	Growth Factor Receptor Bound Protein 10
GSH	Glutathione—Reduced Form
GSK3β	Glycogen Synthase Kinase 3 Beta
HbA1c	Haemoglobin A1c
HUVECs	Human Umbilical Vein Endothelial Cells
IDE	Insulin-Degrading Enzyme
IGF-1	Insulin-Like Growth Factor-1
IRS	Insulin Receptor Substrate
IPP	Isopentenyl Diphosphate
IL	Interleukin
JNK	c-Jun N-terminal kinase
LDH	Lactate Dehydrogenase
LDL	Low Density Lipoproteins-Cholesterol
MCP-1	Monocyte Chemoattractant Protein-1
MDA	Malondialdehyde
MMP	Mitochondrial Membrane Potential
NF-κB	Nuclear Factor Kappa-Light-Chain-Enhancer of Activated B Cells
Pck1	Phosphoenolpyruvate Carboxykinase 1
PI3K	Phosphatidylinositide 3-Kinase
PDK1	3-Phosphoinositide-Dependent Protein Kinase-1
PDX-1	Pancreatic and Duodenal Homeobox 1 or Insulin Promoter Factor 1
PEPCK	Phosphoenolpyruvate Carboxykinase
PPAR	Peroxisome Proliferator-Activated Receptor
ROS	Reactive Oxygen Species
SOD	Superoxide Dismutase
T1D	Type-1 Diabetes
T2D	Type-2 Diabetes
TC	Total Cholesterol
TG	Triglycerides
TNF	Tumor Necrosis Factor
VLDL	Very Low Density Lipoproteins

References

1. WHO. Diabetes. Available online: http://www.who.int/mediacentre/factsheets/fs312/en/ (accessed on 15 August 2017).
2. WHO. Obesity and Overweight. Available online: http://www.who.int/mediacentre/factsheets/fs311/en/ (accessed on 15 August 2017).

3. McCall, A.L.; Farhy, L.S. Treating type 1 diabetes: From strategies for insulin delivery to dual hormonal control. *Minerva Endocrinol.* **2013**, *38*, 145–163. [PubMed]

4. Swinnen, S.G.; Hoekstra, J.B.; DeVries, J.H. Insulin therapy for Type 2 diabetes. *Diabetes Care* **2009**, *32*, S253–S259. [CrossRef] [PubMed]

5. Upadhyay, J.; Polyzos, S.A.; Perakakis, N.; Thakkar, B.; Paschou, S.A.; Katsiki, N.; Underwood, P.; Park, K.-H.; Seufert, J.; Kang, E.S.; et al. Pharmacotherapy of type 2 diabetes: An update. *Metabolism* **2018**, *78*, 13–42. [CrossRef] [PubMed]

6. Chaudhury, A.; Duvoor, C.; Dendi, V.S.R.; Kraleti, S.; Chada, A.; Ravilla, R.; Marco, A.; Shekhawat, S.N.; Montales, M.T.; Kuriakose, K.; et al. Clinical review of antidiabetic drugs: Implications for type 2 diabetes mellitus management. *Front. Endocrinol.* **2017**, *8*, 6. [CrossRef] [PubMed]

7. Stein, S.A.; Lamos, E.M.; Davis, S.N. A review of the efficacy and safety of oral antidiabetic drugs. *Expert Opin. Drug. Saf.* **2013**, *12*, 153–175. [CrossRef] [PubMed]

8. Habtemariam, S. Going back to the good old days: The merit of crude plant drug mixtures in the 21st century. *Int. J. Complement. Altern. Med.* **2017**, *6*, 1–5. [CrossRef]

9. Habtemariam, S.; Varghese, G.K. Antioxidant, anti-α-glucosidase and pancreatic β-cell protective effects of methanolic extract of *Ensete superbum* Cheesm seeds. *Asian Pac. J. Trop. Biomed.* **2017**, *7*, 121–125. [CrossRef]

10. Habtemariam, S.; Lentini, G. The therapeutic potential of rutin for diabetes: An update. *Mini Rev. Med. Chem.* **2015**, *15*, 524–528. [CrossRef] [PubMed]

11. Habtemariam, S.; Varghese, G.K. The antidiabetic therapeutic potential of dietary polyphenols. *Curr. Pharm. Biotechnol.* **2014**, *15*, 391–400. [CrossRef] [PubMed]

12. Tholl, D. Biosynthesis and biological functions of terpenoids in plants. *Adv. Biochem. Eng. Biotechnol.* **2015**, *148*, 63–106. [CrossRef] [PubMed]

13. Dewick, P.M. The biosynthesis of C5–C25 terpenoid compounds. *Nat. Prod. Rep.* **2002**, *19*, 181–222. [CrossRef] [PubMed]

14. Rehman, R.; Hanif, M.A.; Mushtaq, Z.; Al-Sadi, A.M. Biosynthesis of essential oils in aromatic plants: A review. *Food Rev. Int.* **2016**, *32*, 117–160. [CrossRef]

15. Miziorko, H.M. Enzymes of the mevalonate pathway of isoprenoid biosynthesis. *Arch. Biochem. Biophys.* **2011**, *505*, 131–143. [CrossRef] [PubMed]

16. Dinda, B.; Debnath, S.; Banik, R. Naturally occurring iridoids and secoiridoids. An updated review, Part 4. *Chem. Pharm. Bull.* **2011**, *59*, 803–833. [CrossRef] [PubMed]

17. Pichersky, E.; Raguso, R.A. Why do plants produce so many terpenoid compounds? *New Phytol.* **2016**, in press. [CrossRef] [PubMed]

18. Llusià, J.; Estiarte, M.; Peñuelas, J. Terpenoids and plant communication. *Butll. Inst. Catalana d'Hist. Nat.* **1996**, *64*, 125–133.

19. Singh, B.; Sharma, R.A. Plant terpenes: Defense responses, phylogenetic analysis, regulation and clinical applications. *Biotechnology* **2015**, *5*, 129–151. [CrossRef] [PubMed]

20. Thoppil, R.J.; Bishayee, A. Terpenoids as potential chemopreventive and therapeutic agents in liver cancer. *World J. Hepatol.* **2011**, *3*, 228–249. [CrossRef] [PubMed]

21. Wang, G.; Tang, W.; Bidigare, R.R. Terpenoids as therapeutic drugs and pharmaceutical agents. In *Natural Products*; Zhang, L., Demain, A.L., Eds.; ACS Publications: Washington, DC, USA, 2005; pp. 197–227.

22. Dambolena, J.S.; Zunino, M.P.; Herrera, J.M.; Pizzolitto, R.P.; Areco, V.A.; Zygadlo, J.A. Terpenes: Natural products for controlling insects of importance to human Health—A structure-activity relationship study. *Psyche J. Entomol.* **2016**, *2016*, 4595823. [CrossRef]

23. Nazzaro, F.; Fratianni, F.; de Martino, L.; Coppola, R.; de Feo, V. Effect of essential oils on pathogenic bacteria. *Pharmaceuticals* **2013**, *6*, 1451–1474. [CrossRef] [PubMed]

24. Boskabady, M.H.; Jandaghi, P. Relaxant effects of carvacrol on guinea pig tracheal chains and its possible mechanisms. *Pharmazie* **2003**, *58*, 661–663. [PubMed]

25. Eccles, R.; Jones, A.S. The effect of menthol on nasal resistance to air flow. *J. Laryngol. Otol.* **1983**, *97*, 705–709. [CrossRef] [PubMed]

26. Burrow, A.; Eccles, R.; Jones, A.S. The effects of camphor, eucalyptus and menthol vapour on nasal resistance to airflow and nasal sensation. *Acta Otolaryngol.* **1983**, *96*, 157–161. [CrossRef] [PubMed]

27. Brum, L.F.S.; Elisabetsky, E.; Souza, D. Effects of linalool on [^{3}H]MK801 and [^{3}H] muscimol binding in mouse cortical membranes. *Phytother. Res.* **2001**, *15*, 422–425. [CrossRef] [PubMed]

28. Park, T.J.; Park, Y.S.; Lee, T.G.; Ha, H.; Kim, KT. Inhibition of acetylcholine-mediated effects by borneol. *Biochem. Pharmacol.* **2003**, *65*, 83–90. [CrossRef]
29. Dagli Gul, A.S.; Fadillioglu, E.; Karabulut, I.; Yesilyurt, A.; Delibasi, T. The effects of oral carvacrol treatment against H_2O_2 induced injury on isolated pancreas islet cells of rats. *Islets* **2013**, *5*, 149–155. [CrossRef] [PubMed]
30. Joglekar, M.M.; Panaskar, S.N.; Arvindekar, A.U. Inhibition of advanced glycation end product formation by cymene—A common food constituent. *J. Funct. Foods* **2014**, *6*, 107–115. [CrossRef]
31. Ma, C.J.; Nie, A.F.; Zhang, Z.J.; Zhang, Z.G.; Du, L.; Li, X.Y.; Ning, G. Genipin stimulates glucose transport in C_2C_{12} myotubes via an IRS-1 and calcium-dependent mechanism. *J. Endocrinol.* **2013**, *216*, 353–362. [CrossRef] [PubMed]
32. Hao, Y.; Liu, C.; Yin, F.; Zhang, Y.; Liu, J. 5′-AMP-activated protein kinase plays an essential role in geniposide-regulated glucose-stimulated insulin secretion in rat pancreatic INS-1 β cells. *J. Nat. Med.* **2017**, *71*, 123–130. [CrossRef] [PubMed]
33. Liu, C.Y.; Hao, Y.N.; Yin, F.; Zhang, Y.L.; Liu, J.H. Geniposide accelerates proteasome degradation of Txnip to inhibit insulin secretion in pancreatic β-cells. *J. Endocrinol. Investig.* **2017**, *40*, 505–512. [CrossRef] [PubMed]
34. Zhang, Y.; Ding, Y.; Zhong, X.; Guo, Q.; Wang, H.; Gao, J.; Bai, T.; Ren, L.; Guo, Y.; Jiao, X.; et al. Geniposide acutely stimulates insulin secretion in pancreatic β-cells by regulating GLP-1 receptor/cAMP signaling and ion channels. *Mol. Cell Endocrinol.* **2016**, *430*, 89–96. [CrossRef] [PubMed]
35. Zhang, Y.; Yin, F.; Liu, J.; Liu, Z.; Guo, L.; Xia, Z.; Zidichouski, J. Geniposide attenuates insulin-deficiency-induced acceleration of β-amyloidosis in an APP/PS1 transgenic model of Alzheimer's disease. *Neurochem. Int.* **2015**, *89*, 7–16. [CrossRef] [PubMed]
36. Guo, L.X.; Liu, J.H.; Yin, F. Regulation of insulin secretion by geniposide: Possible involvement of phosphatidylinositol 3-phosphate kinase. *Eur. Rev. Med. Pharmacol. Sci.* **2014**, *18*, 1287–1294. [PubMed]
37. Liu, J.; Zhang, Y.; Deng, X.; Yin, F. Geniposide decreases the level of Aβ1-42 in the hippocampus of streptozotocin-induced diabetic rats. *Acta Biochim. Biophys. Sin.* **2013**, *45*, 787–791. [CrossRef] [PubMed]
38. Liu, J.; Yin, F.; Xiao, H.; Guo, L.; Gao, X. Glucagon-like peptide 1 receptor plays an essential role in geniposide attenuating lipotoxicity-induced β-cell apoptosis. *Toxicol. In Vitro* **2012**, *26*, 1093–1097. [CrossRef] [PubMed]
39. Guo, L.X.; Xia, Z.N.; Gao, X.; Yin, F.; Liu, J.H. Glucagon-like peptide 1 receptor plays a critical role in geniposide-regulated insulin secretion in INS-1 cells. *Acta Pharmacol. Sin.* **2012**, *33*, 237–241. [CrossRef] [PubMed]
40. Kojima, K.; Shimada, T.; Nagareda, Y.; Watanabe, M.; Ishizaki, J.; Sai, Y.; Miyamoto, K.; Aburada, M. Preventive effect of geniposide on metabolic disease status in spontaneously obese type 2 diabetic mice and free fatty acid-treated HepG2 cells. *Biol. Pharm. Bull.* **2011**, *34*, 1613–1618. [CrossRef] [PubMed]
41. Huang, X.J.; Li, J.; Mei, Z.Y.; Chen, G. Gentiopicroside and sweroside from *Veratrilla baillonii* Franch. induce phosphorylation of Akt and suppress Pck1 expression in hepatoma cells. *Biochem. Cell Biol.* **2016**, *94*, 270–278. [CrossRef] [PubMed]
42. Kong, P.; Chi, R.; Zhang, L.; Wang, N.; Lu, Y. Effects of paeoniflorin on tumor necrosis factor-α-induced insulin resistance and changes of adipokines in 3T3-L1 adipocytes. *Fitoterapia* **2013**, *91*, 44–50. [CrossRef] [PubMed]
43. Jiang, B.; Qiao, J.; Yang, Y.; Lu, Y. Inhibitory effect of paeoniflorin on the inflammatory vicious cycle between adipocytes and macrophages. *J. Cell. Biochem.* **2012**, *113*, 2560–2566. [CrossRef] [PubMed]
44. Ha, D.T.; Trung, T.N.; Hien, T.T.; Dao, T.T.; Yim, N.; Ngoc, T.M.; Oh, W.K.; Bae, K. Selected compounds derived from MoutanCortex stimulated glucose uptake and glycogen synthesis via AMPK activation in human HepG2 cells. *J. Ethnopharmacol.* **2010**, *131*, 417–424. [CrossRef] [PubMed]
45. Tan, X.C.; Chua, K.H.; Ravishankar-Ram, M.; Kuppusamy, U.R. Monoterpenes: Novel insights into their biological effects and roles on glucose uptake and lipid metabolism in 3T3-L1 adipocytes. *Food Chem.* **2016**, *196*, 242–250. [CrossRef] [PubMed]
46. Malmir, M.; Gohari, A.R.; Saeidnia, S.; Silva, O. A new bioactive monoterpene–flavonoid from *Satureja khuzistanica*. *Fitoterapia* **2015**, *105*, 107–112. [CrossRef] [PubMed]
47. Patel, T.P.; Rawal, K.; Soni, S.; Gupta, S. Swertiamarin ameliorates oleic acid induced lipid accumulation and oxidative stress by attenuating gluconeogenesis and lipogenesis in hepatic steatosis. *Biomed. Pharmacother.* **2016**, *83*, 785–791. [CrossRef] [PubMed]

48. Alkhateeb, H.; Bonen, A. Thujone, a component of medicinal herbs, rescues palmitate-induced insulin resistance in skeletal muscle. *Am. J. Physiol. Regul. Integr. Comp. Physiol.* **2010**, *299*, R804–R812. [CrossRef] [PubMed]

49. Habtemariam, S.; Varghese, G.K. A novel diterpene skeleton: Identification of a highly aromatic, cytotoxic and antioxidant 5-methyl-10-demethyl-abietane-type diterpene from *Premna serratifolia*. *Phyther. Res.* **2015**, *29*, 80–85. [CrossRef] [PubMed]

50. Habtemariam, S. Investigation into the antioxidant and antidiabetic potential of *Moringa stenopetala*: Identification of the active principles. *Nat. Prod. Commun.* **2015**, *10*, 475–478. [PubMed]

51. Habtemariam, S.; Varghese, G.K. Extractability of rutin in herbal tea preparations of *Moringa stenopetala* leaves. *Beverages* **2015**, *1*, 169–182. [CrossRef]

52. Roselli, M.; Lentini, G.; Habtemariam, S. Phytochemical, antioxidant and anti-α-glucosidase activity evaluations of *Bergenia cordifolia*. *Phyther. Res.* **2012**, *26*, 908–914. [CrossRef] [PubMed]

53. Habtemariam, S.; Cowley, R.A. Antioxidant and anti-α-glucosidase ccompounds from the rhizome of *Peltiphyllum peltatum* (Torr.) Engl. *Phytother. Res.* **2012**, *26*, 1656–1660. [CrossRef] [PubMed]

54. Habtemariam, S. Methyl-3-*O*-methyl gallate and gallic acid from the leaves of *Peltiphyllum peltatum*: Isolation and comparative antioxidant, prooxidant, and cytotoxic effects in neuronal cells. *J. Med. Food* **2011**, *14*, 1412–1418. [CrossRef] [PubMed]

55. Juan-Badaturuge, M.; Habtemariam, S.; Thomas, M.J.K. Antioxidant compounds from a South Asian beverage and medicinal plant, *Cassia auriculata*. *Food Chem.* **2011**, *125*, 221–225. [CrossRef]

56. Juan-Badaturugea, M.; Habtemariam, S.; Jackson, C.; Thomas, M.J.K. Antioxidant principles of *Tanacetum vulgare* L. aerial part. *Nat. Prod. Commun.* **2009**, *4*, 1561–1564.

57. Habtemariam, S.; Dagne, E. Comparative antioxidant, prooxidant, and cytotoxic activity of sigmoidin A and eriodictyol. *Planta Med.* **2010**, *76*, 589–594. [CrossRef] [PubMed]

58. Habtemariam, S. Activity-guided isolation and identification of free Radical-scavenging components from ethanolic extract of Boneset (Leaves of *Eupatorium perfoliatum*). *Nat. Prod. Commun.* **2008**, *3*, 1317–1320.

59. Habtemariam, S.; Jackson, C. Antioxidant and cytoprotective activity of leaves of *Peltiphyllum peltatum* (Torr.) Engl. *Food Chem.* **2007**, *105*, 498–503. [CrossRef]

60. Habtemariam, S. Flavonoids as inhibitors or enhancers of the cytotoxicity of tumor necrosis factor-α in L-929 tumor cells. *J. Nat. Prod.* **1997**, *60*, 775–778. [CrossRef] [PubMed]

61. Habtemariam, S. Modulation of tumour necrosis factor-α-induced cytotoxicity by polyphenols. *Phyther. Res.* **1997**, *11*, 277–280. [CrossRef]

62. Habtemariam, S. Catechols and quercetin reduce MTT through iron ions: A possible artefact in cell viability assay. *Phyther. Res.* **1995**, *9*, 603–605. [CrossRef]

63. Varghese, G.K.; Bose, L.V.; Habtemariam, S. Antidiabetic components of *Cassia alata* leaves: Identification through α-glucosidase inhibition studies. *Pharm. Biol.* **2013**, *51*, 345–349. [CrossRef] [PubMed]

64. Habtemariam, S. Antihyperlipidemic components of *Cassia auriculata* aerial parts: Identification through in vitro studies. *Phytother. Res.* **2013**, *27*, 152–155. [CrossRef] [PubMed]

65. Habtemariam, S. α-Glucosidase inhibitory activity of kaempferol-3-*O*-rutinoside. *Nat. Prod. Commun.* **2011**, *6*, 201–203. [PubMed]

66. Jin, L.; Xue, H.Y.; Jin, L.J.; Li, S.Y.; Xu, Y.P. Antioxidant and pancreas-protective effect of aucubin on rats with streptozotocin-induced diabetes. *Eur. J. Pharmacol.* **2008**, *582*, 162–167. [CrossRef] [PubMed]

67. Kodikonda, M.; Naik, P.R. Ameliorative effect of borneol, a natural bycyclic monoterpene against hyperglycemia, hyperlipidemia and oxidative stress in streptozotocin-induced diabetic Wistar rats. *Biomed. Pharmacother.* **2017**, *96*, 336–347. [CrossRef]

68. Ezhumalai, M.; Ashokkumar, N.; Pugalendi, K.V. Combination of carvacrol and rosiglitazone ameliorates high fat diet induced changes in lipids and inflammatory markers in C57BL/6J mice. *Biochimie* **2015**, *110*, 129–136. [CrossRef] [PubMed]

69. Ezhumalai, M.; Radhiga, T.; Pugalendi, K.V. Antihyperglycemic effect of carvacrol in combination with rosiglitazone in high-fat diet-induced type 2 diabetic C57BL/6J mice. *Mol. Cell. Biochem.* **2014**, *385*, 23–31. [CrossRef] [PubMed]

70. Deng, W.; Lu, H.; Teng, J. Carvacrol attenuates diabetes-associated cognitive deficits in rats. *J. Mol. Neurosci.* **2013**, *51*, 813–819. [CrossRef] [PubMed]

71. Samarghandian, S.; Farkhondeh, T.; Samini, F.; Borji, A. Protective Effects of Carvacrol against Oxidative Stress Induced by Chronic Stress in Rat's Brain, Liver, and Kidney. *Biochem. Res. Int.* **2016**, *2016*, 2645237. [CrossRef] [PubMed]

72. Bayramoglu, G.; Senturk, H.; Bayramoglu, A.; Uyanoglu, M.; Colak, S.; Ozmen, A.; Kolankaya, D. Carvacrol partially reverses symptoms of diabetes in STZ-induced diabetic rats. *Cytotechnology* **2014**, *66*, 251–257. [CrossRef] [PubMed]

73. Muruganathan, U.; Srinivasan, S. Beneficial effect of carvone, a dietary monoterpene ameliorates hyperglycemia by regulating the key enzymes activities of carbohydrate metabolism in streptozotocin-induced diabetic rats. *Biomed. Pharmacother.* **2016**, *84*, 1558–1567. [CrossRef] [PubMed]

74. Yang, S.; Deng, H.; Zhang, Q.; Xie, J.; Zeng, H.; Jin, X.; Ling, Z.; Shan, Q.; Liu, M.; Ma, Y.; et al. Amelioration of diabetic mouse nephropathy by catalpol correlates with down-regulation of Grb10 expression and activation of insulin-like growth factor 1/insulin-like growth factor 1 receptor signaling. *PLoS ONE* **2016**, *11*. [CrossRef] [PubMed]

75. Zhou, J.; Xu, G.; Ma, S.; Li, F.; Yuan, M.; Xu, H.; Huang, K. Catalpol ameliorates high-fat diet-induced insulin resistance and adipose tissue inflammation by suppressing the JNK and NF-κB pathways. *Biochem. Biophys. Res. Commun.* **2015**, *467*, 853–858. [CrossRef] [PubMed]

76. Huang, W.J.; Niu, H.S.; Lin, M.H.; Cheng, J.T.; Hsu, F.L. Antihyperglycemic effect of catalpol in streptozotocin-induced diabetic rats. *J. Nat. Prod.* **2010**, *73*, 1170–1172. [CrossRef] [PubMed]

77. Wang, C.F.; Li, D.Q.; Xue, H.Y.; Hu, B. Oral supplementation of catalpol ameliorates diabetic encephalopathy in rats. *Brain Res.* **2010**, *1307*, 158–165. [CrossRef] [PubMed]

78. Srinivasan, S.; Muruganathan, U. Antidiabetic efficacy of citronellol, a citrus monoterpene by ameliorating the hepatic key enzymes of carbohydrate metabolism in streptozotocin-induced diabetic rats. *Chem. Biol. Interact.* **2016**, *250*, 38–46. [CrossRef] [PubMed]

79. Guan, L.; Feng, H.; Gong, D.; Zhao, X.; Cai, L.; Wu, Q.; Yuan, B.; Yang, M.; Zhao, J.; Zou, Y. Genipin ameliorates age-related insulin resistance through inhibiting hepatic oxidative stress and mitochondrial dysfunction. *Exp. Gerontol.* **2013**, *48*, 1387–1394. [CrossRef] [PubMed]

80. Zhang, Y.; Yin, F.; Liu, J.; Liu, Z. Geniposide Attenuates the phosphorylation of tau protein in cellular and insulin-deficient APP/PS1 transgenic mouse model of Alzheimer's disease. *Chem. Biol. Drug Des.* **2016**, *87*, 409–418. [CrossRef] [PubMed]

81. Gao, C.; Liu, Y.; Jiang, Y.; Ding, J.; Li, L. Geniposide ameliorates learning memory deficits, reduces tau phosphorylation and decreases apoptosis via GSK3β pathway in streptozotocin-induced Alzheimer rat model. *Brain Pathol.* **2014**, *24*, 261–269. [CrossRef] [PubMed]

82. Zhang, L.; Yang, B.; Yu, B. Paeoniflorin protects against nonalcoholic fatty liver disease induced by a high-fat diet in mice. *Biol. Pharm. Bull.* **2015**, *38*, 1005–1011. [CrossRef] [PubMed]

83. Ma, T.; Huang, C.; Zong, G.; Zha, D.; Meng, X.; Li, J.; Tang, W. Hepatoprotective effects of geniposide in a rat model of nonalcoholic steatohepatitis. *J. Pharm. Pharmacol.* **2011**, *63*, 587–593. [CrossRef] [PubMed]

84. Wu, S.Y.; Wang, G.F.; Liu, Z.Q.; Rao, J.J.; Lü, L.; Xu, W.; Wu, S.G.; Zhang, J.J. Effect of geniposide, a hypoglycemic glucoside, on hepatic regulating enzymes in diabetic mice induced by a high-fat diet and streptozotocin. *Acta Pharmacol. Sin.* **2009**, *30*, 202–208. [CrossRef] [PubMed]

85. Babukumar, S.; Vinothkumar, V.; Sankaranarayanan, C.; Srinivasan, S. Geraniol, a natural monoterpene, ameliorates hyperglycemia by attenuating the key enzymes of carbohydrate metabolism in streptozotocin-induced diabetic rats. *Pharm. Biol.* **2017**, *55*, 1442–1449. [CrossRef] [PubMed]

86. Murali, R.; Saravanan, R. Antidiabetic effect of D-limonene, a monoterpene in streptozotocin-induced diabetic rats. *Biomed. Prevent. Nutr.* **2012**, *2*, 269–275. [CrossRef]

87. Bacanlı, M.; Anlar, H.G.; Aydın, S.; Çal, T.; Arı, N.; Ündeğer Bucurgat, Ü.; Başaran, A.A.; Başaran, N. D-limonene ameliorates diabetes and its complications in streptozotocin-induced diabetic rats. *Food Chem. Toxicol.* **2017**, *110*, 434–442. [CrossRef] [PubMed]

88. Liu, K.; Xu, H.; Lv, G.; Liu, B.; Lee, M.K.; Lu, C.; Lv, X.; Wu, Y. Loganin attenuates diabetic nephropathy in C57BL/6J mice with diabetes induced by streptozotocin and fed with diets containing high level of advanced glycation end products. *Life Sci.* **2015**, *123*, 78–85. [CrossRef] [PubMed]

89. Muruganathan, U.; Srinivasan, S.; Vinothkumar, V. Antidiabetogenic efficiency of menthol, improves glucose homeostasis and attenuates pancreatic β-cell apoptosis in streptozotocin-nicotinamide induced experimental rats through ameliorating glucose metabolic enzymes. *Biomed Pharmacother.* **2017**, *92*, 229–239. [CrossRef] [PubMed]

90. Rathinam, A.; Pari, L. Myrtenal ameliorates hyperglycemia by enhancing GLUT2 through Akt in the skeletal muscle and liver of diabetic rats. *Chem. Biol. Interact.* **2016**, *256*, 161–166. [CrossRef] [PubMed]

91. Rathinam, A.; Pari, L.; Chandramohan, R.; Sheikh, B.A. Histopathological findings of the pancreas, liver, and carbohydrate metabolizing enzymes in STZ-induced diabetic rats improved by administration of myrtenal. *J. Physiol. Biochem.* **2014**, *70*, 935–946. [CrossRef] [PubMed]

92. Ayyasamy, R.; Leelavinothan, P. Myrtenal alleviates hyperglycaemia, hyperlipidaemia and improves pancreatic insulin level in STZ-induced diabetic rats. *Pharm. Biol.* **2016**, *54*, 2521–2527. [CrossRef] [PubMed]

93. Sun, X.; Li, S.; Xu, L.; Wang, H.; Ma, Z.; Fu, Q.; Qu, R.; Ma, S. Paeoniflorin ameliorates cognitive dysfunction via regulating SOCS2/IRS-1 pathway in diabetic rats. *Physiol. Behav.* **2017**, *174*, 162–169. [CrossRef] [PubMed]

94. Vaidya, H.; Prajapati, A.; Rajani, M.; Sudarsanam, V.; Padh, H.; Goyal, R.K. Beneficial Effects of Swertiamarin on dyslipidaemia in streptozotocin-induced Type 2 diabetic rats. *Phytother. Res.* **2012**, *26*, 1259–1261. [CrossRef] [PubMed]

95. Saravanan, S.; Pari, L. Role of thymol on hyperglycemia and hyperlipidemia in high fat diet-induced type 2 diabetic C57BL/6J mice. *Eur. J. Pharmacol.* **2015**, *761*, 279–287. [CrossRef] [PubMed]

96. Saravanan, S.; Pari, L. Protective effect of thymol on high fat diet induced diabetic nephropathy in C57BL/6J mice. *Chem. Biol. Interact.* **2016**, *245*, 1–11. [CrossRef] [PubMed]

97. Fang, F.; Li, H.; Qin, T.; Li, M.; Ma, S. Thymol improves high-fat diet-induced cognitive deficits in mice via ameliorating brain insulin resistance and upregulating NRF2/HO-1 pathway. *Metab. Brain Dis.* **2017**, *32*, 385–393. [CrossRef] [PubMed]

98. Haque, M.R.; Ansari, S.H.; Najmi, A.K.; Ahmad, M.A. Monoterpene phenolic compound thymol prevents high fat diet induced obesity in murine model. *Toxicol. Mech. Methods* **2014**, *24*, 116–123. [CrossRef] [PubMed]

99. Liu, Z.Q.; Jiang, Z.H.; Liu, L.; Hu, M. Mechanisms responsible for poor oral bioavailability of paeoniflorin: Role of intestinal disposition and interactions with sinomenine. *Pharm. Res.* **2006**, *23*, 2768–2780. [CrossRef] [PubMed]

100. Cheng, C.; Lin, J.-Z.; Li, L.; Yang, J.-L.; Jia, W.-W.; Huang, W.-H.; Du, F.-F.; Wang, F.-Q.; Li, M.-J.; Li, Y.F.; et al. Pharmacokinetics and disposition of monoterpene glycosides derived from *Paeonia lactiflora* roots (Chishao) after intravenous dosing of antiseptic XueBiJing injection in human subjects and rats. *Acta Pharmacol. Sin.* **2016**, *37*, 530–544. [CrossRef] [PubMed]

101. Martey, O.N.K.; Shi, X.; He, X. Advance in pre-clinical pharmacokinetics of paeoniflorin, a major monoterpene glucoside from the root of *Paeonia lactiflora*. *Pharmacol. Pharm.* **2013**, *4*, 4–14. [CrossRef]

102. Austgulen, L.T.; Solheim, E.; Scheline, R.R. Metabolism in rats of *p*-cymene derivatives: Carvacrol and thymol. *Pharmacol. Toxicol.* **1987**, *61*, 98–102. [CrossRef] [PubMed]

103. Dong, R.H.; Fang, Z.Z.; Zhu, L.L.; Liang, S.C.; Ge, G.B.; Liu, Z.Y. Investigation of UDP-glucuronosyltransferases (UGTs) inhibitory properties of carvacrol. *Phytother. Res.* **2012**, *26*, 86–90. [CrossRef] [PubMed]

104. Kohlert, C.; Schindler, G.; März, R.W.; Abel, G.; Brinkhaus, B.; Derendorf, H.; Gräfe, E.U.; Veit, M. Systemic availability and pharmacokinetics of thymol in humans. *J. Clin. Pharmacol.* **2002**, *42*, 731–737. [CrossRef] [PubMed]

105. Miller, J.A.; Lang, J.E.; Ley, M.; Nagle, R.; Hsu, C.H.; Thompson, P.A.; Cordova, C.; Waer, A.; Chow, H.H. Human breast tissue disposition and bioactivity of limonene in women with early stage breast cancer. *Cancer Prev. Res.* **2013**, *6*, 577–584. [CrossRef] [PubMed]

106. Miller, J.A.; Hakim, I.A.; Chew, W.; Thompson, P.; Thomson, C.A.; Chow, H.H. Adipose tissue accumulation of D-limonene with the consumption of a lemonade preparation rich in D-limonene content. *Nutr. Cancer* **2010**, *62*, 783–788. [CrossRef] [PubMed]

107. Li, H.L.; He, J.C.; Bai, M.; Song, Q.Y.; Feng, E.F.; Rao, G.X.; Xu, G.L. Determination of the plasma pharmacokinetic and tissue distributions of swertiamarin in rats by liquid chromatography with tandem mass spectrometry. *Arzneimittelforschung* **2012**, *62*, 138–144. [CrossRef] [PubMed]

108. Cheng, S.; Lin, L.C.; Lin, C.H.; Tsai, T.H. Comparative oral bioavailability of geniposide following oral administration of geniposide, *Gardenia jasminoides* Ellis fruits extracts and Gardenia herbal formulation in rats. *J. Pharm. Pharmacol.* **2014**, *66*, 705–712. [CrossRef] [PubMed]

109. Ahmadian, M.; Suh, J.M.; Hah, N.; Liddle, C.; Atkins, A.R.; Downes, M.; Evans, R.M. PPARγ signaling and metabolism: The good, the bad and the future. *Nat. Med.* **2013**, *19*, 557–566. [CrossRef] [PubMed]

110. Eldor, R.; DeFronzo, R.A.; Abdul-Ghani, M. In vivo actions of peroxisome proliferator-activated receptors: Glycemic control, insulin sensitivity, and insulin secretion. *Diabetes Care* **2013**, *36* (Suppl. 2), S162–S174. [CrossRef] [PubMed]

111. Soccio, R.E.; Chen, E.R.; Rajapurkar, S.R.; Safabakhsh, P.; Marinis, J.M.; Dispirito, J.R.; Emmett, M.J.; Briggs, E.R.; Fang, B.; Everett, L.J.; et al. Genetic variation determines PPARγ function and anti-diabetic drug response in vivo. *Cell* **2015**, *162*, 33–44. [CrossRef] [PubMed]

112. Delea, T.E.; Edelsberg, J.S.; Hagiwara, M.; Oster, G.; Phillips, L.S. Use of thiazolidinediones and risk of heart failure in people with type 2 diabetes: A retrospective cohort study. *Diabetes Care* **2003**, *26*, 2983–2989. [CrossRef] [PubMed]

113. Chi, C.; Snaith, J.; Gunton, J.E. Diabetes Medications and Cardiovascular Outcomes in Type 2 Diabetes. *Heart Lung Circ.* **2017**, *26*, 1133–1141. [CrossRef] [PubMed]

114. Wang, L.; Waltenberger, B.; Pferschy-Wenzig, E.M.; Blunder, M.; Liu, X.; Malainer, C.; Blazevic, T.; Schwaiger, S.; Rollinger, J.M.; Heiss, E.H.; et al. Natural product agonists of peroxisome proliferator-activated receptor γ (PPARγ): A review. *Biochem. Pharmacol.* **2014**, *92*, 73–89. [CrossRef] [PubMed]

115. Atanasov, A.G.; Wang, J.N.; Gu, S.P.; Bu, J.; Kramer, M.P.; Baumgartner, L.; Fakhrudin, N.; Ladurner, A.; Malainer, C.; Vuorinen, A.; et al. Honokiol: A nonadipogenic PPARγ agonist from nature. *Biochim. Biophys. Acta* **2013**, *1830*, 4813–4819. [CrossRef] [PubMed]

116. Chopra, I.; Li, H.F.; Wang, H.; Webster, K.A. Phosphorylation of the insulin receptor by AMP-activated protein kinase (AMPK) promotes ligand-independent activation of the insulin signalling pathway in rodent muscle. *Diabetologia* **2012**, *55*, 783–794. [CrossRef] [PubMed]

117. Meng, S.; Cao, J.; He, Q.; Xiong, L.; Chang, E.; Radovick, S.; Wondisford, F.E.; He, L. Metformin activates AMP-activated protein kinase by promoting formation of the αβγ heterotrimeric complex. *J. Biol. Chem.* **2015**, *290*, 3793–3802. [CrossRef] [PubMed]

118. Gaidhu, M.P.; Fediuc, S.; Anthony, N.M.; So, M.; Mirpourian, M.; Perry, R.L.; Ceddia, R.B. Prolonged AICAR-induced AMP-kinase activation promotes energy dissipation in white adipocytes: Novel mechanisms integrating HSL and ATGL. *J. Lipid Res.* **2009**, *50*, 704–715. [CrossRef] [PubMed]

119. Kim, J.W.; You, Y.H.; Ham, D.S.; Yang, H.K.; Yoon, K.H. The paradoxical effects of AMPK on insulin gene expression and glucose-induced insulin secretion. *J. Cell Biochem.* **2016**, *117*, 239–246. [CrossRef] [PubMed]

120. Yavari, A.; Stocker, C.J.; Ghaffari, S.; Wargent, E.T.; Steeples, V.; Czibik, G.; Pinter, K.; Bellahcene, M.; Woods, A.; Martínez de Morentin, P.B.; et al. Chronic activation of γ2 AMPK induces obesity and reduces β cell function. *Cell Metab.* **2016**, *23*, 821–836. [CrossRef] [PubMed]

121. Li, J.; Zhong, L.; Wang, F.; Zhu, G. Dissecting the role of AMP-activated protein kinase in human diseases. *Acta Pharm. Sin. B* **2017**, *7*, 249–259. [CrossRef] [PubMed]

122. Belwal, T.; Nabavi, S.F.; Nabavi, S.M.; Habtemariam, S. Dietary anthocyanins and insulin resistance: When food becomes a medicine. *Nutrients* **2017**, *9*, 1111. [CrossRef] [PubMed]

123. Atkinson, B.J.; Griesel, B.A.; King, C.D.; Josey, M.A.; Olson, A.L. Moderate GLUT4 overexpression improves insulin sensitivity and fasting triglyceridemia in high-fat diet–fed transgenic mice. *Diabetes* **2013**, *62*, 2249–2258. [CrossRef] [PubMed]

124. Huang, S.; Czech, M.P. The GLUT4 Glucose Transporter. *Cell Metab.* **2007**, *5*, 237–252. [CrossRef] [PubMed]

125. Morgan, B.J.; Chai, S.Y.; Albiston, A.L. GLUT4 associated proteins as therapeutic targets for diabetes. *Recent Pat. Endocr. Metab. Immune Drug Discov.* **2011**, *5*, 25–32. [CrossRef] [PubMed]

126. Olson, A.L. Regulation of GLUT4 and insulin-dependent glucose flux. *Int. Sch. Res. Not.* **2012**, *2012*, 856987. [CrossRef] [PubMed]

127. Lee, J.O.; Lee, S.K.; Kim, J.H.; Kim, N.; You, G.Y.; Moon, J.W.; Kim, S.J.; Park, S.H.; Kim, H.S. Metformin regulates glucose transporter 4 (GLUT4) translocation through AMP-activated protein kinase (AMPK)-mediated Cbl/CAP signaling in 3T3-L1 preadipocyte cells. *J. Biol. Chem.* **2012**, *287*, 44121–44129. [CrossRef] [PubMed]

128. Choi, K.H.; Lee, H.A.; Park, M.H.; Han, J.S. Mulberry (*Morus alba* L.) fruit extract containing anthocyanins improves glycemic control and insulin sensitivity via activation of AMP-activated protein kinase in diabetic C57BL/Ksj-db/db mice. *J. Med. Food* **2016**, *19*, 737–745. [CrossRef] [PubMed]

129. Talagavadi, V.; Rapisarda, P.; Galvano, F.; Pelicci, P.; Giorgio, M. Cyanidin-3-*O*-β-glucoside and protocatechuic acid activate AMPK/mTOR/S6K pathway and improve glucose homeostasis in mice. *J. Funct. Foods* **2016**, *21*, 338–348. [CrossRef]

130. Huang, B.; Wang, Z.; Park, J.H.; Ryu, O.H.; Choi, M.K.; Lee, J.Y. Anti-diabetic effect of purple corn extract on C57BL/KsJ db/db mice. *Nutr. Res. Pract.* **2015**, *9*, 22–29. [CrossRef] [PubMed]

131. Yan, F.; Zheng, X. Anthocyanin-rich mulberry fruit improves insulin resistance and protects hepatocytes against oxidative stress during hyperglycemia by regulating AMPK/ACC/mTOR pathway. *J. Funct. Foods* **2017**, *30*, 270–281. [CrossRef]

132. Zhu, W.; Jia, Q.; Wang, Y.; Zhang, Y.; Xia, M. The anthocyanin cyanidin-3-*O*-β-glucoside, a flavonoid, increases hepatic glutathione synthesis and protects hepatocytes against reactive oxygen species during hyperglycemia: Involvement of a cAMP-PKA-dependent signaling pathway. *Free Rad. Biol. Med.* **2012**, *52*, 314–327. [CrossRef] [PubMed]

133. Barf, T. Intervention of hepatic glucose production. Small molecule regulators of potential targets for type 2 diabetes therapy. *Mini Rev. Med. Chem.* **2004**, *4*, 897–908. [CrossRef] [PubMed]

134. Agius, L. New hepatic targets for glycaemic control in diabetes. *Best Pract. Res. Clin. Endocrinol. Metab.* **2007**, *21*, 587–605. [CrossRef] [PubMed]

135. Kurukulasuriya, R.; Link, J.T.; Madar, D.J.; Pei, Z.; Richards, S.J.; Rohde, J.J.; Souers, A.J.; Szczepankiewicz, B.G. Potential drug targets and progress towards pharmacologic inhibition of hepatic glucose production. *Curr. Med. Chem.* **2003**, *10*, 123–153. [CrossRef] [PubMed]

136. McCormack, J.G.; Westergaard, N.; Kristiansen, M.; Brand, C.L.; Lau, J. Pharmacological approaches to inhibit endogenous glucose production as a means of anti-diabetic therapy. *Curr. Pharm. Des.* **2001**, *7*, 1451–1474. [CrossRef] [PubMed]

137. Luft, V.C.; Schmidt, M.I.; Pankow, J.S.; Couper, D.; Ballantyne, C.M.; Young, J.H.; Duncan, B.B. Chronic inflammation role in the obesity-diabetes association: A case-cohort study. *Diabetol. Metab. Syndr.* **2013**, *5*, 31. [CrossRef] [PubMed]

138. Pereira, S.S.; Alvarez-Leite, J.I. Low-grade inflammation, obesity, and diabetes. *Curr. Obes. Rep.* **2014**, *3*, 422–431. [CrossRef] [PubMed]

139. Edirisinghe, I.; Banaszewski, K.; Cappozzo, J. Strawberry anthocyanin and its association with postprandial inflammation and insulin. *Br. J. Nutr.* **2011**, *106*, 913–922. [CrossRef] [PubMed]

140. Donath, M.Y.; Dalmas, É.; Sauter, N.S.; Böni-Schnetzler, M. Inflammation in obesity and diabetes: Islet dysfunction and therapeutic opportunity. *Cell Metab.* **2013**, *17*, 860–872. [CrossRef] [PubMed]

141. Zeidan, M.; Rayan, M.; Zeidan, N.; Falah, M.; Rayan, A. Indexing Natural Products for Their Potential Anti-Diabetic Activity: Filtering and Mapping Discriminative Physicochemical Properties. *Molecules* **2017**, *22*, 1563. [CrossRef] [PubMed]

International Journal of
Molecular Sciences

MDPI

Article

Chemical Composition and Bioactivity of Essential Oil from *Blepharocalyx salicifolius*

Fabiana Barcelos Furtado [1], Bruna Cristina Borges [2], Thaise Lara Teixeira [2], Hans Garcia Garces [3], Luiz Domingues de Almeida Junior [4], Fernanda Cristina Bérgamo Alves [1], Claudio Vieira da Silva [2] and Ary Fernandes Junior [1,*]

[1] Institute of Biosciences of Botucatu, Laboratory of Bacteriology, Unesp-São Paulo State University, Botucatu CEP 18618-000, Brazil; fabisbarcelos@hotmail.com (F.B.F.); fe.bergamo@gmail.com (F.C.B.A.)
[2] Institute of Biomedical Sciences, Trypanosomatids Laboratory (LATRI), UFU-Federal University of Uberlândia, Uberlândia CEP 38400-902, Brazil; brunacb90@gmail.com (B.C.B.); thaiselara@yahoo.com.br (T.L.T.); silva_cv@yahoo.com.br (C.V.d.S.)
[3] Institute of Biosciences of Botucatu, Laboratory of Fungi Biology, Unesp-São Paulo State University, Botucatu CEP 18618-000, Brazil; atiweb@gmail.com
[4] Institute of Biosciences of Botucatu, Laboratory of Phytomedicines, Pharmacology and Biotechnology (PhytoPharmaTech), Unesp-São Paulo State University, Botucatu CEP 18618-000, Brazil; domingues_luiz@hotmail.com
* Correspondence: ary@ibb.unesp.br; Tel.: +55-143-880-0412

Received: 6 November 2017; Accepted: 13 December 2017; Published: 4 January 2018

Abstract: Natural products represent a source of biologically active molecules that have an important role in drug discovery. The aromatic plant *Blepharocalyx salicifolius* has a diverse chemical constitution but the biological activities of its essential oils have not been thoroughly investigated. The aims of this paper were to evaluate in vitro cytotoxic, antifungal and antibacterial activities of an essential oil from leaves of *B. salicifolius* and to identify its main chemical constituents. The essential oil was extracted by steam distillation, chemical composition was determined by gas chromatography/mass spectrometry, and biological activities were performed by a microdilution broth method. The yield of essential oil was 0.86% (w/w), and the main constituents identified were bicyclogermacrene (17.50%), globulol (14.13%), viridiflorol (8.83%), γ-eudesmol (7.89%) and α-eudesmol (6.88%). The essential oil was cytotoxic against the MDA-MB-231 (46.60 μg·mL^{-1}) breast cancer cell line, being more selective for this cell type compared to the normal breast cell line MCF-10A (314.44 μg·mL^{-1}). Flow cytometry and cytotoxicity results showed that this oil does not act by inducing cell death, but rather by impairment of cellular metabolism specifically of the cancer cells. Furthermore, it presented antifungal activity against *Paracoccidioides brasiliensis* (156.25 μg·mL^{-1}) but was inactive against other fungi and bacteria. Essential oil from *B. salicifolius* showed promising biological activities and is therefore a source of molecules to be exploited in medicine or by the pharmaceutical industry.

Keywords: bicyclogermacrene; globulol; viridiflorol; eudesmol; breast cancer; MDA-MB-231; MCF-10A; *Paracoccidioides brasiliensis*; flow cytometry

1. Introduction

Blepharocalix salicifolius (Kunth) O. Berg is an aromatic species that belongs to the Myrtaceae family, widely distributed in South America. It was reported in Paraguay, Uruguay, Argentina, Bolivia, Ecuador and Brazil [1,2]. In Brazil, it is extensively distributed, being found in the North, Southeast and South regions associated with several climatic conditions (humidity and temperature) [1]. *B. salicifolius* is a tree 10–20 m high and 20–40 cm in diameter. It grows in a straight and cylindrical way, barely tortuous, and with a dark brown bark [3]. Several synonyms are attributed to this species, among them *B. tweediei* and *B. giganteus* [4].

In Brazil, it is known with the name of "Murta", and is popularly used to treat respiratory diseases, coughs, colds, hypotension, rheumatism, hypoglycemia, diarrhea, leukorrhea, urethritis and bladder diseases [5–8]. Extracts of this species showed antiparasitic, antifungal, antibacterial, allelopathic, cytotoxic and insecticide effects [2,3,9–15].

The chemical composition of the essential oil in leaves extracted from this species has been reported and mainly includes 1,8-cineol, α-pinene, limonene, β-pinene, α-terpineol, (E)-caryophyllene and linalool [16–27]. However, few studies relating to the biological activities of the essential oil from this plant are available. Until now, only fungistatic activity against *Phyllosticta citricarpa* and antitussive, antispasmodic, bronchodilating and cardiac inotropic effects have been described for the essential oil from leaves of this plant [27,28].

Breast cancer has become a great concern, as it is a leading cause of cancer-related deaths for women worldwide [29]. According to the World Health Organization, more than one million cases of breast cancer are reported worldwide annually [30]. High cost, increasing drug resistance, and side effects of current therapeutic approaches are forcing researchers to explore alternative procedures as options to find new chemical entities for cancer treatments [31]. Likewise, microbial pathogens, such as bacteria and fungi, are a major cause of human disease and are among the most concerning threats to public health due to drug-resistant strains [32,33]. A review estimated that continued rise in microbial resistance could lead to 10 million deaths every year by 2050 [34]. Thus, there is a need for the development of new drugs and, due to broad biological action, essential oils or their isolated compounds can inspire novel discoveries. Therefore, the aim of this study was to evaluate in vitro cytotoxic, antifungal and antibacterial activities of essential oils from the leaves of *B. salicifolius*, and to identify their main chemical constituents.

2. Results and Discussion

2.1. Yield and Chemical Composition of the Essential Oil

The average yield of the essential oil from *B. salicifolius* was 0.86% (*w*/*w*). This is similar when compared to the percentage yield of plants already explored commercially in essential oil production, such as *Ocimum basilicum* (1.24%) and *Thymus vulgaris* (1.05%) [35] or *Eucalyptus citriodora* (1.05%) [36].

According to chemical analysis, 29 compounds were identified from the essential oil of the leaves from this species, and sesquiterpene hydrocarbons were predominant (Table 1). Major compounds identified were bicyclogermacrene (17.50%), globulol (14.13%), viridiflorol (8.83%), γ-eudesmol (7.89%), α-eudesmol (6.88%), rosifoliol (4.54%), (E)-caryophyllene (4.49%), cubeban-11-ol (4.40%), palustrol (3.83%) and α-pinene (3.22%).

Table 1. Chemical composition of the essential oil of the leaves from *Blepharocalyx salicifolius*.

Compound	Arithmetic Retention Index (AI) Reference	Arithmetic Retention Index (AI) Calculated	%
Monoterpene hydrocarbons			
α-pinene	932 [a]	925	3.22
β-pinene	974 [a]	967	0.75
(E)-β-ocimene	1044 [a]	1039	0.26
Oxygenated monoterpenes			
linalool	1095 [a]	1092	0.14
Sesquiterpene hydrocarbons			
δ-elemene	1335 [a]	1329	0.20
α-ylangene	1373 [a]	1365	0.13
isoledene	1374 [a]	1370	0.08
β-elemene	1389 [a]	1384	1.34

Table 1. *Cont.*

Compound	Arithmetic Retention Index (AI) Reference	Arithmetic Retention Index (AI) Calculated	%
α-gurjunene	1409 [a]	1401	0.41
(E)-caryophyllene	1417 [a]	1411	4.49
NI	-	1419	0.09
NI	-	1426	0.10
α-guaiene	1437 [a]	1430	0.57
β-barbatene	1440 [a]	1435	0.12
α-humulene	1452 [a]	1445	0.47
allo-aromadendrene	1458 [a]	1453	0.79
NI	-	1467	0.23
germacrene D	1480 [a]	1473	0.59
4(14),11-eudesmadiene	1487 [a]	1477	0.24
NI	-	1480	0.19
bicyclogermacrene	1500 [a]	1488	17.50
germacrene A	1508 [a]	1496	0.65
γ-cadinene	1513 [a]	1506	0.09
δ-cadinene	1522 [a]	1515	0.40
Oxygenated sesquiterpenes			
elemol	1548 [a]	1541	1.39
NI	-	1543	0.40
NI	-	1550	0.59
palustrol	1567 [a]	1559	3.83
globulol	1582 [b]	1575	14.13
viridiflorol	1592 [a]	1583	8.83
cubeban-11-ol	1595 [a]	1585	4.40
rosifoliol	1600 [a]	1593	4.54
NI	-	1614	2.27
γ-eudesmol	1630 [a]	1622	7.89
β-eudesmol	1649 [a]	1641	3.05
α-eudesmol	1652 [a]	1645	6.88
			Total (%): 91.26

NI = not identified. [a] Adams mass-spectral retention index library [37]; [b] NIST: Standard Reference Data [38].

These results for major compounds are similar to those reported in the literature regarding globulol [17,18], γ-eudesmol and α-eudesmol [22], (E)-caryophyllene [21,23] and α-pinene [17,18,20,21,27]. Other terpenes, such as bicyclogermacrene, viridiflorol, rosifoliol, cubeban-11-ol and palustrol, are being described here for the first time as major components of the essential oil in leaves of *B. salicifolius*. Previous studies showed a diverse chemical constitution for the essential oil of this same species, probably because many factors, such as soil characteristics, extraction techniques, climatic characteristics, and others, can affect the quantitative and qualitative constitution of the essential oil [17].

2.2. Cytotoxic Activity

The cytotoxic effect of the essential oil was tested in estrogen-receptor positive MCF-7 and estrogen-receptor negative MDA-MB-231 human breast cancer cell lines. Additionally, the human normal breast cell line, MCF-10A, was evaluated. The results of the cytotoxicity assays are shown in Table 2.

Table 2. Cytotoxic concentration (CC_{50} µg·mL^{-1}) of essential oil against breast cancer and normal breast cell lines.

Cell Line	*B. salicifolius*	Selectivity index (SI)
MCF-7	>512	−0.21
MDA-MB-231	46.60 ± 8.22 [a]	0.83
MCF-10A	314.44 ± 60.12 [b]	-

Six replicates; values are mean ± standard deviation. MCF-7: human metastatic adenocarcinoma; MDA-MB-231: human metastatic adenocarcinoma; MCF-10A: human mammary epithelial cells. Different letters in a column represent significant differences when $p < 0.05$.

The essential oil was toxic against MDA-MB-231 (46.60 µg·mL^{-1}) cancer cell lines. MDA-MB-231 cells, which represent an aggressive phenotype, responded more favorably to the cytotoxic effect of the essential oil than the less-aggressive MCF-7 breast cancer cell line. Several difficulties are associated with the treatment of estrogen-receptor negative breast cancer cells because these usually have poor prognosis and hormone therapies are not effective [39], which makes the results found here considered of great relevance. Moreover, this essential oil presented similar results compared to other plant essential oils considered active and with relevant action against this cell type, such as *Decatropis bicolor*, which had a CC_{50} of 53.81 µg·mL^{-1} [40] and *Hedychium spicatum* with values around 65 µg·mL^{-1} [41].

The normal breast cell line, MCF-10A (314.44 µg·mL^{-1}), was around sevenfold more resistant to essential oil action than MDA-MB-231 cancer cells. A relationship between normal breast cell lines and breast cancer cell lines was established through the selectivity index (SI) (Table 2) in order to check specificity. A positive SI value is desirable, and represents more selectivity against cancer cells than toxicity to normal cells. The negative value of SI for MCF-7 (−0.21) showed selectivity by normal cells, but the SI for the MDA-MB-231 cells was 0.83, which means selectivity by cancer cells and confirms the potential of this essential oil as a good candidate as a cytotoxic agent.

Against MCF-7, the essential oil did not present cytotoxic action at the tested concentrations. Methanolic extract, chalcones and ursolic acid isolated from the leaves of this species presented cytotoxic activity against such cellular types [9,42]. This suggests that differential content or concentration, especially of polar compounds in *B. salicifolius*, may be related to cytotoxic action against MCF-7 cells.

2.3. Flow Cytometry Analysis

Flow cytometry analysis was performed to investigate if the cytotoxicity effect induced by *B. salicifolius* essential oil is related to pathways that activate cell death. As shown in Figure 1, the treatment with essential oil did not induce cell death in MDA-MB-231 and MCF-10A at the doses of 46 µg·mL^{-1} and 314 µg·mL^{-1}, respectively, when compared to control cells. Therefore, the cytotoxicity of this essential oil is not involved with the activation of cell death mechanisms, but with the reduction of the cellular metabolic capacity. The resazurin reduction to verify cell viability may be indicating that there was impairment in cellular metabolism and not necessarily an interruption of electron transport and mitochondrial dysfunction [43].

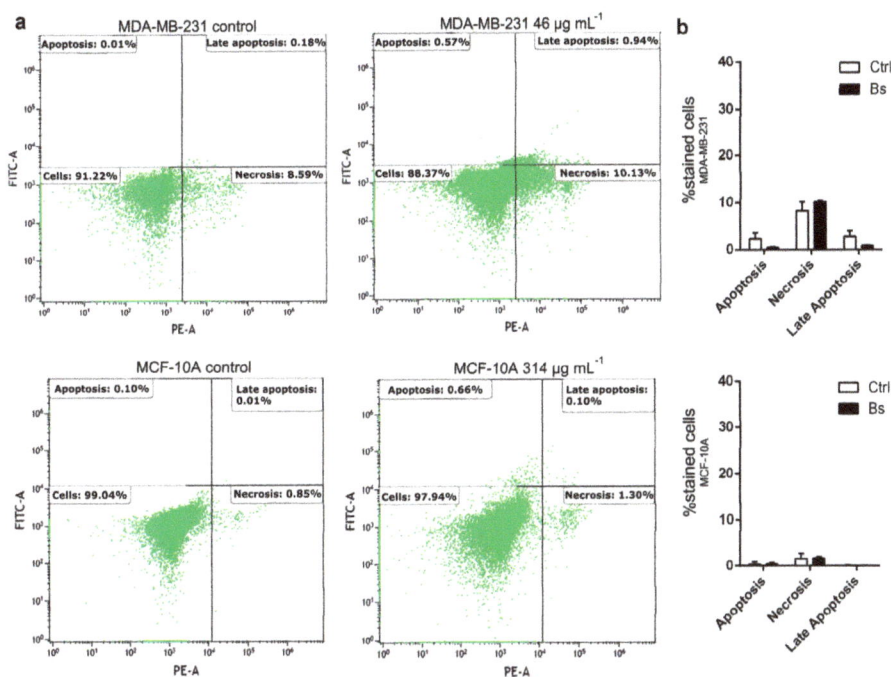

Figure 1. (**a**) Flow cytometry analysis of MDA-MB-231 and MCF-10A cells under action of *B. salicifolius* essential oil; (**b**) Bar diagram demonstrates that there is no difference between control cells and treated cells at the concentration of essential oil determined by cytotoxic activity assays for both cells lines. Ctrl: control; Bs: treatment with *B. salicifolius* essential oil. Data are expressed as mean ± standard deviation for experiments carried out in triplicate. Comparison between bars was performed by two-way ANOVA test followed by Sidak's test using GraphPad Prism 6.01.

2.4. Antifungal Activity

As shown in Table 3, antifungal activity was not observed against *Candida* spp., *Cryptococcus neoformans* and *Microsporum canis*. A weak action against *Trichophyton mentagrophytes* was found, and *Paracoccidioides brasiliensis* demonstrated a moderate degree of sensitivity to the essential oil based on the minimal inhibitory concentration (MIC) classification adopted by Holetz et al. (2002) [44].

Table 3. Minimal inhibitory concentration (μg·mL^{-1}) for tested fungi.

Fungi Strains	*B. salicifolius*	Amphotericin B
Candida krusei	>5000	2
Candida albicans	>5000	2
Candida guilliermondii	>5000	2
Candida parapsilosis	>5000	1
Candida orthopsilosis	>5000	0.5
Candida metapsilosis	>5000	0.25
Cryptococcus neoformans	2500 [a],*	0.5
Paracoccidioides brasiliensis	156 [b],*	0.5
Microsporum canis	2500 [a],*	2
Trichophyton mentagrophytes	625 [c],*	1

Six replicates; amphotericin B: positive control. Different letters in a column represent significant differences in the minimum inhibitory concentration between microorganism for the same treatment; * differences when minimal inhibitory concentration value is compared to positive control; $p < 0.05$.

Essential oils with strong antimicrobial activity against yeasts are characterized by a high content of thymol, carvacrol, cymene, linalool or α-pinene [45]. Among the compounds found in the essential oil from *B. salicifolius*, linalool (0.14%) and α-pinene (3.22%) may have exerted some antifungal action against yeasts of *P. brasiliensis*, the causative agent of paracoccidioidomycosis. Few studies related to the action of essential oils against this fungus are available, possibly because it is a mycoses restricted to some countries of Latin America. In addition to previously published studies, this work provides an indication that some terpenes from *B. salicifolius* can be related to inhibition of *P. brasiliensis* growth and this essential oil is therefore a relevant target for future investigation regarding its isolated compounds. Essential oil from *B. salicifolius* showed inhibition with an MIC value lower than already found for other essential oils or isolated compounds tested against this fungus [46–48].

Until now, with regard to the antifungal activities of essential oil of the leaves from *B. salicifolius*, only a fungistatic effect against *Phyllosticta citricarpa* was described [28]. Against fungi that are pathogenic to humans, the essential oil of this species had not yet been evaluated. Except for *P. brasiliensis*, the results found for the other fungi indicate that this essential oil was not capable of acting in any mechanism that could provide the fungistatic or fungicidal activity, and therefore, the present terpenes in *B. salicifolius* were not efficient against these species.

2.5. Antibacterial Activity

Antibacterial activity of the essential oil from leaves of *B. salicifolius* was determined against some microorganisms of clinical importance, as shown in Table 4.

Table 4. Minimal inhibitory concentration 90% (μg·mL^{-1}) for tested bacteria.

Gram Staining Reaction	Bacterial Strains	*B. salicifolius*	Polymyxin B	Cephalothin
Gram Positive	MRSA	4125 [a],*	-	2
	MSSA	2062 [b],*	-	0.5
Gram Negative	*E. coli*	5000 [c],*	1	-
	P. aeruginosa	>5000	1	-

n = 7:1 ATCC and 6 clinical isolates; MRSA: methicillin-resistant *Staphylococcus aureus*; MSSA: methicillin-sensitive *Staphylococcus aureus*. Polymyxin B and cephalothin: positive controls. Different letters in a column represent significant differences in the minimum inhibitory concentration between microorganism for the same treatment; * differences when MIC value is compared to its respective positive control; *p* < 0.05.

Essential oils from the leaves of this species has not been investigated previously against any bacterial organism. This study shows that, against the tested bacteria, the essential oil did not have antibacterial activity. However, some studies [2,23] showed activity of extracts from this species against *Staphylococcus aureus* and *Escherichia coli*, which leads to a conclusion that, regarding metabolites from this species, antibacterial activity is dependent on the presence of polar compounds not present in the essential oil. (E)-caryophyllene and α-pinene may be responsible for a certain effect against the MRSA and MSSA strains, since other authors have reported antibacterial activity against *S. aureus* when these compounds are in a majority in essential oils [49–55]. It was observed that the essential oil had greater difficulty in acting against the Gram-negative bacteria. In general, essential oils are more effective against Gram-positive than Gram-negative bacterial strains, because the outer membrane surrounding the cell wall restricts the diffusion of hydrophobic compounds through the lipopolysaccharide covering [56]. Furthermore, the species *Pseudomonas aeruginosa* is generally less susceptible to a diverse range of antimicrobial compounds, including essential oils [57]. This reduced susceptibility has been attributed to the outer membrane and associated properties, such as drug efflux [58,59].

3. Materials and Methods

3.1. Plant Material and Extraction of Essential Oil

Leaves of *B. salicifolius* were collected in Botucatu City, Brazil (22°57′55,90″ N 48°24′16,99″ W), on a morning during December 2014. For the purpose of scientific research of the species under study, an authorization to access samples of components of the genetic patrimony (N° 010621/2015-6) was obtained from the National Council of Scientific and Technological Development (CNPq). The plant was identified by a specialist, and a voucher specimen was deposited in the Herbarium of the Federal University of Uberlândia, under number HUFU 71037. The essential oil was obtained by steam distillation of fresh leaves for 2 h using a distiller designed for essential oil production (model MA480—Marconi). The essential oil was separated from water by decantation, filtered (membrane filter—0.2 μm), stored at low temperature (−10 °C) and protected from light until analysis. The percentage yield was calculated relative to the dried mass of the initial sample and this analysis was performed in triplicate.

3.2. Gas Chromatography/Mass Spectrometry (GC/MS) Analysis

Essential oils from *B. salicifolius* leaves were analyzed using a gas chromatograph coupled to a mass spectrometer, model FOCUS ISQ 230ST (Thermo Scientific, Austin, TX, USA), equipped with a TraceGOLD TG-5MS capillary column (Thermo Scientific, 30 m × 0.25 mm × 0.25 μm film thickness). The carrier gas was helium, at a flow rate of 1 mL·min^{-1}. Injector and detector temperatures were 220 °C and 240 °C, respectively; the injection volume was 1 μL and split ratio was 1:20. The oven temperature was programmed from 60 °C to 246 °C, at 3 °C/min. The electron impact energy was set at 70 eV and fragments from 40 to 415 m/z were collected [37].

The identification of the essential oil components was carried out by comparison of the mass spectrum obtained with that stored in the software library (Nist08), and by comparing the calculated arithmetic retention index (AI) according to the equation proposed by Van Den Dool and Kratz (1963) [60] with arithmetic retention index reported in the literature [37,38]. AI calculation was based on retention times of linear alkane standards (C8-C40, Sigma-Aldrich, St. Louis, MO, USA) run under the same operating conditions as previously described.

Quantitative analysis was carried out in triplicate and the amounts of volatile compounds were calculated using the internal standard method, taking into account the relative response factor (RRF) according to IOFI Working Group on Methods of Analysis [61]. For this purpose, heptanal (Sigma-Aldrich; 0.0092 mg·mL^{-1}) was selected as the internal standard. Phellandrene, linalool, caryophylene and nerolidol (Sigma-Aldrich) were selected to represent the response factors for monoterpene hydrocarbons, oxygenated monoterpenes, sesquiterpene hydrocarbons and oxygenated sesquiterpenes, respectively. Analyses for the RRF calculation were completed five times at four concentrations (0.0104, 0.0146, 0.0187 and 0.0229 mg·mL^{-1}) [62] and the average response factors were obtained.

3.3. Cytotoxic Activity

A sample of the essential oil was solubilized in methanol (Synth, São Paulo, Brazil) and diluted in supplemented Dulbecco's Modified Eagle Medium (DMEM, Sigma-Aldrich) to form a stock solution of 640 μg·mL^{-1}. Cell viability was tested with cell lines from American Type Culture Collection (ATCC, Rockville, MD, USA). MCF-7 (human metastatic adenocarcinoma-estrogen-receptor positive; ATCC HTB-22) and MDA-MB-231 (human metastatic adenocarcinoma-estrogen-receptor negative; ATCC HTB-26) breast cancer cell lines, and the MCF-10A (human mammary epithelial cells; ATCC CRL-10317) normal breast cell line, were selected for cytotoxicity assays. A solution containing 1×10^4 cells in 100 μL of supplemented DMEM was pipetted into each well and the plate was incubated overnight at 37 °C, with a humidified atmosphere and 5% CO_2, allowing cell adhesion in wells. Once attached and after removal of the culture medium, the stock solution of essential oil was added to the microplate and a serial dilution was performed to achieve concentrations ranging from 4 to 512 $\mu g \cdot mL^{-1}$. For this analysis, the controls of cell growth, solvent (methanol 3%), samples and negative control (100% lysed cells) were performed. Microplates were incubated for 48 h at 37 °C with a humidified atmosphere and 5% CO_2. Next, a revealing solution of resazurin (3 mM) diluted in phosphate-buffered saline (PBS) was added to each well [63] and the plate was incubated again for 24 h under the same conditions. Readings of absorbance at 595 nm were performed using a Multiskan GO Microplate Spectrophotometer (Thermo Scientific). Assays were performed in six replicates and the results of absorbance were calculated according to the growth control. The cytotoxic concentration at which 50% of the cells are viable (CC_{50}) was calculated by a dose-response graph of nonlinear regression.

In order to check specificity, a relationship between normal breast and breast cancer cell lines was established by the selectivity index (SI) and calculated according to the Equation (1) adapted from Case et al. (2006) [64]:

$$SI = \log (CC_{50} \text{ normal cell line}/CC_{50} \text{ cancer cell line}). \tag{1}$$

3.4. Flow Cytometry Analysis

For this analysis, MDA-MB-231 and MCF-10A cells were selected based on results from cytotoxic activity assays. Cells were seeded (5×10^6) in 6-well dishes and treated with essential oil for 48 h at 37 °C at the dose of CC_{50} previously determined by the cytotoxic activity assay. After treatment, the cells were harvested by trypsinisation, washed with PBS and stained with annexin V and propidium iodide (PI) using a FITC Annexin V Apoptosis Detection Kit I (BD Biosciences, San Jose, CA, USA), following the manufacturer's instructions. The fluorescence was measured in a CytoFLEX Flow Cytometer (Beckman Coulter Inc., Miami, FL, USA) and data analyzed in Kaluza Analysis Software (Beckman Coulter Inc., Miami, FL, USA).

3.5. Antifungal Activity

Tested strains were obtained from ATCC and the Laboratory of Fungi Biology, Department of Microbiology and Immunology, São Paulo State University (UNESP-Botucatu Campus). No ATCC strains were molecularly identified and deposited at GenBank database of the National Center for Biotechnology Information (NCBI). The following microorganisms were used in the evaluation of antifungal activity of the essential oil: *Candida krusei* (ATCC 6258), *Candida albicans* (ATCC 36801), *Candida guilliermondii* (ATCC 22017), *Candida parapsilosis* (ATCC 90018), *Candida orthopsilosis* (ATCC 96141), *Candida metapsilosis* (ATCC 96142), *Cryptococcus neoformans* (ATCC 90012), *Paracoccidioides brasiliensis* (strain TLM17LM2, GenBank accession number KX774393, yeast form), *Microsporum canis* (strain RS5, GenBank accession number KT443098) and *Trichophyton mentagrophytes* (ATCC 11480). A microdilution broth susceptibility assay for yeasts and filamentous fungi was performed, as recommended by Clinical and Laboratory Standards Institute [65,66], with some adaptations. Assays were performed on 96-well microplates (Costar, Cambridge, MA, USA) with RPMI-1640 supplemented with glutamine without sodium bicarbonate (Cultilab, Campinas, Brazil). A sample of essential oil was dissolved in dimethyl sulfoxide (DMSO, Synth, São Paulo) and a serial dilution was performed to achieve concentrations ranging from 2.44 to 5000 $\mu g \cdot mL^{-1}$. The final DMSO concentration was 3% (*v/v*). For inoculums, adjustment of yeast (*Candida* spp., *C. neoformans*, *P. brasiliensis*) was performed using microscopic enumeration with a cell-counting hematocytometer (Neubauer chamber, Knittel Gläser, Braunschweig, Germany) from the initial growth at RPMI-1640 medium. *Candida* spp. was grown at 37 °C for 48 h, *C. neoformans* at 37 °C for 72 h and *P. brasiliensis* at 35 °C for 10 days. After counting, the inoculums were adjusted to a final concentration of 1.2×10^3 cell mL^{-1}. Filamentous fungi, *M. canis* and *T. mentagrophytes*, were grown on potato dextrose agar (PDA; Oxoid, Basingstoke, Hampshire, UK) for 7 days at 25 °C before counting. Cell counting in a Neubauer chamber was also performed, and microplates were inoculated with a final concentration of 2×10^4 cell mL^{-1}. The tested systems were incubated in a humid atmosphere, with agitation, at 37 °C for 48 h for *Candida* spp. and

72 h for *C. neoformans*. *P. brasiliensis* was incubated for 14 days at 35 °C with agitation and *M. canis* and *T. mentagrophytes* for 10 days at 25 °C without agitation. The experiment was performed in six replicates and the growth inhibition was determined by measuring turbidity of the cultured medium at 530 nm using a spectrophotometer (BioTek, Winooski, VT, USA). Amphotericin B (Sigma-Aldrich) was used as positive control. Controls of sterility of the broth, control culture (inoculum), essential oil and DMSO (3%) were performed.

3.6. Antibacterial Activity

Experiments were performed with standard strains from ATCC and clinical isolates ($n = 6$) obtained from the Laboratory of Microbiology, Department of Microbiology and Immunology, São Paulo State University (UNESP-Botucatu Campus). The study was conducted in accordance with the Declaration of Helsinki and the use of isolated microorganisms was approved by the Institutional Committee on Human Research (number 47186415.0.0000.5411-8 March, 2015), according to the ethical principles for medical research involving human subjects, and the strains were submitted to biochemical tests for phenotypic confirmation of the species [67]. The following microorganisms were used: methicillin-resistant *Staphylococcus aureus* (MRSA) (ATCC 33591), methicillin-sensitive *Staphylococcus aureus* (MSSA) (ATCC 25923), *Escherichia coli* (ATCC 43895) and *Pseudomonas aeruginosa* (ATCC 27853). Susceptibility assays were performed according to the Clinical and Laboratory Standards Institute [68] on 96-well microplates (Costar) with Mueller Hinton broth (Difco, Detroit, MI, USA) culture medium. The sample of essential oil was dissolved in DMSO (Synth, 160,000 $\mu g \cdot mL^{-1}$) and serial dilution was performed to achieve concentrations ranging from 2.44 to 5000 $\mu g \cdot mL^{-1}$. The final DMSO concentration was 3% (v/v). Bacterial strains were grown (37 °C for 18–24 h) in brain heart infusion (BHI; Difco) and, after standardization by 0.5 McFarland scale, were inoculated (around 10^5 cfu·mL^{-1}) in wells at concentrations previously prepared. After incubation (37 °C for 18–24 h), the minimal inhibitory concentration (MIC) of each strain was visually recorded after addition of 50 μL of resazurin (0.01%) in respective wells. Polymyxin B (Latinofarma, Cotia, Brazil) was used as a positive control for Gram-negative bacteria and Cephalothin (Latinofarma) for Gram-positive bacteria. Controls of sterility of the broth, culture control (inoculum), essential oil and DMSO (3%) were performed.

3.7. Statistical Analysis

Statistical analysis of the data was performed by analysis of variance (ANOVA) followed by Tukey's test for analysis of cytotoxic activity and the Kruskal–Wallis test followed by the Student–Newman–Keuls test or Dunn's test for analysis of minimal inhibitory concentration against microorganisms using SigmaPlot 11.0 software (Erkrath, North Rhine-Westphalia, Germany). For flow cytometry analysis, two-way ANOVA followed by Sidak's test was performed using the software GraphPad Prism 6.01. Probability values $p < 0.05$ were considered to denote a statistically significant difference.

4. Conclusions

In this study, we report cytotoxic and antimicrobial activities of essential oils from the leaves of *B. salicifolius*. This essential oil is a promising cytotoxic agent by its action and selectivity demonstrated here, and is a diverse source of molecules that may have an important role in drug discovery for the treatment of estrogen-receptor negative tumors, which has been a challenge in oncology. Moreover, the antifungal action of this essential oil against *Paracoccidioides brasiliensis*, the causative agent of paracoccidioidomycosis, shows potential to be used in the treatment of this mycosis. Thus, relevant biological activities and high yield of the essential oil from *B. salicifolius* means it is a source of molecules to be exploited for medicine and by the pharmaceutical industry.

Acknowledgments: This work was supported by São Paulo Research Foundation—FAPESP (2015/14278-6). The authors thank the botanist Inara Regiane Moreira Coneglian for plant identification; our collaborators Mário Machado Martins, Eduardo Bagagli, Vera Lúcia Mores Rall and Luiz Claudio Di Stasi for analysis tools, materials and helpful suggestions; Coordination for the Improvement of Higher Education Personnel—CAPES for the scholarship granted to Fabiana Barcelos Furtado.

Author Contributions: Fabiana Barcelos Furtado, Ary Fernandes Junior: idea and concept of the paper; Fabiana Barcelos Furtado, Bruna Cristina Borges, Hans Garcia Garces, Luiz Domingues de Almeida Junior, Thaise Lara Teixeira: performed the experiments; Fabiana Barcelos Furtado, Fernanda Cristina Bérgamo Alves, Ary Fernandes Junior: analyzed the data; Claudio Vieira da Silva, Ary Fernandes Junior: contributed reagents/materials/analysis tools; all authors: article preparation.

Conflicts of Interest: The authors declare no conflict of interest.

References

1. Lima, J.S.S.; Castro, J.M.C.; Sabino, L.B.S.; Lima, A.C.S.; Torres, L.B.V. Physicochemical properties of Gabiroba (*Campomanesia lineatifolia*) and Myrtle (*Blepharocalyx salicifolius*) native to the mountainous. *Rev. Caatinga* **2016**, *29*, 753–757. [CrossRef]
2. Vivot, E.P.; Sánchez, C.; Cacik, F.; Sequin, C. Actividad antibacteriana en plantas medicinales de la flora de Entre Ríos (Argentina). *Cienc. Docencia y Tecnol.* **2012**, *45*, 165–185.
3. Poncio, S. Bioatividade de Inseticidas Botânicos Sobre *Microtheca ochroloma* Stal (Coleoptera: Chrysomelidae). Master's Thesis, Universidade Federal de Santa Maria, Santa Maria, RS, Brazil, 2010.
4. Tropicos—*Blepharocalyx salicifolius*. Available online: http://tropicos.org/Name/22102059?tab=synonyms (accessed on 17 November 2017).
5. Alice, C.B.; Vargas, V.M.F.; Silva, G.A.A.B.; de Siqueira, N.C.S.; Schapoval, E.E.S.; Gleye, J.; Henriques, J.A.P.; Henriques, A.T. Screening of plants used in south Brazilian folk medicine. *J. Ethnopharmacol.* **1991**, *35*, 165–171. [CrossRef]
6. Mors, W.B.; Rizzini, C.T.; Pereira, N.A.; De Filipps, R.A. *Medicinal Plants of Brazil*; Reference Publications: Detroit, MI, USA, 2000.
7. Piva, M.G. *O Caminho das Plantas Medicinais: Estudo Etnobotânico*; Rio de Janeiro: Mondrian, Brazil, 2002.
8. Ratera, E.L.; Ratera, M.O. *Plantas de la Flora Argentina Empleadas en la Medicina Popular*; Hemisferio Sur: Buenos Aires, Argentina, 1980.
9. Calderón, Á.I.; Vázquez, Y.; Solís, P.N.; Caballero-George, C.; Zacchino, S.; Gimenez, A.; Pinzón, R.; Cáceres, A.; Tamayo, G.; Correa, M.; et al. Screening of Latin American Plants for Cytotoxic Activity. *Pharm. Biol.* **2006**, *44*, 130–140. [CrossRef]
10. Charneau, S.; de Mesquita, M.L.; Bastos, I.M.D.; Santana, J.M.; de Paula, J.E.; Grellier, P.; Espindola, L.S. In vitro investigation of Brazilian Cerrado plant extract activity against *Plasmodium falciparum*, *Trypanosoma cruzi* and *T. brucei* gambiense. *Nat. Prod. Res.* **2016**, *30*, 1320–1326. [CrossRef] [PubMed]
11. Luján, M.C.; Pérez Corral, C. Cribado para evaluar actividad antibacteriana y antimicótica en plantas utilizadas en medicina popular de Argentina. *Rev. Cuba. Farm.* **2008**, *42*, 1–10.
12. Freixa, B.; Vila, R.; Vargas, L.; Lozano, N.; Adzet, T.; Cañigueral, S. Screening for antifungal activity of nineteen Latin American plants. *Phyther. Res.* **1998**, *12*, 427–430. [CrossRef]
13. Vivot Lupi, E.P.; Sánchez Brizuela, C.I.; Jeifetz, F.C.; Sequin Acosta, C.J. Screening of antifungal activity of extracts present in Entre Ríos flora species [Tamizaje de la actividad antifúngica de extractos de especies de la flora de Entre Ríos]. *Rev. Cuba. Farm.* **2009**, *43*, 74–84.
14. Imatomi, M. Estudo Alelopático de Espécies da Família Myrtaceae do Cerrado. Ph.D. Thesis, Universidade Federal de São Carlos, San Carlos, SP, Brazil, 2010.
15. Noal, C.B.; Monteiro, D.U.; de Brum, T.F.; Emmanouilidis, J.; Zanette, R.A.; Morel, A.F.; de Cassia Stefanon, E.B.; Frosi, M.; la Rue, M.L.D. In vitro effects of *Blepharocalyx salicifolius* (H.B.K.) O. Berg on the viability of *Echinococcus ortleppi* protoscoleces. *Rev. Inst. Med. Trop.* **2017**, *59*, 1–5. [CrossRef] [PubMed]
16. Castelo, A.V.M.; Del Menezzi, C.H.S.; Resck, I.S. Yield and spectroscopic analysis (1H, 13C NMR; IR) of essential oils from four plants of the Brazilian Savannah. *Cerne* **2010**, *16*, 573–584. [CrossRef]
17. Castelo, A.V.M.; Menezzi, C.H.S.D.; Resck, I.S. Seasonal Variation in the Yield and the Chemical Composition of Essential Oils from Two Brazilian Native Arbustive Species. *J. Appl. Sci.* **2012**, *12*, 753–760. [CrossRef]

18. Costa, O.B.D.; Del Menezzi, C.H.S.; Benedito, L.E.C.; Resck, I.S.; Vieira, R.F.; Ribeiro Bizzo, H.; Sabioni, S.; Vieira, R.F.; Ribeiro Bizzo, H. Essential oil constituents and yields from leaves of *Blepharocalyx salicifolius* (Kunt) O. Berg and *Myracrodruon urundeuva* (Allemão) collected during daytime. *Int. J. For. Res.* **2014**, *2014*, 982576.

19. Dellacassa, E.; Lorenzo, D.; Mondello, L.; Dugo, P. Uruguayan essential oils. Part IX. Composition of leaf oil of *Blepharocalyx tweediei* (Hook. et Arn.) Berg var. tweediei (Myrtaceae). *J. Essent. Oil Res.* **1997**, *9*, 673–676. [CrossRef]

20. Furlán, R.; Zacchino, S.; Gattuso, M.; Bradessi, P.; Casanova, J.; Vila, R.; Cañigueral, S. Constituents of the essential oil from leaves and seeds of *Blepharocalyx tweediei* (Hook, et Arn.) Berg and *B. gigantea* Lillo. *J. Essent. Oil Res.* **2002**, *14*, 175–178. [CrossRef]

21. Garneau, F.X.; Collin, G.J.; Jean, F.I.; Gagnon, H.; Arze, J.B.L. Essential oils from Bolivia. XIII. Myrtaceae: *Blepharocalyx salicifolius* (Kunth.) O. Berg. *J. Essent. Oil Res.* **2013**, *25*, 166–170. [CrossRef]

22. Godinho, W.M.; Farnezi, M.M.; Pereira, I.M.; Gregório, L.E.; Grael, C.F.F. Volatile constituents from leaves of *Blepharocalyx salicifolius* (Kunth) O. Berg (Myrtaceae). *Bol. Latinoam. y del Caribe Plantas Med. y Aromát.* **2014**, *13*, 249–253.

23. Limberger, R.P.; Sobral, M.E.G.; Zuanazzi, J.A.S.; Moreno, P.R.H.; Schapoval, E.E.S.; Henriques, A.T. Biological activities and essential oil composition of leaves of *Blepharocalyx salicifolius*. *Pharm. Biol.* **2001**, *39*, 308–311. [CrossRef]

24. Moreira, J.J.S.; Pereira, C.S.; Ethur, E.M.; Machado, E.C.S.; Morel, A.F.; König, W.A. Volatile constituents composition of *Blepharocalyx salicifolius* leaf oil. *J. Essent. Oil Res.* **1999**, *11*, 45–48. [CrossRef]

25. Talenti, E.C.; Taher, H.A.; Ubiergo, G.O. Constituents of the essential oil of *Blepharocalyx tweediei*. *J. Nat. Prod.* **1984**, *47*, 905–906. [CrossRef] [PubMed]

26. Tucker, A.O.; Maciarello, M.J.; Landrum, L.R. Volatile leaf oils of american myrtaceae. I. *Blepharocalyx cruckshanksii* (Hook. & Arn.) Niedenzu of Chile and *B. salicifolius* (Humb., Bonpl. & Kunth) Berg of Argentina. *J. Essent. Oil Res.* **1993**, *5*, 333–335.

27. Hernández, J.J.; Ragone, M.I.; Bonazzola, P.; Bandoni, A.L.; Consolini, A.E. Antitussive, antispasmodic, bronchodilating and cardiac inotropic effects of the essential oil from *Blepharocalyx salicifolius* leaves. *J. Ethnopharmacol.* **2017**, *210*, 107–117. [CrossRef] [PubMed]

28. Lombardo, P.; Guimaraens, A.; Franco, J.; Dellacassa, E.; Pérez, E. Postharvest biology and technology effectiveness of essential oils for postharvest control of *Phyllosticta citricarpa* (citrus black spot) on citrus fruit. *Postharvest Biol. Technol.* **2016**, *121*, 1–8. [CrossRef]

29. Qu, Y.; Han, B.; Yu, Y.; Yao, W.; Bose, S.; Karlan, B.Y.; Giuliano, A.E.; Cui, X. Evaluation of MCF-10A as a reliable model for normal human mammary epithelial cells. *PLoS ONE* **2015**, *10*, e0131285. [CrossRef] [PubMed]

30. WHO | Global Cancer Rates Could Increase by 50% to 15 Million by 2020. Available online: http://www.who.int/mediacentre/news/releases/2003/pr27/en/ (accessed on 24 November 2017).

31. Gautam, N.; Mantha, A.K.; Mittal, S. Essential oils and their constituents as anticancer agents: A mechanistic view. *BioMed Res. Int.* **2014**, *2014*, 154106. [CrossRef] [PubMed]

32. Bajpai, V.K.; Shukla, S.; Sharma, A. Essential Oils as Antimicrobial Agents. In *Natural Products*; Springer: Berlin/Heidelberg, Germany, 2013; pp. 3975–3988.

33. Solórzano-Santos, F.; Miranda-Novales, M.G. Essential oils from aromatic herbs as antimicrobial agents. *Curr. Opin. Biotechnol.* **2012**, *23*, 136–141. [CrossRef] [PubMed]

34. O'Neill, J. Antimicrobial Resistance: Tackling a crisis for the health and wealth of nations. *Rev. Antimicrob. Resist.* **2014**, 1–18.

35. Lee, S.J.; Umano, K.; Shibamoto, T.; Lee, K.G. Identification of volatile components in basil (*Ocimum basilicum* L.) and thyme leaves (*Thymus vulgaris* L.) and their antioxidant properties. *Food Chem.* **2005**, *91*, 131–137. [CrossRef]

36. Zrira, S.S.; Benjilali, B.B.; Fechtal, M.M.; Richard, H.H. Essential oils of twenty-seven eucalyptus species grown in Morocco. *J. Essent. Oil Res.* **1992**, *4*, 259–264. [CrossRef]

37. Adams, R.P. *Identification of Essential Oil Components by Gas Chromatography/Mass Spectrometry*, 4th ed.; Allured Publishing Corporation: Carol Stream, IL, USA, 2007; p. 803.

38. NIST Standard Reference Data. Available online: http://webbook.nist.gov/chemistry/name-ser.html (accessed on 25 January 2016).

39. Nakagawa-Goto, K.; Chen, J.Y.; Cheng, Y.T.; Lee, W.L.; Takeya, M.; Saito, Y.; Lee, K.H.; Shyur, L.F. Novel sesquiterpene lactone analogues as potent anti-breast cancer agents. *Mol. Oncol.* **2016**, *10*, 921–937. [CrossRef] [PubMed]

40. Gómez, C.C.E.; Carreño, A.A.; Ishiwara, D.G.P.; Martínez, E.S.M.; López, J.M.; Hernández, N.P.; García, M.C.G. *Decatropis bicolor* (Zucc.) Radlk essential oil induces apoptosis of the MDA-MB-231 breast cancer cell line. *BMC Complement. Altern. Med.* **2016**, *16*, 1–11.

41. Mishra, T.; Pal, M.; Meena, S.; Datta, D.; Dixit, P.; Kumar, A.; Meena, B.; Rana, T.S.; Upreti, D.K. Composition and in vitro cytotoxic activities of essential oil of *Hedychium spicatum* from different geographical regions of western Himalaya by principal components analysis. *Nat. Prod. Res.* **2016**, *30*, 1224–1227. [CrossRef] [PubMed]

42. Siqueira, E.P.; Oliveira, D.M.; Johann, S.; Cisalpino, P.S.; Cota, B.B.; Rabello, A.; Alves, T.M.A.; Zani, C.L. Bioactivity of the compounds isolated from *Blepharocalyx salicifolius*. *Rev. Bras. Farmacogn.* **2011**, *21*, 645–651. [CrossRef]

43. Bols, N.C.; Dayeh, V.R.; Lee, L.E.J.; Schirmer, K. Chapter 2: Use of fish cell lines in the toxicology and ecotoxicology of fish. Piscine cell lines in environmental toxicology. *Biochem. Mol. Biol. Fishes* **2005**, *6*, 43–84.

44. Holetz, F.B.; Pessini, G.L.; Sanches, N.R.; Cortez, D.A.G.; Nakamura, C.V.; Filho, B.P.D. Screening of some plants used in the Brazilian folk medicine for the treatment of infectious diseases. *Memórias Do Instituto Oswaldo Cruz* **2002**, *97*, 1027–1031. [CrossRef] [PubMed]

45. Lang, G.; Buchbauer, G. A review on recent research results (2008–2010) on essential oils as antimicrobials and antifungals. A review. *Flavour Fragr. J.* **2012**, *27*, 13–39. [CrossRef]

46. Costa, D.P.; Filho, E.G.A.; Silva, L.M.A.; Santos, S.C.; Passos, X.S.; Do Rosário, R.; Silva, M.; Seraphin, J.C.; Ferri, P.H. Influence of fruit biotypes on the chemical composition and antifungal activity of the essential oils of *Eugenia uniflora* leaves. *J. Braz. Chem. Soc.* **2010**, *21*, 851–858. [CrossRef]

47. Johann, S.; Oliveira, F.B.; Siqueira, E.P.; Cisalpino, P.S.; Rosa, C.A.; Alves, T.M.; Zani, C.L.; Cota, B.B. Activity of compounds isolated from *Baccharis dracunculifolia* D.C. (Asteraceae) against *Paracoccidioides brasiliensis*. *Med. Mycol.* **2012**, *50*, 843–851. [CrossRef] [PubMed]

48. Passos, X.S.; Castro, A.C.M.; Pires, J.S.; Garcia, A.C.F.; Campos, F.C.; Orionalda, F.L.; Paula, J.R.; Ferreira, H.D.; Santos, S.C.; Ferri, P.H.; et al. Composition and antifungal activity of the essential oils of *Caryocar brasiliensis*. *Pharm. Biol.* **2003**, *41*, 319–324. [CrossRef]

49. Ahmadi, F.; Sadeghi, S.; Modarresi, M.; Abiri, R.; Mikaeli, A. Chemical composition, in vitro anti-microbial, antifungal and antioxidant activities of the essential oil and methanolic extract of *Hymenocrater longiflorus* Benth., of Iran. *Food Chem. Toxicol.* **2010**, *48*, 1137–1144. [CrossRef] [PubMed]

50. Ebrahimabadi, A.H.; Mazoochi, A.; Kashi, F.J.; Djafari-Bidgoli, Z.; Batooli, H. Essential oil composition and antioxidant and antimicrobial properties of the aerial parts of *Salvia eremophila* Boiss. from Iran. *Food Chem. Toxicol.* **2010**, *48*, 1371–1376. [CrossRef] [PubMed]

51. El-Massrry, K.F.; El-Ghorab, A.H.; Shaaban, H.A.; Shibamoto, T. Chemical compositions and antioxidant/antimicrobial activities of various samples prepared from *Schinus terebinthifolius* leaves cultivated in Egypt. *J. Agric. Food Chem.* **2009**, *57*, 5265–5270. [CrossRef] [PubMed]

52. Formisano, C.; Napolitano, F.; Rigano, D.; Arnold, N.A.; Piozzi, F.; Senatore, F. Essential oil composition of *Teucrium divaricatum* Sieb. ssp. *villosum* (Celak.) Rech. fil. Growing Wild in Lebanon. *J. Med. Food* **2010**, *13*, 1281–1285. [PubMed]

53. Joshi, S.C.; Verma, A.R.; Mathela, C.S. Antioxidant and antibacterial activities of the leaf essential oils of Himalayan Lauraceae species. *Food Chem. Toxicol.* **2010**, *48*, 37–40. [CrossRef] [PubMed]

54. Kirmizibekmez, H.; Demirci, B.; Yeşilada, E.; Başer, K.H.C.; Demirci, F. Chemical composition and antimicrobial activity of the essential oils of *Lavandula stoechas* L. ssp. growing wild in Turkey. *Nat. Prod. Commun.* **2009**, *4*, 1001–1006. [PubMed]

55. Özek, G.; Demirci, F.; Özek, T.; Tabanca, N.; Wedge, D.E.; Khan, S.I.; Başer, K.H.C.; Duran, A.; Hamzaoglu, E. Gas chromatographic-mass spectrometric analysis of volatiles obtained by four different techniques from *Salvia rosifolia* Sm., and evaluation for biological activity. *J. Chromatogr. A* **2010**, *1217*, 741–748. [CrossRef] [PubMed]

56. Vergis, J.; Gokulakrishnan, P.; Agarwal, R.K.; Kumar, A. Essential oils as natural food antimicrobial agents: A review. *Crit. Rev. Food Sci. Nutr.* **2015**, *55*, 1320–1323. [CrossRef] [PubMed]

57. Thormar, H. *Lipids and Essential Oils as Antimicrobial Agents*; John Wiley & Sons, Ltd.: Chichester, UK, 2011.

58. Griffin, S.G.; Wyllie, S.G.; Markham, J.L. Role of the outer membrane of *Escherichia coli* AG100 and *Pseudomonas aeruginosa* NCTC 6749 and Resistance/Susceptibility to Monoterpenes of Similar Chemical Structure. *J. Essent. Oil Res.* **2001**, *13*, 380–386. [CrossRef]

59. Longbottom, C.J.; Carson, C.F.; Hammer, K.A.; Mee, B.J.; Riley, T.V. Tolerance of *Pseudomonas aeruginosa* to *Melaleuca alternifolia* (tea tree) oil is associated with the outer membrane and energy-dependent cellular processes. *J. Antimicrob. Chemother.* **2004**, *54*, 386–392. [CrossRef] [PubMed]

60. Van Den Dool, H.; Kratz, P. A generalization of the retention index system including linear temperature programmed gas-liquid partition chromatography. *J. Chromatogr. A* **1963**, *11*, 463–471. [CrossRef]

61. IOFI Working Group on Methods of Analysis. Guidelines for the quantitative gas chromatography of volatile flavouring substances, from the Working Group on Methods of Analysis of the International Organization of the Flavor Industry (IOFI). *Flavour Fragr. J.* **2011**, *26*, 297–299.

62. Ligiero, C.B.P.; dos Reis, L.A.; Parrilha, G.L.; Filho, M.B.; Canela, M.C. Comparação entre métodos de quantificação em cromatografia gasosa: Um experimento para cursos de química. *Quim. Nova* **2009**, *32*, 1338–1341. [CrossRef]

63. Rolón, M.; Vega, C.; Escario, J.A.; Gómez-Barrio, A. Development of resazurin microtiter assay for drug sensibility testing of *Trypanosoma cruzi* epimastigotes. *Parasitol. Res.* **2006**, *99*, 103–107. [CrossRef] [PubMed]

64. Case, R.J.; Franzblau, S.G.; Wang, Y.; Cho, S.H.; Soejarto, D.D.; Pauli, G.F. Ethnopharmacological evaluation of the informant consensus model on anti-tuberculosis claims among the Manus. *J. Ethnopharmacol.* **2006**, *106*, 82–89. [CrossRef] [PubMed]

65. Clinical and Laboratory Standards Institute (CLSI). *M27-A3 Reference Method for Broth Dilution Antifungal Susceptibility Testing of Yeasts; Approved Standard*, 3rd ed.; CLSI: Wayne, PA, USA, 2008.

66. Clinical and Laboratory Standards Institute (CLSI). *M38-A2 Reference Method for Broth Dilution Antifungal Susceptibility Testing of Filamentous Fungi; Approved Standard*, 2nd ed.; CLSI: Wayne, PA, USA, 2008.

67. Koneman, E.W.; Allen, S.D.; Janda, W.M.; Schreckenberger, P.C.; Winn, W.C., Jr. *Color Atlas and Textbook of Diagnostic Microbiology*, 6th ed.; Lippincott: Philadelphia, PA, USA, 2005.

68. Clinical and Laboratory Standards Institute (CLSI). *M100-S24 Performance Standards for Antimicrobial Susceptibility Testing; Twenty-Fourth Informational Supplement*; CLSI: Wayne, PA, USA, 2014.

International Journal of
Molecular Sciences

MDPI

Article

Dietary Resveratrol Does Not Affect Life Span, Body Composition, Stress Response, and Longevity-Related Gene Expression in *Drosophila melanogaster*

Stefanie Staats [1,*], Anika E. Wagner [2], Bianca Kowalewski [1], Florian T. Rieck [3], Sebastian T. Soukup [3], Sabine E. Kulling [3] and Gerald Rimbach [1]

[1] Institute of Human Nutrition and Food Science, University of Kiel, Hermann-Rodewald-Strasse 6, D-24118 Kiel, Germany; b-i-b-i@o2online.de (B.K.); rimbach@foodsci.uni-kiel.de (G.R.)

[2] Institute of Nutritional Medicine, University of Lübeck, Ratzeburger Allee 160, D-23538 Lübeck, Germany; Anika.Wagner@uksh.de

[3] Department of Safety and Quality of Fruit and Vegetables, Max Rubner Institute, Haid-und-Neu-Strasse 9, D-76131 Karlsruhe, Germany; florian@rieck.ru (F.T.R.); sebastian.soukup@mri.bund.de (S.T.S.); sabine.kulling@mri.bund.de (S.E.K.)

* Correspondence: staats@foodsci.uni-kiel.de; Tel.: +49-431-880-5313

Received: 30 November 2017; Accepted: 5 January 2018; Published: 11 January 2018

Abstract: In this study, we tested the effect of the stilbene resveratrol on life span, body composition, locomotor activity, stress response, and the expression of genes encoding proteins centrally involved in ageing pathways in the model organism *Drosophila melanogaster*. Male and female w^{1118} *D. melanogaster* were fed diets based on sucrose, corn meal, and yeast. Flies either received a control diet or a diet supplemented with 500 µmol/L resveratrol. Dietary resveratrol did not affect mean, median, and maximal life span of male and female flies. Furthermore, body composition remained largely unchanged following the resveratrol supplementation. Locomotor activity, as determined by the climbing index, was not significantly different between control and resveratrol-supplemented flies. Resveratrol-fed flies did not exhibit an improved stress response towards hydrogen peroxide as compared to controls. Resveratrol did not change mRNA steady levels of antioxidant (*catalase, glutathione-S-transferase, NADH dehydrogenase, glutathione peroxidase, superoxide dismutase* 2) and longevity-related genes, including *sirtuin 2, spargel,* and *I'm Not Dead Yet*. Collectively, present data suggest that resveratrol does not affect life span, body composition, locomotor activity, stress response, and longevity-associated gene expression in w^{1118} *D. melanogaster*.

Keywords: resveratrol; *Drosophila*; healthy ageing; life span; longevity

1. Introduction

Diet is an important determinant of health and disease prevention. Epidemiological data on the consumption of foods rich in fruits and vegetables suggests that secondary plant metabolites may favour health and successful ageing [1].

The traditional Asian and the Mediterranean diets are rich in fruits and vegetables [2]. There are specific plant bioactives, which predominantly occur in the Mediterranean (e.g., resveratrol from red wine, hydroxytyrosol from olives) and in the Asian diets (e.g., isoflavones from soybean and epigallocatechin gallate from green tea). In this context, we have recently introduced the concept of the so-called "MediterrAsian" diet combining foods of the traditional Asian as well as Mediterranean diet as a promising dietary strategy in chronic disease prevention [2].

Drosophila melanogaster is widely used as a model organism in ageing studies. *Drosophila* exhibits a relatively short life span of 60 to 90 days, which makes it particularly attractive for life span studies [3,4].

Furthermore, in recent years, the fruit fly has also been increasingly recognised as a model organism in nutrition research. Feed intake, body composition, locomotor activity, gut function, composition of the microbiota, ageing, as well as life span can be systematically determined in *Drosophila* in response to dietary factors [5–11]. Moreover, diet-induced pathophysiological mechanisms including both intestinal and systemic inflammatory processes [12–17], and stress response against various triggers like reactive oxygen species, alcohol, acids, or heat [6,12,18,19] may be evaluated in the fruit fly under defined experimental conditions. We have recently shown that secondary plant metabolites including isoflavones [5], green tea catechins [10], and isothiocyanates [20] are capable of improving health status and survival in male *D. melanogaster*.

The stilbene *trans*-resveratrol (3,4′,5-trihydroxystilbene) has been widely suggested as a putative "anti-ageing" molecule, e.g., in *Saccharomyces cerevisiae* [21–23], *Caenorhabditis elegans* [24–26], and killifish [27–29]. However, literature is contradictory regarding the life span modulating properties of resveratrol in *D. melanogaster* [24,30–32]. Furthermore, resveratrol mostly failed to improve life span in studies conducted in mice [33]. Several mechanisms, including induction of autophagy and sirtuins [34–39], modulation of IGF signalling [26,40,41], improvement of stress response [42–46], endogenous antioxidant defence [43,47,48], mitochondrial function [41,49–51], as well as anti-inflammatory properties [52–59], have been suggested by which resveratrol may counteract the ageing process. Moreover, there is literature data indicating that resveratrol may affect body weight [60–65], body composition [62,64,66], and metabolism [65–68] in different species—however, data are partly contradicting.

Although resveratrol has been shown to increase the life span in short-lived species like worms (*C. elegans*) [24,25] and killifish [27,28], the role of resveratrol in the fruit fly is less clear. Therefore, the aim of the present study was to systematically investigate the effect of dietary resveratrol on life span, body composition, stress response, and longevity-associated gene expression in *D. melanogaster*.

2. Results

Since feed intake may affect body composition, metabolism, locomotor activity, and life span, we monitored the feed intake of *D. melanogaster* in the presence and absence of resveratrol by the food-dye-based "sulforhodamine B gustatory assay" [5,69]. Under the conditions investigated there were no significant differences in feed intake between resveratrol-supplemented flies and controls both in males ($p = 0.162$) and females ($p = 0.126$) (Figure 1).

Accordingly, resveratrol-supplemented and control-fed flies exhibited similar fat, protein, and glucose contents (Table 1), whereby flies showed a rather heterogeneous response to the dietary resveratrol treatment as revealed by higher standard errors. Solely the protein content was slightly increased in resveratrol-fed males compared to controls. Thus, overall body composition of *D. melanogaster* remained largely unchanged in response to dietary resveratrol supplementation.

Table 1. Changes in body weight and body composition of male and female w^{1118} *D. melanogaster* in dependence of dietary resveratrol (RESV; 500 µmol/L) administration for ten days compared to controls.

Parameter	Male			Female		
	Control	RESV [1]	*p*-Value	Control	RESV [1]	*p*-Value
Body weight (µg/fly)	778 ± 46	774 ± 42	0.953	1222 ± 43	1223 ± 77	0.989
Triglycerides (% control)	100 ± 9.0	165 ± 31	0.083	100 ± 5.1	95.5 ± 13	0.750
Protein (% control)	100 ± 1.8	121 ± 7.5	**0.035**	100 ± 2.0	103 ± 5.3	0.679
Glucose (% control)	100 ± 12	115 ± 21	0.500	100 ± 14	156 ± 34	0.100

All data are shown as means $\pm$ SEM from three independent experiments comprising 178–224 flies/group (body weight) and 20–35 flies/group (body composition). Outliers were removed. Statistics: 2-sided Student's *t*-test and Mann–Whitney U. [1] RESV: resveratrol (500 µmol/L).

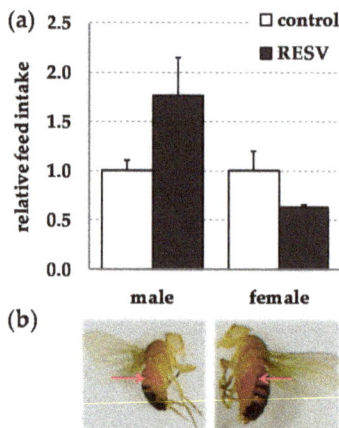

Figure 1. Dietary resveratrol (RESV; 500 µmol/L) does not affect feed intake in w^{1118} *D. melanogaster*. (a) Relative feed intake in male and female flies following a five-day feeding period with a RESV-supplemented or a control diet. Flies were administered the experimental diets for five days prior to administration of a sulforhodamine B (0.2% *w/v*)-supplemented medium for 8 h. Feed intake was quantified via fluorometric measurements. Feed intake of RESV-treated males and females was normalised to the feed intake of their control fed counterparts, respectively. Bars show means ± SEM comprising 60–100 flies/group. Statistics: 2-sided Student's *t*-test and Mann–Whitney U.; (b) Bright-field pictures of male and female w^{1118} flies administered with sulforhodamine B for 8 h. Arrows point to pink-coloured body parts due to the sulforhodamine B ingestion.

Locomotor activity of *D. melanogaster* was determined by calculating the climbing score applying the so-called RING assay [70,71]. Under the conditions investigated, locomotor activity was similar between control and resveratrol fed flies both in males ($p = 0.092$) and females ($p = 0.743$) as shown in Figure 2.

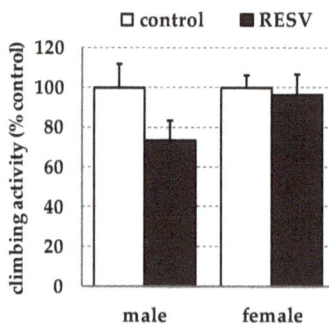

Figure 2. Dietary resveratrol (RESV; 500 µmol/L) does not affect locomotor activity in w^{1118} *D. melanogaster*. Relative climbing activity of male and female flies following a thirty-day feeding period in the presence or absence of resveratrol. Locomotor activity was quantified via the rapid iterative negative geotaxis (RING) assay. Bars show means ± SEM from 20 measurements/group derived from two independent experiments. Statistics: Mann–Whitney U. RESV: resveratrol (500 µmol/L).

The hydrogen peroxide-based stress resistance assay is well established and suitable to examine both direct and indirect antioxidant effects of secondary plant metabolites in fruit flies [6,72–75]. In order to test flies for stress resistance against reactive oxygen species, male and female w^{1118} were challenged with hydrogen peroxide (10% w/v diluted in a 5% w/v sucrose solution) following a ten-day feeding period with a resveratrol-supplemented or a control diet. The hydrogen peroxide administration dramatically increased mortality of both male and female *D. melanogaster* as reported in the literature. However, there was no significant advantage for survival when flies received dietary resveratrol prior to hydrogen peroxide challenge as compared to controls (Figure 3). Both male and female flies did not benefit from dietary resveratrol supplementation or even displayed slightly reduced mean and median survival rates compared to their control-fed counterparts.

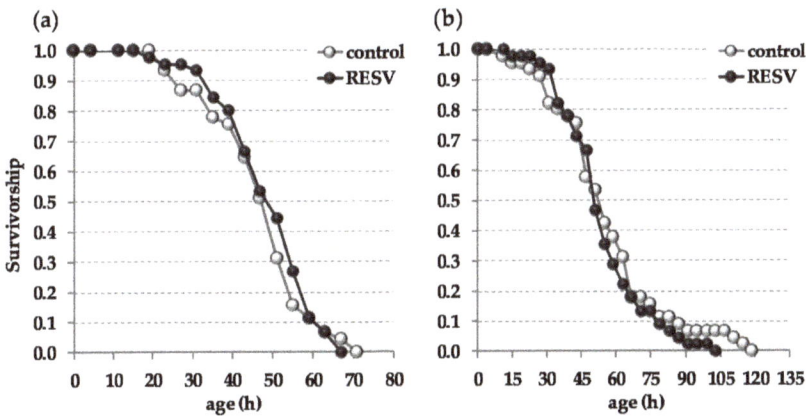

Figure 3. Dietary resveratrol (RESV; 500 µmol/L) does not improve stress resistance of w^{1118} *D. melanogaster* against reactive oxygen species. Flies received a resveratrol-supplemented or a control diet for ten days prior to the exposure to hydrogen peroxide (10% w/v diluted in a 5% w/v sucrose solution). Dead flies were steadily counted every four hours. (**a**) Survival curve of male and (**b**) female flies. The stress resistance experiment was independently performed three times with 45 flies/group each revealing similar results. Statistics: Log-Rank.

Accordingly, mRNA expression levels of genes encoding antioxidant enzymes were not significantly modulated by resveratrol ingestion (Table 2).

Table 2. Relative mRNA expression of antioxidant enzymes in female w^{1118} *D. melanogaster* in dependence of resveratrol (RESV; 500 µmol/L) administration for ten days compared to controls.

Target	mRNA Expression Level vs. Control	p-Value [6]
Cat [1]	0.87 ± 0.15	0.700
GstD2 [2]	1.02 ± 0.22	0.948
ND-75 [3]	0.99 ± 0.22	0.963
PHGPx [4]	0.89 ± 0.17	0.741
Sod2 [5]	1.03 ± 0.18	0.930

Data are shown as means $\pm$ SEM comprising 40 flies/group from two independent experiments; α-Tubulin at 84B (αTub84B) and Ribosomal protein L32 (RpL32) served as the housekeeping genes. Statistics: 2-sided Student's *t*-test and Mann–Whitney U. [1] *Cat*: Catalase; [2] *GstD2*: Glutathione S transferase D2; [3] *ND-75*: NADH dehydrogenase (ubiquinone) 75 kDa subunit; [4] *PHGPx*: PHGPx with glutathione peroxidase activity; [5] *Sod2*: Superoxide dismutase 2 (Mn). [6] *p*-value indicates differences in relative mRNA expression in whole body RNA extracts of resveratrol treated flies compared to controls.

Several studies suggest that resveratrol may affect the life span of model organisms. Therefore, we determined mean, median, and maximum life span of flies in response to the resveratrol treatment. However, dietary resveratrol did not change mean, median, and maximum life span of male and female

w^{1118} *D. melanogaster* in general (Table 3). Resveratrol ingestion rather decreased than increased mean, medium, and maximum life span in both males and females compared to the respective control groups in the majority of the assays performed within this study (Table 3), although resveratrol-dependent differences in life span remained mostly non-significant.

Table 3. Differences in mean, median, and maximum survival rates (%) of male and female w^{1118} *D. melanogaster* in dependence of life-long dietary resveratrol (RESV; 500 µmol/L) administration compared to controls.

Trial	Males				Females			
	Mean	Median	Max. [1]	p-Value [2]	Mean	Median	Max. [1]	p-Value [2]
No. 1	+15.2	+2.50	+1.03	0.307	−8.20	−7.50	−14.1	0.066
No. 2	−11.8	−19.6	+3.04	0.345	−14.7	−46.2	−1.04	0.099
No. 3	−33.1	−56.3	−9.35	<0.001	+15.9	±0.00	+1.06	0.341

The life span experiment was independently performed three times (Trial No. 1–3) comprising 150 flies/group in each experiment. Changes in mean, medium, and maximum life span (% control) of RESV treated males and females in comparison to the respective control group are shown for each single experiment. RESV did not improve the life span of male and female w^{1118} *D. melanogaster*. Statistics: Log-Rank. RESV: resveratrol (500 µmol/L); data are shown as means. [1] max.: maximum lifespan defined as the mean survival of the longest-lived 10%; [2] *p*-value indicates differences in overall survival benefit of resveratrol treated flies compared to controls.

Furthermore, we monitored mRNA steady levels of various genes encoding proteins that have been reported to be related to ageing and longevity in the fruit fly. Thus, mRNA levels of the health and life span associated genes *I'm Not Dead Yet* (INDY) [76], *sirtuin* (Sir2) [77], and *spargel* (srl) [78] were determined by quantitative real time PCR in whole body homogenates. Similar to our life span data, none of these genes were significantly regulated on the transcript level following dietary resveratrol supplementation. Relative transcript levels are summarized in Table 4.

Table 4. Relative mRNA expression of longevity-associated genes in male and female w^{1118} *D. melanogaster* in dependence of resveratrol (RESV; 500 µmol/L) administration for ten days compared to controls.

Sex	Transcript					
	Sir2 [1]	p-Value [4]	INDY [2]	p-Value [4]	Srl [3]	p-Value [4]
Male	0.90 ± 0.19	0.700	1.13 ± 0.15	0.437	1.77 ± 0.74	0.375
Female	0.83 ± 0.12	0.405	1.02 ± 0.14	0.913	1.12 ± 0.19	0.593

Data are shown as means ± SEM comprising 40 flies/group in total from two independent experiments; *α-Tubulin at 84B* (αTub84B) and *Ribosomal protein L32* (RpL32) served as the housekeeping genes. Statistics: 2-sided Student's *t*-test and Mann–Whitney U. [1] *Sir2*: *Sirtuin 2*; [2] *INDY*: *I'm Not Dead Yet*; [3] *srl*: *spargel*; [4] *p*-value indicates differences in relative mRNA expression in whole body RNA extracts of resveratrol-treated flies compared to controls.

Resveratrol displays a rather low bioavailability; hence tissue distribution may affect its efficacy to modulate metabolism and life span [79]. Therefore, we determined resveratrol concentrations in the whole body homogenates of male and female flies. We recovered substantial quantities of resveratrol in the whole body homogenates of resveratrol-treated flies, whereas no resveratrol was detected in homogenates of controls (limit of detection 0.49 nmol/g flies). These data suggest that resveratrol was readily taken up from the diet. Interestingly, female flies fed the resveratrol-supplemented diet exhibited a 2.2-fold higher resveratrol body concentration as compared to male flies (20.98 ± 0.44 vs. 9.60 ± 2.80 nmol/g fly; means ± SD; N = 54 flies/group following a ten-day feeding period).

3. Discussion

Several studies in lemurs, mice, honey bees, and humans suggest that resveratrol may affect food intake due to its bitter taste [65,80–83], which may in turn affect stress response and metabolism. However, in the current fly study, we did not observe significant differences in feed intake between groups (Figure 1). Although literature data suggest that resveratrol may affect body weight and body composition [60–66,84], we did not find any significant changes in these parameters, except for the slight increase in protein content in males, following dietary resveratrol supplementation (Table 1).

Furthermore, we did not observe changes in both locomotor activity and life span in resveratrol-fed flies as compared to controls (Figure 2; Table 3). One limitation of our present study is that only one dietary resveratrol concentration has been investigated and no dose-response analysis has been performed. Thus, we cannot fully exclude the possibility that higher or lower dietary resveratrol concentrations may modulate overall health status and life span in *D. melanogaster* as indicated by other but rather contradictory studies [24,30,31,85,86]. Therefore, further studies may include additional resveratrol concentrations and treatment periods for body weight and composition measurements, transcript analyses, and oxidative stress resistance evaluation. Due to comparable high variances in the treatment groups, the inclusion of further fly strains should be also considered.

We have recently conducted a systematic literature review concerning the effect of resveratrol on life span in various model organisms. Interestingly, resveratrol supplementation has been reported to increase life span in approximately 60% of the studies conducted in model organisms [33]. However, literature data is rather inconsistent, suggesting that the life span effects of resveratrol vary in relation to the model organism. Furthermore, it should be considered that other factors such as dose, gender, genetic background, and diet composition [35,87] may contribute to the high variance in the life span prolonging ability of resveratrol in the fruit fly.

Several plant bioactives, including lutein, epicatechin, epigallocatechin gallate, and apple polyphenols, have been reported to counteract reactive oxygen species induced mortality in *D. melanogaster* [6,88,89]. It is suggested that the protective action of these plant bioactives may be partly mediated via the induction of the transcription factor cnc, which represents an ortholog of mammalian Nrf2 in the fruit fly [90], and its target genes exhibiting antioxidant activity [89,91,92]. However, under the conditions investigated resveratrol did not improve the survival of flies treated with hydrogen peroxide as compared to controls (Figure 3). Accordingly, the relative transcript levels of cnc target genes and further antioxidant enzymes such as *glutathione-S-transferase*, *glutathione peroxidase*, *NADH dehydrogenase*, *superoxide dismutase*, and *catalase* remained largely unchanged in response to the resveratrol treatment (Table 2).

Cell culture as well as in vivo studies suggest that resveratrol may modulate the expression of genes encoding proteins that are centrally involved in ageing related pathways and suppresses DNA damage, thereby affecting the ageing process [93–96]. Among others, sirtuins seem to be important molecular targets of resveratrol [49,97,98]. Sir2 is an important modulator of the ageing process and locomotor activity in the fruit fly [77,86,99]. Moderate induction of Sir2 expression is associated with life span elongation in *D. melanogaster* in response to both nutrient composition as well as dietary plant bioactives [5,77,99–102]. In contrast, mutant fly strains with diminished or absent Sir2 expression display a markedly shortened life span and lack responsiveness to life span extending dietary interventions [31,99,101]. However, in the present study, Sir2 mRNA levels were not affected in whole body homogenates of *D. melanogaster*. The same holds true for other genes previously reported to affect ageing and life span in the fruit fly, including INDY and srl (Table 4), being in accordance with the ineffectiveness of resveratrol to prolong life span in w^{1118} males and females (Table 3). It should be considered that INDY expression is not inevitably associated with longevity in the w^{1118} *D. melanogaster* strain [103]. Moreover, it needs to be taken into account that, in the present study, gene expression was determined in whole body homogenates and not in specific tissues and organs of *D. melanogaster*. Thus, it could be possible that resveratrol may have affected gene expression in distinct tissues of the fly, which was however beyond the scope of the current study.

Unlike in the present study in w^{1118} *D. melanogaster*, resveratrol was shown to increase life span in the short-lived killifish [27,28] and in laboratory mice in dependence of their genetic background [41,104]. Besides differences in the dietary resveratrol concentrations and durations of the experimental trials, there may be also species-specific differences in resveratrol metabolism [33].

Importantly, resveratrol whole body concentrations in both male and female flies were elevated in response to dietary resveratrol supplementation. Interestingly, female flies exhibited a whole body resveratrol concentration more than twice as high as male flies, which warrants further investigations into the underlying mechanisms. Studies by Wang et al. and Zou et al. [32,87] suggested that female but not male *D. melanogaster* may benefit from dietary resveratrol supplementation in terms of life span prolongation in dependence of the macronutrient composition of the diet. This may be partly related to a higher resveratrol body concentration in female vs. male flies as observed in the present study. The storage of fat diverges in male and female fruit flies starting in the early adulthood while the extent of the sex-specific deviation depends on the fly strain [105,106]. Thus, difference in whole body resveratrol content between male and female flies may also partly depend on the higher fat content in w^{1118} females compared to males (triglyceride level in ten-day-old males: 13.2 ± 0.7 ng/μg fly; females: 32.7 ± 2.8 ng/μg fly; $p < 0.001$) resulting in an increased accumulation of the primarily fat-soluble resveratrol.

In conclusion, present data suggest that dietary resveratrol, although sufficiently absorbed, does not affect life span, body composition, stress response, and longevity-related gene expression in male and female w^{1118} *D. melanogaster*.

4. Materials and Methods

4.1. Fly Strain and Husbandry

The *Drosophila melanogaster* strain w^{1118} (Bloomington Drosophila Stock Center #5905, Indiana University, Bloomington, IN, USA) was used for all experiments. w^{1118} flies were maintained on a standard diet at 25 °C and 60% humidity with a 12/12 h light–dark cycle. Standard medium was prepared as described previously [5]. The experimental diet consisted of sucrose (5% *w/v*; Carl Roth, Karlsruhe, Germany), Agar Type II (0.5% *w/v*; Dutscher Scientific, Grays, UK), corn meal (8.6% *w/v*; Dutscher Scientific), and inactive dry yeast (5% *w/v*; Dutscher Scientific), while Tegosept (0.3% *w/v*; Dutscher Scientific) and propionic acid (0.3% *v/v*; Carl Roth) served as preservatives. Resveratrol was purchased from Carl Roth and was dissolved in dimethyl sulfoxide (DMSO; Carl Roth) at a concentration of 100 mmol/L. This resveratrol stock solution was stored at −80 °C until supplementation of the experimental diet at a final concentration of 500 μmol/L. This concentration was chosen based on contradictory literature data reporting both beneficial [30,35,85] and detrimental [24,32] effects of oral resveratrol administration on the fruit fly that were observed in a concentration range between 100 and 1000 μmol/L. To ensure sufficient supply and absorption of resveratrol, 500 μmol/L were orally applied in this study. DMSO (0.5% *v/v*) served as the vehicle control. For all experiments, flies received either a resveratrol-supplemented (500 μmol/L) or a control diet and were reared on the experimental diets at a population density of 25 flies/vial.

4.2. Life Span Analyses

Newly enclosed synchronized flies were permitted to mate for two days as described previously [5]. Sexual activity may act as a confounding variable in life span experiments [107,108] as frequent mating shortens life expectancy in *D. melanogaster* [107,109]. Therefore, flies were separated according to sex two days past enclosure and were housed in single-sex cohorts. The flies were fed the resveratrol-supplemented or the control diet life-long while transferred to fresh medium three times a week. The number of dead flies was recorded at each transfer until all flies were dead. The experiment was independently performed three times with 150 flies/group, respectively.

4.3. Gustatory Assay

Sulforhodamine B (0.2% *w/v*; Sigma-Aldrich Chemie GmbH, Taufkirchen, Germany) was used to determine food intake in both male and female flies fed a resveratrol-supplemented or a control diet. Flies were reared on the experimental diet for five days prior performing the food-dye-based gustatory assay according to the protocol described elsewhere [5,69]. This feeding period was chosen as food consumption in older flies remarkably declines in comparison to early-life feeding levels [110]. The gustatory assay was independently performed two times.

4.4. Body Weight and Body Composition

Flies received a resveratrol-supplemented or a control diet for ten days. Body weight was estimated as described in [5], the same applies to the quantification of whole body triglyceride, protein and glucose levels. Average body weight of male and female w^{1118} was calculated from two independent experiments with 51–137 flies/group, respectively (178–224 flies/group in total). Body composition data were collected from three independent experiments comprising 5–15 flies/group, respectively (20–35 flies/group in total).

4.5. Locomotor Activity

Locomotor activity can be easily evaluated in *D. melanogaster* by applying the rapid iterative negative geotaxis (RING) assay [70,71] indicating the health status of fruit flies. Climbing speed was determined in both male and female flies by performing the RING assay as previously described in detail in [5] following oral administration of a resveratrol-supplemented or a control diet for thirty days. As locomotor activity declines with age [11] and senescence is functionally associated with the flies' climbing ability reflecting the functional status of muscle and locomotor function [111], flies were pre-treated for thirty days to investigate a putative anti-ageing effect of resveratrol in *D. melanogaster*. The experiment was performed two times comprising ten repetitive measurements each (N = 100).

4.6. Oxidative Stress Resistance

Resistance against reactive oxygen species was investigated by applying the hydrogen peroxide stress test as described previously [12]. As H_2O_2 generates hydroxyl radicals in the presence of metal ions [6], it was used to assess the resistance of male and female w^{1118} flies that were pre-fed with a resveratrol (500 µmol/L)-supplemented or a control diet for ten days. This pre-feeding period was chosen as flies exhibited adequate whole body resveratrol concentrations that allowed the investigation of putative resveratrol-dependent antioxidant effects in vivo. The experiment was independently performed three times including 45 flies/group, respectively.

4.7. RNA Isolation and qRT-PCR

Flies were orally administered a resveratrol (500 µmol/L)-supplemented or a control diet for ten days. Whole flies (ten per sample) were homogenised in a TissueLyser II (Qiagen, Hilden, Germany) prior to total RNA isolation with the help of TriFast reagent (peqlab, Erlangen, Germany) according to the manufacturer's protocol. RNA concentration was determined via NanoDrop measurements (NanoDrop2000c; ThermoScientific, Waltham, MA, USA). mRNA expression was quantified by qRT-PCR measurements using the SensiFast SYBR No-ROX One-Step Kit (Bioline, London, UK) and a Rotor-Gene 6000 real-time PCR cycler (Corbett/Qiagen). Relative mRNA concentrations were calculated using a standard curve. The expression of the longevity associated genes (*INDY*, F: GATTGGTTGTGTTCCTGGTG, R: CGTCACATAGAGAGGCAAGG; *Sir2*, F: CCGTTACTGAGGAGGAGCTG, R: GTAGATCGCACACGTCCTTG; *srl*, F: CTCTTGGAGTCCGAG ATCCGCAA, R: GGGACCGCGAGCTGATGGTT [78]), and of the antioxidant enzymes (*Cat*, F: CCTCT GATTCCTGTGGGCAA, R: GACGACCATGCAGCATCTTG; *GstD2*, F: GTCTACTTCGCAGGCATC AC, R: CTTCTCGATCCAGGCCTTGA; *ND-75*, F: CGGACATTAACTACACGGGC, R: CAATCTCGG AGGCGAAACG; *PHGPx*, F: CCTCAACTTCCCGTGCAATC, R: CCATTCACATCGACCTTGGC; *Sod2*, F: AACGCAGATATGTTCGTGGC, R: GGTGATGCAGCTCCATGATC) was normalized to the expression of the housekeeping genes α-*Tubulin at 84B* (α*Tub84B*) and *Ribosomal protein L32* (*RpL32*), respectively. Experiment was independently performed two times with 20 flies/group, respectively.

4.8. Whole Body Resveratrol Concentration

Newly hatched two-day-old flies were orally administered a resveratrol (500 µmol/L)-supplemented or a control diet for ten days. Whole flies were stored at −80 °C until analysis. For analysis, whole flies (27 flies/sample) were mixed with 200 µL ice-cold aqueous triethylammonium acetate solution (1 mol/L, pH = 7.0) followed by adding 1 mL of an ice-cold acetonitril–methanol mixture (1:1, *v*/*v*). Afterwards, 1-mm silica spheres were added and flies were treated with a FastPrep homogenizer (3 × 15 s, 4 m/s) with intermediate cooling periods on ice (FastPrep-24, MP Biomedicals, Solon, OH, USA). Homogenates were transferred to a new sample tube. The residual silica spheres were washed with 100 µL of an ice-cold acetonitrile–methanol mixture (1:1, *v*/*v*) and this washing fraction was combined with the transferred homogenates. Next, homogenates were first stirred on a vortex mixer for 10 s and then treated in an ultrasonic bath for 10 min. Once again, homogenates were stirred on a vortex mixer for 10 s followed by a centrifugation of the samples (23,100× *g*; 10 min; 0 °C). Supernatants were transferred to new sample tubes und evaporated to dryness using a SpeedVac (SPD131; Thermo Electron LED GmbH, Langenselbold, Germany). Residues were first dissolved in 70 µL of 30% (*v*/*v*) methanol in water and then centrifuged at 23,100× *g* for 10 min. Supernatants were transferred to a HPLC vial and analysed by LC-MS. A TripleTOF 5600 mass spectrometer (AB Sciex, Darmstadt, Germany) equipped with a 1290 Infinity LC system (Agilent, Waldbronn, Germany) was used. The LC-MS system was controlled by the software Analyst TF 1.6.0. LC separation was carried out on an Agilent Eclipse Plus C18 column (150 mm × 3.0 mm internal diameter, 3.5 µm; Agilent). Eluent A was an aqueous ammonium formiate buffer (25 mmol/L, pH = 3.0) and eluent B was an acetonitrile–methanol mixture (1:1, *v*/*v*). A linear gradient was used with a flow rate of 0.6 mL/min and the following elution profile: 0–1 min, isocratic with 10% B; 1–19 min, from 10 to 28% B; 19–33 min from 28 to 70% B; 33–34 min, from 70 to 95% B; 34–39 min, isocratic with 95% B; 39–40 min, from 95 to 10% B; and 40–45 min, isocratic with initial conditions. The column oven was set to 40 °C and the injection volume was 20 µL. The DuoSpray source was operated in negative ESI mode using the following source parameters: curtain gas, 35 psi; ion spray voltage, −4500 V; ion source gas 1, 60 psi; ion source gas 2, 70 psi; ion source gas 2 temperature, 650 °C. The MS full scans were recorded from *m*/*z* 100 to 800 with an accumulation time of 200 ms, a declustering potential of −110 V and a collision energy voltage of −10 V. The MS/MS spectra (product ion) were recorded from *m*/*z* 50 to 600 in the high-sensitivity mode with an accumulation time of 80 ms, a collision energy voltage of −45 V, and a collision energy spread of 25 V. Nitrogen was used as collision gas. Analysis of data was performed with the MultiQuant 2.1.1 and PeakView 1.2.0.3 software (AB Sciex, Darmstadt, Germany). *trans*-Resveratrol was identified by retention time and MS/MS spectrum. Accurate XICs (10 mDa extraction width) were used to monitor and quantify the analytes. In detail, *trans*-resveratrol was quantified by an external calibration using commercial reference compound (Sigma-Aldrich, Deideshofen, Germany). Therefore, control flies were spiked with 5 µL of the specific standard solution (concentration range between 0.8 and 100 µmol/L *trans*-resveratrol in DMSO) and these spiked samples were worked up as described above. A best fit line was obtained by linear regression using a weighting of 1/*x*. Whole body resveratrol measurements comprised 54 flies/group, respectively.

4.9. Statistics

Life span and survival rates following H_2O_2 treatment were analysed using the DLife software (Winchecker version 3.0 [112]). Values are given as means and were statistically compared via a Log-Rank Test based on R (i386 version 3.1.0). All other data are given as means ± SEM, except otherwise mentioned, and were statistically analysed by applying SPSS (version 24; SPSS Inc., Munich, Germany). Data were tested for normality of distribution (Kolmogorov–Smirnov and Shapiro–Wilk) and mean comparisons were carried out using a 2-sided Student's *t*-test. If the assumption of a normal distribution was violated, the non-parametric Mann–Whitney U. test was used. Significance was accepted at *p*-values < 0.05.

Int. J. Mol. Sci. **2018**, *19*, 223

Acknowledgments: Sources of funding for research work: University of Kiel (federate state Schleswig-Holstein, Germany). For covering the costs to publish in open access: We acknowledge financial support by Land Schleswig-Holstein within the funding programme "Open Access Publikationsfonds".

Author Contributions: Anika E. Wagner and Stefanie Staats conceived and designed the experiments; Bianca Kowalewski performed the experiments; Bianca Kowalewski, Stefanie Staats, Florian T. Rieck, and Sebastian T. Soukup analysed the data; Gerald Rimbach and Sabine E. Kulling contributed reagents, materials, and analysis tools; Gerald Rimbach, Stefanie Staats, Sebastian T. Soukup, and Sabine E. Kulling wrote the paper.

Conflicts of Interest: The authors declare no conflict of interest. The founding sponsors had no role in the design of the study; in the collection, analyses, or interpretation of data; in the writing of the manuscript; and in the decision to publish the results.

Abbreviations

Cat	Catalase
GstD2	Glutathione S transferase D2
INDY	I'm Not Dead Yet
Max.	maximum
ND-75	NADH dehydrogenase (ubiquinone) 75 kDa subunit
PHGPx	PHGPx with glutathione peroxidase activity
RESV	resveratrol
Sir2	Sirtuin 2
Sod2	Superoxide dismutase 2 (Mn)
srl	spargel

References

1. Pallauf, K.; Duckstein, N.; Rimbach, G. A literature review of flavonoids and lifespan in model organisms. *Proc. Nutr. Soc.* **2016**, *76*, 145–162. [CrossRef] [PubMed]

2. Pallauf, K.; Giller, K.; Huebbe, P.; Rimbach, G. Nutrition and healthy ageing: Calorie restriction or polyphenol-rich "MediterrAsian" diet? *Oxid. Med. Cell. Longev.* **2013**, *2013*, 707421. [CrossRef] [PubMed]

3. Grandison, R.C.; Wong, R.; Bass, T.M.; Partridge, L.; Piper, M.D.W. Effect of a Standardised Dietary Restriction Protocol on Multiple Laboratory Strains of *Drosophila melanogaster*. *PLoS ONE* **2009**, *4*, e4067. [CrossRef] [PubMed]

4. Kang, H.-L.; Benzer, S.; Min, K.-T. Life extension in Drosophila by feeding a drug. *Proc. Natl. Acad. Sci. USA* **2002**, *99*, 838–843. [CrossRef] [PubMed]

5. Piegholdt, S.; Rimbach, G.; Wagner, A.E. The phytoestrogen prunetin affects body composition and improves fitness and lifespan in male *Drosophila melanogaster*. *FASEB J.* **2016**, *30*, 948–958. [CrossRef] [PubMed]

6. Zhang, Z.; Han, S.; Wang, H.; Wang, T. Lutein extends the lifespan of *Drosophila melanogaster*. *Arch. Gerontol. Geriatr.* **2014**, *58*, 153–159. [CrossRef] [PubMed]

7. Rera, M.; Clark, R.I.; Walker, D.W. Intestinal barrier dysfunction links metabolic and inflammatory markers of aging to death in Drosophila. *Proc. Natl. Acad. Sci. USA* **2012**, *109*, 21528–21533. [CrossRef] [PubMed]

8. Ja, W.W.; Carvalho, G.B.; Zid, B.M.; Mak, E.M.; Brummel, T.; Benzer, S. Water- and nutrient-dependent effects of dietary restriction on Drosophila lifespan. *Proc. Natl. Acad. Sci. USA* **2009**, *106*, 18633–18637. [CrossRef] [PubMed]

9. Wang, L.; Li, Y.M.; Lei, L.; Liu, Y.; Wang, X.; Ma, K.Y.; Chen, Z.Y. Cranberry anthocyanin extract prolongs lifespan of fruit flies. *Exp. Gerontol.* **2015**, *69*, 189–195. [CrossRef] [PubMed]

10. Wagner, A.E.; Piegholdt, S.; Rabe, D.; Baenas, N.; Schloesser, A.; Eggersdorfer, M.; Stocker, A.; Rimbach, G. Epigallocatechin gallate affects glucose metabolism and increases fitness and lifespan in *Drosophila melanogaster*. *Oncotarget* **2015**, *6*, 30568–30578. [CrossRef] [PubMed]

11. Simon, A.F.; Liang, D.T.; Krantz, D.E. Differential decline in behavioral performance of *Drosophila melanogaster* with age. *Mech. Ageing Dev.* **2006**, *127*, 647–651. [CrossRef] [PubMed]

12. Piegholdt, S.; Rimbach, G.; Wagner, A.E. Effects of the isoflavone prunetin on gut health and stress response in male *Drosophila melanogaster*. *Redox Biol.* **2016**, *8*, 119–126. [CrossRef] [PubMed]

13. Stramer, B.; Wood, W. Inflammation and wound healing in Drosophila. *Methods Mol. Biol.* **2009**, *571*, 137–149. [CrossRef] [PubMed]

14. Vlisidou, I.; Wood, W. Drosophila blood cells and their role in immune responses. *FEBS J.* **2015**, *282*, 1368–1382. [CrossRef] [PubMed]

15. Panayidou, S.; Apidianakis, Y. Regenerative inflammation: Lessons from Drosophila intestinal epithelium in health and disease. *Pathogens* **2013**, *2*, 209–231. [CrossRef] [PubMed]
16. Apidianakis, Y.; Rahme, L.G. *Drosophila melanogaster* as a model for human intestinal infection and pathology. *Dis. Model. Mech.* **2011**, *4*, 21–30. [CrossRef] [PubMed]
17. Silverman, N.; Maniatis, T. NF-kappaB signaling pathways in mammalian and insect innate immunity. *Genes Dev.* **2001**, *15*, 2321–2342. [CrossRef] [PubMed]
18. Clark, A.G.; Fucito, C.D. Stress tolerance and metabolic response to stress in *Drosophila melanogaster*. *Heredity* **1998**, *81*, 514–527. [CrossRef] [PubMed]
19. Sorensen, J.G.; Nielsen, M.M.; Kruhoffer, M.; Justesen, J.; Loeschcke, V. Full genome gene expression analysis of the heat stress response in *Drosophila melanogaster*. *Cell Stress Chaperones* **2005**, *10*, 312–328. [CrossRef] [PubMed]
20. Baenas, N.; Piegholdt, S.; Schloesser, A.; Moreno, D.A.; García-Viguera, C.; Rimbach, G.; Wagner, A.E. Metabolic Activity of Radish Sprouts Derived Isothiocyanates in *Drosophila melanogaster*. *Int. J. Mol. Sci.* **2016**, *17*, 251. [CrossRef] [PubMed]
21. Howitz, K.T.; Bitterman, K.J.; Cohen, H.Y.; Lamming, D.W.; Lavu, S.; Wood, J.G.; Zipkin, R.E.; Chung, P.; Kisielewski, A.; Zhang, L.L.; et al. Small molecule activators of sirtuins extend *Saccharomyces cerevisiae* lifespan. *Nature* **2003**, *425*, 191–196. [CrossRef] [PubMed]
22. Yang, H.; Baur, J.A.; Chen, A.; Miller, C.; Sinclair, D.A. Design and synthesis of compounds that extend yeast replicative lifespan. *Aging Cell* **2007**, *6*, 35–43. [CrossRef] [PubMed]
23. Jarolim, S.; Millen, J.; Heeren, G.; Laun, P.; Goldfarb, D.S.; Breitenbach, M. A novel assay for replicative lifespan in *Saccharomyces cerevisiae*. *FEMS Yeast Res.* **2004**, *5*, 169–177. [CrossRef] [PubMed]
24. Bass, T.M.; Weinkove, D.; Houthoofd, K.; Gems, D.; Partridge, L. Effects of resveratrol on lifespan in *Drosophila melanogaster* and *Caenorhabditis elegans*. *Mech. Ageing Dev.* **2007**, *128*, 546–552. [CrossRef] [PubMed]
25. Lee, J.; Kwon, G.; Park, J.; Kim, J.-K.; Lim, Y.-H. Brief Communication: SIR-2.1-dependent lifespan extension of *Caenorhabditis elegans* by oxyresveratrol and resveratrol. *Exp. Biol. Med.* **2016**, *241*, 1757–1763. [CrossRef] [PubMed]
26. Viswanathan, M.; Kim, S.K.; Berdichevsky, A.; Guarente, L. A role for SIR-2.1 regulation of ER stress response genes in determining *C. elegans* life span. *Dev. Cell* **2005**, *9*, 605–615. [CrossRef] [PubMed]
27. Genade, T.; Lang, D.M. Resveratrol extends lifespan and preserves glia but not neurons of the *Nothobranchius guentheri* optic tectum. *Exp. Gerontol.* **2013**, *48*, 202–212. [CrossRef] [PubMed]
28. Valenzano, D.R.; Terzibasi, E.; Genade, T.; Cattaneo, A.; Domenici, L.; Cellerino, A. Resveratrol Prolongs Lifespan and Retards the Onset of Age-Related Markers in a Short-Lived Vertebrate. *Curr. Biol.* **2006**, *16*, 296–300. [CrossRef] [PubMed]
29. Valenzano, D.R.; Cellerino, A. Resveratrol and the Pharmacology of Aging: A New Vertebrate Model to Validate an Old Molecule. *Cell Cycle* **2006**, *5*, 1027–1032. [CrossRef] [PubMed]
30. Balasubramani, S.P.; Mohan, J.; Chatterjee, A.; Patnaik, E.; Kukkupuni, S.K.; Nongthomba, U.; Venkatasubramanian, P. Pomegranate Juice Enhances Healthy Lifespan in *Drosophila melanogaster*: An Exploratory Study. *Front. Public Health* **2014**, *2*, 245. [CrossRef] [PubMed]
31. Kayashima, Y.; Katayanagi, Y.; Tanaka, K.; Fukutomi, R.; Hiramoto, S.; Imai, S. Alkylresorcinols activate SIRT1 and delay ageing in *Drosophila melanogaster*. *Sci. Rep.* **2017**, *7*, 43679. [CrossRef] [PubMed]
32. Wang, C.; Wheeler, C.T.; Alberico, T.; Sun, X.; Seeberger, J.; Laslo, M.; Spangler, E.; Kern, B.; de Cabo, R.; Zou, S. The effect of resveratrol on lifespan depends on both gender and dietary nutrient composition in *Drosophila melanogaster*. *Age* **2013**, *35*, 69–81. [CrossRef] [PubMed]
33. Pallauf, K.; Rimbach, G.; Rupp, P.M.; Chin, D.; Wolf, I.M. Resveratrol and Lifespan in Model Organisms. *Curr. Med. Chem.* **2016**, *23*, 4639–4680. [CrossRef] [PubMed]
34. Pallauf, K.; Rimbach, G. Autophagy, polyphenols and healthy ageing. *Ageing Res. Rev.* **2013**, *12*, 237–252. [CrossRef] [PubMed]
35. Wood, J.G.; Rogina, B.; Lavu, S.; Howitz, K.; Helfand, S.L.; Tatar, M.; Sinclair, D. Sirtuin activators mimic caloric restriction and delay ageing in metazoans. *Nature* **2004**, *430*, 686–689. [CrossRef] [PubMed]
36. Morselli, E.; Maiuri, M.C.; Markaki, M.; Megalou, E.; Pasparaki, A.; Palikaras, K.; Criollo, A.; Galluzzi, L.; Malik, S.A.; Vitale, I.; et al. The life span-prolonging effect of sirtuin-1 is mediated by autophagy. *Autophagy* **2010**, *6*, 186–188. [CrossRef] [PubMed]

37. Park, D.; Jeong, H.; Lee, M.N.; Koh, A.; Kwon, O.; Yang, Y.R.; Noh, J.; Suh, P.-G.; Park, H.; Ryu, S.H. Resveratrol induces autophagy by directly inhibiting mTOR through ATP competition. *Sci. Rep.* **2016**, *6*, 21772. [CrossRef] [PubMed]

38. Morselli, E.; Maiuri, M.C.; Markaki, M.; Megalou, E.; Pasparaki, A.; Palikaras, K.; Criollo, A.; Galluzzi, L.; Malik, S.A.; Vitale, I.; et al. Caloric restriction and resveratrol promote longevity through the Sirtuin-1-dependent induction of autophagy. *Cell Death Dis.* **2010**, *1*, e10. [CrossRef] [PubMed]

39. Ding, S.; Jiang, J.; Zhang, G.; Bu, Y.; Zhang, G.; Zhao, X. Resveratrol and caloric restriction prevent hepatic steatosis by regulating SIRT1-autophagy pathway and alleviating endoplasmic reticulum stress in high-fat diet-fed rats. *PLoS ONE* **2017**, *12*, e0183541. [CrossRef] [PubMed]

40. Saini, A.; Al-Shanti, N.; Sharples, A.P.; Stewart, C.E. Sirtuin 1 regulates skeletal myoblast survival and enhances differentiation in the presence of resveratrol. *Exp. Physiol.* **2012**, *97*, 400–418. [CrossRef] [PubMed]

41. Baur, J.A.; Pearson, K.J.; Price, N.L.; Jamieson, H.A.; Lerin, C.; Kalra, A.; Prabhu, V.V.; Allard, J.S.; Lopez-Lluch, G.; Lewis, K.; et al. Resveratrol improves health and survival of mice on a high-calorie diet. *Nature* **2006**, *444*, 337–342. [CrossRef] [PubMed]

42. Bosutti, A.; Degens, H. The impact of resveratrol and hydrogen peroxide on muscle cell plasticity shows a dose-dependent interaction. *Sci. Rep.* **2015**, *5*, 8093. [CrossRef]

43. Ghanim, H.; Sia, C.L.; Korzeniewski, K.; Lohano, T.; Abuaysheh, S.; Marumganti, A.; Chaudhuri, A.; Dandona, P. A Resveratrol and Polyphenol Preparation Suppresses Oxidative and Inflammatory Stress Response to a High-Fat, High-Carbohydrate Meal. *J. Clin. Endocrinol. Metab.* **2011**, *96*, 1409–1414. [CrossRef] [PubMed]

44. Kristl, J.; Teskač, K.; Caddeo, C.; Abramović, Z.; Šentjurc, M. Improvements of cellular stress response on resveratrol in liposomes. *Eur. J. Pharm. Biopharm.* **2009**, *73*, 253–259. [CrossRef] [PubMed]

45. Xiang, L.; Nakamura, Y.; Lim, Y.M.; Yamasaki, Y.; Kurokawa-Nose, Y.; Maruyama, W.; Osawa, T.; Matsuura, A.; Motoyama, N.; Tsuda, L. Tetrahydrocurcumin extends life span and inhibits the oxidative stress response by regulating the FOXO forkhead transcription factor. *Aging* **2011**, *3*, 1098–1109. [CrossRef] [PubMed]

46. Danilov, A.; Shaposhnikov, M.; Shevchenko, O.; Zemskaya, N.; Zhavoronkov, A.; Moskalev, A. Influence of non-steroidal anti-inflammatory drugs on *Drosophila melanogaster* longevity. *Oncotarget* **2015**, *6*, 19428–19444. [CrossRef] [PubMed]

47. Lee, L.-C.; Weng, Y.-T.; Wu, Y.-R.; Soong, B.-W.; Tseng, Y.-C.; Chen, C.-M.; Lee-Chen, G.-J. Downregulation of proteins involved in the endoplasmic reticulum stress response and Nrf2-ARE signaling in lymphoblastoid cells of spinocerebellar ataxia type 17. *J. Neural Transm.* **2014**, *121*, 601–610. [CrossRef] [PubMed]

48. Robb, E.L.; Winkelmolen, L.; Visanji, N.; Brotchie, J.; Stuart, J.A. Dietary resveratrol administration increases MnSOD expression and activity in mouse brain. *Biochem. Biophys. Res. Commun.* **2008**, *372*, 254–259. [CrossRef] [PubMed]

49. Lagouge, M.; Argmann, C.; Gerhart-Hines, Z.; Meziane, H.; Lerin, C.; Daussin, F.; Messadeq, N.; Milne, J.; Lambert, P.; Elliott, P.; et al. Resveratrol Improves Mitochondrial Function and Protects against Metabolic Disease by Activating SIRT1 and PGC-1α. *Cell* **2006**, *127*, 1109–1122. [CrossRef] [PubMed]

50. Hui, Y.; Lu, M.; Han, Y.; Zhou, H.; Liu, W.; Li, L.; Jin, R. Resveratrol improves mitochondrial function in the remnant kidney from 5/6 nephrectomized rats. *Acta Histochem.* **2017**, *119*, 392–399. [CrossRef] [PubMed]

51. Zhang, B.; Xu, L.; Zhuo, N.; Shen, J. Resveratrol protects against mitochondrial dysfunction through autophagy activation in human nucleus pulposus cells. *Biochem. Biophys. Res. Commun.* **2017**, *493*, 373–381. [CrossRef] [PubMed]

52. Karuppagounder, V.; Arumugam, S.; Thandavarayan, R.A.; Pitchaimani, V.; Sreedhar, R.; Afrin, R.; Harima, M.; Suzuki, H.; Nomoto, M.; Miyashita, S.; et al. Resveratrol attenuates HMGB1 signaling and inflammation in house dust mite-induced atopic dermatitis in mice. *Int. Immunopharmacol.* **2014**, *23*, 617–623. [CrossRef] [PubMed]

53. Kjær, T.N.; Thorsen, K.; Jessen, N.; Stenderup, K.; Pedersen, S.B. Resveratrol Ameliorates Imiquimod-Induced Psoriasis-Like Skin Inflammation in Mice. *PLoS ONE* **2015**, *10*, e0126599. [CrossRef] [PubMed]

54. Bereswill, S.; Muñoz, M.; Fischer, A.; Plickert, R.; Haag, L.-M.; Otto, B.; Kühl, A.A.; Loddenkemper, C.; Göbel, U.B.; Heimesaat, M.M. Anti-Inflammatory Effects of Resveratrol, Curcumin and Simvastatin in Acute Small Intestinal Inflammation. *PLoS ONE* **2010**, *5*, e15099. [CrossRef] [PubMed]

55. Larrosa, M.; Yañéz-Gascón, M.J.; Selma, M.V.; González-Sarrías, A.; Toti, S.; Cerón, J.J.; Tomás-Barberán, F.; Dolara, P.; Espín, J.C. Effect of a Low Dose of Dietary Resveratrol on Colon Microbiota, Inflammation and Tissue Damage in a DSS-Induced Colitis Rat Model. *J. Agric. Food Chem.* **2009**, *57*, 2211–2220. [CrossRef] [PubMed]

56. Jeong, S.I.; Shin, J.A.; Cho, S.; Kim, H.W.; Lee, J.Y.; Kang, J.L.; Park, E.-M. Resveratrol attenuates peripheral and brain inflammation and reduces ischemic brain injury in aged female mice. *Neurobiol. Aging* **2016**, *44*, 74–84. [CrossRef] [PubMed]

57. Tran, H.T.; Liong, S.; Lim, R.; Barker, G.; Lappas, M. Resveratrol ameliorates the chemical and microbial induction of inflammation and insulin resistance in human placenta, adipose tissue and skeletal muscle. *PLoS ONE* **2017**, *12*, e0173373. [CrossRef] [PubMed]

58. Zhao, W.; Li, A.; Feng, X.; Hou, T.; Liu, K.; Liu, B.; Zhang, N. Metformin and resveratrol ameliorate muscle insulin resistance through preventing lipolysis and inflammation in hypoxic adipose tissue. *Cell. Signal.* **2016**, *28*, 1401–1411. [CrossRef] [PubMed]

59. Bariani, M.V.; Correa, F.; Leishman, E.; Domínguez Rubio, A.P.; Arias, A.; Stern, A.; Bradshaw, H.B.; Franchi, A.M. Resveratrol protects from lipopolysaccharide-induced inflammation in the uterus and prevents experimental preterm birth. *MHR Basic Sci. Reprod. Med.* **2017**, *23*, 571–581. [CrossRef] [PubMed]

60. Kim, S.; Jin, Y.; Choi, Y.; Park, T. Resveratrol exerts anti-obesity effects via mechanisms involving down-regulation of adipogenic and inflammatory processes in mice. *Biochem. Pharmacol.* **2011**, *81*, 1343–1351. [CrossRef] [PubMed]

61. Chang, C.C.; Lin, K.Y.; Peng, K.Y.; Day, Y.J.; Hung, L.M. Resveratrol exerts anti-obesity effects in high-fat diet obese mice and displays differential dosage effects on cytotoxicity, differentiation, and lipolysis in 3T3-L1 cells. *Endocr. J.* **2016**, *63*, 169–178. [CrossRef] [PubMed]

62. Mendez-del Villar, M.; Gonzalez-Ortiz, M.; Martinez-Abundis, E.; Perez-Rubio, K.G.; Lizarraga-Valdez, R. Effect of resveratrol administration on metabolic syndrome, insulin sensitivity, and insulin secretion. *Metab. Syndr. Relat. Disord.* **2014**, *12*, 497–501. [CrossRef] [PubMed]

63. Jeon, S.-M.; Lee, S.-A.; Choi, M.-S. Antiobesity and Vasoprotective Effects of Resveratrol in ApoE-Deficient Mice. *J. Med. Food* **2014**, *17*, 310–316. [CrossRef] [PubMed]

64. Qiao, Y.; Sun, J.; Xia, S.; Tang, X.; Shi, Y.; Le, G. Effects of resveratrol on gut microbiota and fat storage in a mouse model with high-fat-induced obesity. *Food Funct.* **2014**, *5*, 1241–1249. [CrossRef] [PubMed]

65. Dal-Pan, A.; Blanc, S.; Aujard, F. Resveratrol suppresses body mass gain in a seasonal non-human primate model of obesity. *BMC Physiol.* **2010**, *10*, 11. [CrossRef] [PubMed]

66. Um, J.-H.; Park, S.-J.; Kang, H.; Yang, S.; Foretz, M.; McBurney, M.W.; Kim, M.K.; Viollet, B.; Chung, J.H. AMP-Activated Protein Kinase–Deficient Mice Are Resistant to the Metabolic Effects of Resveratrol. *Diabetes* **2010**, *59*, 554–563. [CrossRef] [PubMed]

67. Ran, G.; Ying, L.; Li, L.; Yan, Q.; Yi, W.; Ying, C.; Wu, H.; Ye, X. Resveratrol ameliorates diet-induced dysregulation of lipid metabolism in zebrafish (*Danio rerio*). *PLoS ONE* **2017**, *12*, e0180865. [CrossRef] [PubMed]

68. Jimenez-Gomez, Y.; Mattison, J.A.; Pearson, K.J.; Martin-Montalvo, A.; Palacios, H.H.; Sossong, A.M.; Ward, T.M.; Younts, C.M.; Lewis, K.; Allard, J.S.; et al. Resveratrol improves adipose insulin signaling and reduces the inflammatory response in adipose tissue of rhesus monkeys on a high-fat, high-sugar diet. *Cell Metab.* **2013**, *18*, 533–545. [CrossRef] [PubMed]

69. Bahadorani, S.; Bahadorani, P.; Phillips, J.P.; Hilliker, A.J. The effects of vitamin supplementation on Drosophila life span under normoxia and under oxidative stress. *J. Gerontol. A Biol. Sci. Med. Sci.* **2008**, *63*, 35–42. [CrossRef] [PubMed]

70. Bazzell, B.; Ginzberg, S.; Healy, L.; Wessells, R.J. Dietary composition regulates Drosophila mobility and cardiac physiology. *J. Exp. Biol.* **2013**, *216*, 859–868. [CrossRef] [PubMed]

71. Gargano, J.W.; Martin, I.; Bhandari, P.; Grotewiel, M.S. Rapid iterative negative geotaxis (RING): A new method for assessing age-related locomotor decline in Drosophila. *Exp. Gerontol.* **2005**, *40*, 386–395. [CrossRef] [PubMed]

72. Zou, Y.; Liu, Y.; Ruan, M.; Feng, X.; Wang, J.; Chu, Z.; Zhang, Z. *Cordyceps sinensis* oral liquid prolongs the lifespan of the fruit fly, *Drosophila melanogaster*, by inhibiting oxidative stress. *Int. J. Mol. Med.* **2015**, *36*, 939–946. [CrossRef] [PubMed]

73. Schriner, S.E.; Katoozi, N.S.; Pham, K.Q.; Gazarian, M.; Zarban, A.; Jafari, M. Extension of Drosophila lifespan by Rosa damascena associated with an increased sensitivity to heat. *Biogerontology* **2012**, *13*, 105–117. [CrossRef] [PubMed]

74. Minois, N.; Carmona-Gutierrez, D.; Bauer, M.A.; Rockenfeller, P.; Eisenberg, T.; Brandhorst, S.; Sigrist, S.J.; Kroemer, G.; Madeo, F. Spermidine promotes stress resistance in *Drosophila melanogaster* through autophagy-dependent and -independent pathways. *Cell Death Dis.* **2012**, *3*, e401. [CrossRef] [PubMed]

75. Zou, Y.X.; Ruan, M.H.; Luan, J.; Feng, X.; Chen, S.; Chu, Z.Y. Anti-Aging Effect of Riboflavin via Endogenous Antioxidant in Fruit fly Drosophila Melanogaster. *J. Nutr. Health Aging* **2017**, *21*, 314–319. [CrossRef] [PubMed]

76. Rogina, B.; Reenan, R.A.; Nilsen, S.P.; Helfand, S.L. Extended life-span conferred by cotransporter gene mutations in Drosophila. *Science* **2000**, *290*, 2137–2140. [CrossRef] [PubMed]

77. Whitaker, R.; Faulkner, S.; Miyokawa, R.; Burhenn, L.; Henriksen, M.; Wood, J.G.; Helfand, S.L. Increased expression of Drosophila Sir2 extends life span in a dose-dependent manner. *Aging* **2013**, *5*, 682–691. [CrossRef] [PubMed]

78. Tinkerhess, M.J.; Healy, L.; Morgan, M.; Sujkowski, A.; Matthys, E.; Zheng, L.; Wessells, R.J. The Drosophila PGC-1α Homolog spargel Modulates the Physiological Effects of Endurance Exercise. *PLoS ONE* **2012**, *7*, e31633. [CrossRef] [PubMed]

79. Gambini, J.; Inglés, M.; Olaso, G.; Lopez-Grueso, R.; Bonet-Costa, V.; Gimeno-Mallench, L.; Mas-Bargues, C.; Abdelaziz, K.M.; Gomez-Cabrera, M.C.; Vina, J.; et al. Properties of Resveratrol: In Vitro and In Vivo Studies about Metabolism, Bioavailability, and Biological Effects in Animal Models and Humans. *Oxid. Med. Cell. Longev.* **2015**, *2015*, 837042. [CrossRef] [PubMed]

80. Kim, S.J.; Lee, Y.H.; Han, M.D.; Mar, W.; Kim, W.K.; Nam, K.W. Resveratrol, purified from the stem of Vitis coignetiae Pulliat, inhibits food intake in C57BL/6J Mice. *Arch. Pharm. Res.* **2010**, *33*, 775–780. [CrossRef] [PubMed]

81. Rascon, B.; Hubbard, B.P.; Sinclair, D.A.; Amdam, G.V. The lifespan extension effects of resveratrol are conserved in the honey bee and may be driven by a mechanism related to caloric restriction. *Aging* **2012**, *4*, 499–508. [CrossRef] [PubMed]

82. Koga, C.C.; Becraft, A.R.; Lee, Y.; Lee, S.-Y. Taste Detection Thresholds of Resveratrol. *J. Food Sci.* **2015**, *80*, S2064–S2070. [CrossRef] [PubMed]

83. Gaudette, N.J.; Pickering, G.J. Sensory and chemical characteristics of trans-resveratrol-fortified wine. *Aust. J. Grape Wine Res.* **2011**, *17*, 249–257. [CrossRef]

84. Chandrashekara, K.T.; Shakarad, M.N. Aloe vera or resveratrol supplementation in larval diet delays adult aging in the fruit fly, *Drosophila melanogaster*. *J. Gerontol. A Biol. Sci. Med. Sci.* **2011**, *66*, 965–971. [CrossRef] [PubMed]

85. Bauer, J.H.; Goupil, S.; Garber, G.B.; Helfand, S.L. An accelerated assay for the identification of lifespan-extending interventions in *Drosophila melanogaster*. *Proc. Natl. Acad. Sci. USA* **2004**, *101*, 12980–12985. [CrossRef] [PubMed]

86. Parashar, V.; Rogina, B. dSir2 mediates the increased spontaneous physical activity in flies on calorie restriction. *Aging* **2009**, *1*, 529–541. [CrossRef] [PubMed]

87. Zou, S.; Carey, J.R.; Liedo, P.; Ingram, D.K.; Müller, H.-G.; Wang, J.-L.; Yao, F.; Yu, B.; Zhou, A. The Prolongevity Effect of Resveratrol Depends on Dietary Composition and Calorie Intake in a Tephritid Fruit Fly. *Exp. Gerontol.* **2009**, *44*, 472–476. [CrossRef] [PubMed]

88. Jimenez-Del-Rio, M.; Guzman-Martinez, C.; Velez-Pardo, C. The Effects of Polyphenols on Survival and Locomotor Activity in *Drosophila melanogaster* Exposed to Iron and Paraquat. *Neurochem. Res.* **2010**, *35*, 227–238. [CrossRef] [PubMed]

89. Peng, C.; Chan, H.Y.E.; Huang, Y.; Yu, H.; Chen, Z.-Y. Apple Polyphenols Extend the Mean Lifespan of *Drosophila melanogaster*. *J. Agric. Food Chem.* **2011**, *59*, 2097–2106. [CrossRef] [PubMed]

90. Fuse, Y.; Kobayashi, M. Conservation of the Keap1-Nrf2 System: An Evolutionary Journey through Stressful Space and Time. *Molecules* **2017**, *22*, 436. [CrossRef] [PubMed]

91. Sykiotis, G.P.; Bohmann, D. Keap1/Nrf2 signaling regulates oxidative stress tolerance and lifespan in Drosophila. *Dev. Cell* **2008**, *14*, 76–85. [CrossRef] [PubMed]

92. Misra, J.R.; Horner, M.A.; Lam, G.; Thummel, C.S. Transcriptional regulation of xenobiotic detoxification in Drosophila. *Genes Dev.* **2011**, *25*, 1796–1806. [CrossRef] [PubMed]

93. Abharzanjani, F.; Afshar, M.; Hemmati, M.; Moossavi, M. Short-term High Dose of Quercetin and Resveratrol Alters Aging Markers in Human Kidney Cells. *Int. J. Prev. Med.* **2017**, *8*, 64. [CrossRef]

94. Hsu, S.-C.; Huang, S.-M.; Chen, A.; Sun, C.-Y.; Lin, S.-H.; Chen, J.-S.; Liu, S.-T.; Hsu, Y.-J. Resveratrol increases anti-aging Klotho gene expression via the activating transcription factor 3/c-Jun complex-mediated signaling pathway. *Int. J. Biochem. Cell Biol.* **2014**, *53*, 361–371. [CrossRef] [PubMed]

95. Halicka, H.D.; Zhao, H.; Li, J.; Lee, Y.S.; Hsieh, T.C.; Wu, J.M.; Darzynkiewicz, Z. Potential anti-aging agents suppress the level of constitutive mTOR- and DNA damage- signaling. *Aging* **2012**, *4*, 952–965. [CrossRef] [PubMed]

96. Chung, H.J.; Lee, H.K.; Kim, H.J.; Baek, S.H.; Hong, S.T. Gene expression profiles and physiological data from mice fed resveratrol-enriched rice DJ526. *Sci. Data* **2016**, *3*, 160114. [CrossRef] [PubMed]

97. Kulkarni, S.S.; Cantó, C. The molecular targets of resveratrol. *Biochim. Biophys. Acta* **2015**, *1852*, 1114–1123. [CrossRef] [PubMed]

98. Bagul, P.K.; Dinda, A.K.; Banerjee, S.K. Effect of resveratrol on sirtuins expression and cardiac complications in diabetes. *Biochem. Biophys. Res. Commun.* **2015**, *468*, 221–227. [CrossRef] [PubMed]

99. Rogina, B.; Helfand, S.L. Sir2 mediates longevity in the fly through a pathway related to calorie restriction. *Proc. Natl. Acad. Sci. USA* **2004**, *101*, 15998–16003. [CrossRef] [PubMed]

100. Hoffmann, J.; Romey, R.; Fink, C.; Yong, L.; Roeder, T. Overexpression of Sir2 in the adult fat body is sufficient to extend lifespan of male and female Drosophila. *Aging* **2013**, *5*, 315–327. [CrossRef] [PubMed]

101. Banerjee, K.K.; Ayyub, C.; Ali, S.Z.; Mandot, V.; Prasad, N.G.; Kolthur-Seetharam, U. dSir2 in the Adult Fat Body, but Not in Muscles, Regulates Life Span in a Diet-Dependent Manner. *Cell Rep.* **2012**, *2*, 1485–1491. [CrossRef] [PubMed]

102. Bauer, J.H.; Morris, S.N.; Chang, C.; Flatt, T.; Wood, J.G.; Helfand, S.L. dSir2 and Dmp53 interact to mediate aspects of CR-dependent lifespan extension in *D. melanogaster*. *Aging* **2009**, *1*, 38–48. [CrossRef] [PubMed]

103. Toivonen, J.M.; Walker, G.A.; Martinez-Diaz, P.; Bjedov, I.; Driege, Y.; Jacobs, H.T.; Gems, D.; Partridge, L. No influence of Indy on lifespan in Drosophila after correction for genetic and cytoplasmic background effects. *PLoS Genet.* **2007**, *3*, e95. [CrossRef] [PubMed]

104. Gerhardt, E.; Graber, S.; Szego, E.M.; Moisoi, N.; Martins, L.M.; Outeiro, T.F.; Kermer, P. Idebenone and resveratrol extend lifespan and improve motor function of HtrA2 knockout mice. *PLoS ONE* **2011**, *6*, e28855. [CrossRef] [PubMed]

105. Fairbanks, L.D.; Burch, G.E. Rate of water loss and water and fat content of adult *Drosophila melanogaster* of different ages. *J. Insect Physiol.* **1970**, *16*, 1429–1436. [CrossRef]

106. Ye, J.; Cui, X.; Loraine, A.; Bynum, K.; Kim, N.C.; White, G.; De Luca, M.; Garfinkel, M.D.; Lu, X.; Ruden, D.M. Methods for Nutrigenomics and Longevity Studies in Drosophila. In *Biological Aging: Methods and Protocols*; Tollefsbol, T.O., Ed.; Humana Press: Totowa, NJ, USA, 2007; pp. 111–141.

107. Partridge, L.; Farquhar, M. Sexual activity reduces lifespan of male fruitflies. *Nature* **1981**, *294*, 580–582. [CrossRef]

108. Piper, M.D.W.; Partridge, L. Dietary Restriction in Drosophila: Delayed Aging or Experimental Artefact? *PLOS Genet.* **2007**, *3*, e57. [CrossRef] [PubMed]

109. Partridge, L.; Green, A.; Fowler, K. Effects of egg-production and of exposure to males on female survival in *Drosophila melanogaster*. *J. Insect Physiol.* **1987**, *33*, 745–749. [CrossRef]

110. Wong, R.; Piper, M.D.W.; Wertheim, B.; Partridge, L. Quantification of Food Intake in Drosophila. *PLoS ONE* **2009**, *4*, e6063. [CrossRef] [PubMed]

111. Iliadi, K.G.; Knight, D.; Boulianne, G.L. Healthy aging—Insights from Drosophila. *Front. Physiol.* **2012**, *3*, 106. [CrossRef] [PubMed]

112. Linford, N.J.; Bilgir, C.; Ro, J.; Pletcher, S.D. Measurement of Lifespan in *Drosophila melanogaster*. *J. Vis. Exp.* **2013**, *71*, 50068. [CrossRef]

International Journal of
Molecular Sciences

MDPI

Article

Caffeic Acid and Metformin Inhibit Invasive Phenotype Induced by TGF-β1 in C-4I and HTB-35/SiHa Human Cervical Squamous Carcinoma Cells by Acting on Different Molecular Targets

Malgorzata Tyszka-Czochara [1,*], Malgorzata Lasota [2] and Marcin Majka [3]

[1] Department of Food Chemistry and Nutrition, Faculty of Pharmacy, Jagiellonian University Medical College, Medyczna 9, 30-688 Krakow, Poland
[2] Chair of Medical Biochemistry, Faculty of Medicine, Jagiellonian University-Medical College, Kopernika 7, 31-034 Krakow, Poland; malgorzata.lasota@uj.edu.pl
[3] Department of Transplantation, Faculty of Medicine, Jagiellonian University Medical College, Wielicka 258, 30-688 Krakow, Poland; mmajka@cm-uj.krakow.pl
* Correspondence: mtyszka@poczta.fm or malgorzata.tyszka-czochara@uj.edu.pl

Received: 13 November 2017; Accepted: 11 January 2018; Published: 16 January 2018

Abstract: During the progression of epithelial cancer, the cells may lose epithelial markers and gain mesenchymal phenotype via Epithelial-Mesenchymal Transition (EMT). Such transformation of epithelial cancer cells to mesenchymal-like characteristic benefits plasticity and supports their ability to migrate. The aim of this study was to evaluate the influence of natural compound Caffeic Acid (CA) alone and in combination with antidiabetic drug Metformin (Met) on metastatic progression of two human cervical squamous cell cancer lines, C-4I and HTB-35/SiHa cells. EMT program was triggered by exposition of both epithelial cell lines to TGF-β1. Gene expression patterns related to epithelial/mesenchymal phenotype were evaluated by Real-Time PCR analysis and the protein amount was detected by western blot. The treatment of human squamous cancer cells with CA and with Met, suppressed the motility of cells and the effect depended on a particular cell line. Both compounds regulated the EMT process in C4-I and HTB-35 cells by interfering with different molecular targets. In TGF-β1-stimulated C4-I cells, CA suppressed the expression of mesenchymal transcription factor SNAI1 which resulted in enhanced expression of epithelial markers E-cadherin, Occludin and Claudin. Additionally, CA blocked *MMP-9* and upregulated *TIMP*-1 expression, a specific inhibitor of MMP-9. In HTB-35 cells stimulated with TGF-β1, Met decreased the expression of Vimentin. By suppressing hypoxia master regulator *HIF-1α*, Met caused downregulation of *CAIX*, an enzyme involved in metastasis of aggressive malignant cells. In this study we showed that CA and Met inhibited EMT process in cancer cells via different mechanisms. However, when applied together, compounds exerted the greater effect on EMT than each compound alone. This is the first report revealing that CA alone and co-treated with Met may reverse mesenchymal phenotype of TGF-β1-treated cervical tumor cells and we believe that the use of the two small molecules may be considered as a potential therapeutic approach for metastatic cervical cancer.

Keywords: caffeic acid; metformin; epithelial-mesenchymal transition; cervical cancer; epithelial markers; mesenchymal markers; metastasis; hypoxia; carbonic anhydrase IX

1. Introduction

Squamous cell carcinoma is the most common cervical cancer in women and accounts for almost 80% of cervical carcinomas in this population [1]. Along with the advent of human papilloma virus (HPV) vaccines, the primary prevention of cervical malignancy has become more successful; however, survival and prognosis are poor, particularly due to cancer metastasis [2]. Considering the short survival period in patients with recurrent or metastatic cancer, there is an urgent need for improvement of existing therapies for cervical malignancy [3].

Malignant cell transformation consists of a series of processes resulting in the ability of cells to migrate and invade other tissues. When polarized epithelial cells gain invasive characteristic, they lose epithelial markers, especially cell-cell adhesion molecules, such as E-cadherin, Occludin, Claudin and β-catenin, and acquire expression of mesenchymal markers, such as Vimentin, which finally results in the activation of Epithelial-Mesenchymal Transition (EMT) program [4]. The conversion of cancer cells function and morphology to mesenchymal-like phenotype benefits the plasticity and facilitates leaving of the primary site and disseminating to the secondary sites through blood or lymph vessels.

Emerging data suggest that targeting EMT with small molecules derived from plants, such as Caffeic Acid (3,4-dihydroxycinnamic acid, CA), may be an effective chemopreventive approach in various cancers. CA is one of the major hydroxycinnamic acids in the human diet. It occurs in various foods, including herbs and beverages such as coffee and red wine. In the human body, CA may be produced by the hydrolyzation of chlorogenic acid, a phenolic acid abundant in plants. The anticancer activity of CA and its derivatives was demonstrated in vivo [5] as well in vitro [6,7], also against gynecological carcinomas [8]. CA and its derivatives may inhibit the migratory capacity of cancer cells via several mechanisms, including inhibition/activation of transcription factors such as nuclear factor κ-light-chain-enhancer of activated B cells (NFκB) and regulation of Akt/mTOR signaling pathway [9,10]. It was reported that CA reversed TGF-β1- induced EMT in human tumor cell lines [11] and alleviated prostate malignant cells aggressiveness via non-canonical Wnt signaling [5]. CA and its derivatives suppressed EMT process via inhibition of Vimentin and upregulation of E-cadherin, as has been shown in pancreatic [3] and malignant keratinocyte cells [7]. However, the effect of this natural compound on the migratory capacity of neoplastic cervical cancer cells is still unknown. Metformin (dimethylbiguanide, Met) used in humans for the treatment of type 2 diabetes, was shown to restrain EMT in highly metastatic cervical HTB-35/SiHa cells [3] Met, similarly to CA, may prevent TGF-β1- induced EMT and cell invasion by controlling of mesenchymal/epithelial markers expression and by regulation of intracellular signaling pathways triggering cell death [12,13]. The experimental approach used in this study is based on the hypothesis that transforming growth factor β1 (TGF-β1) causes a phenotypical transformation of epithelial cervical cancer cells into mesenchymal-like cells, with subsequent increase in motility of the cells. The two human cervical squamous cell cancer lines, C-4I and HTB-35/SiHa, were selected to investigate the in vitro effects of CA and Met. Both cell lines expressed epithelial characteristic, but expression of E-cadherin and Vimentin differed depending on a particular cell line. In this study we aimed to find out, if exposition of cancer cells to CA and Met may suppress TGF-β1-induced EMT phenotype of cancer cells. We tried to evaluate whether both compounds act on EMT process in C-4I and HTB-35 cells via the same or different regulatory proteins and, based on these findings, assess if simultaneous treatment of cells with CA and Met may exert stronger effect than a single drug. A wound healing assay was also conducted to analyze the influence of CA and Met on motility of the cells. In hypoxic conditions, we tested whether drugs may suppress HIF-1α and regulate HIF-1α downstream protein CAIX.

2. Results

2.1. Transforming Growth Factor Beta 1 (TGF-β1) Induces Epithelial-to-Mesenchymal Transition (EMT) in C4-I and HTB-35 Cells

EMT is characterized by a drop in the expression of proteins involved in junctional complexes related to polarized epithelial phenotype, such as E-cadherin, with concomitant upregulation of mesenchymal proteins, especially reorganizing cytoskeleton, such as Vimentin. We exposed cervical cancer squamous cell lines C4-I and HTB-35 to 10 ng/mL of TGF-β1 for 48 h. Unstimulated cells cultured in the same conditions were appropriate controls. Western blot and qPCR analyses were performed to determine the expression of EMT-specific proteins in C4-I and HTB-35 cells (Figure 1A,B). In unstimulated C4-I cells, strong E-cadherin (*CDH1*) expression was detected, whereas Vimentin (*VIM*) was barely expressed. The incubation of C4-I cells with TGF-β1 caused downregulation of E-cadherin ($p < 0.1$ vs. untreated cells) and enhanced expression of Vimentin ($p < 0.1$ vs. untreated cells). As shown in Figure 1C, C4-I cells displayed an epithelial appearance [14]. Following exposure to TGF-β1 for 48 h, the cells started to dissociate from monolayer. The unstimulated HTB-35 cells expressed Vimentin (Figure 1A,B), while in TGF-β1-stimulated HTB-35 cells the expression of Vimentin was further enforced ($p < 0.1$ vs. untreated cells). At the same time, the enhanced scattering in TGF-β1-stimulated HTB-35 cells was observed (Figure 1C). E-cadherin was weakly expressed in HTB-35 cells and the treatment with TGF-β1 caused no distinct alteration of the expression of the protein (Figure 1A,B).

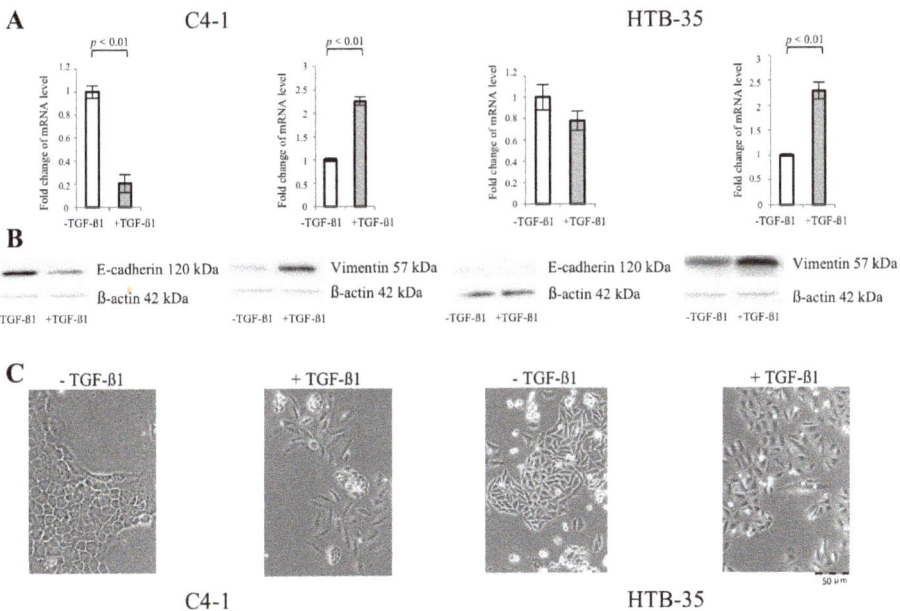

Figure 1. TGF-β1 induces Epithelial-to-Mesenchymal Transition (EMT) in C4-I and HTB-35 cells (**A–C**). The human cervical squamous cell cancer lines, C-4I and HTB-35 cells were incubated for 48 h with the addition of 10 ng/mL of TGF-β1. The untreated cells were grown in the same conditions and used as controls. Real-time PCR analysis revealed significant decrease in E-cadherin transcript level relative to untreated control at $p < 0.01$ in TGF-β1-stimulated C-4I cells, while in HTB-35 the drop in mRNA level for E-cadherin was not statistically significant at $p < 0.05$. Note that TGF-β1 caused significant increase in the expression of Vimentin in both cancer cell lines, as measured using qPCR ((**A**), $p < 0.01$ vs. untreated control for C-4I cells and $p < 0.01$ vs. untreated control for HTB-35 cells) and demonstrated with Western blot analysis ((**B**), 20 µg of total cell lysates were subjected to SDS-PAGE followed by immunoblotting and chemiluminescent detection; β-actin was used as loading control). The experiments were repeated three times with similar results; the Real-Time PCR data were presented as mean values ± SD (**A**), a representative immunoblots were shown (**B**). The incubation of the cells with TGF-β1 for 48 h caused morphological changes in both cell lines, as shown in phase contrast microphotographs (**C**). The enhanced scattering of C-4I and HTB-35 cells was observed following TGF-β1 treatment.

2.2. CA Attenuates the Migratory Capacity of C4-I and Met Inhibits Motility of HTB-35 Cells

Scratch assays were performed to determine the possible influence of CA and Met on the functional effects of EMT in C4-I and HTB-35 human squamous cell cancer lines. The sub-confluent cell cultures were incubated with CA and/or Met for 24 h. In parallel, cultures treated with tested compounds ware exposed to 10 ng/mL of TGF-β1. As shown in Figures 2A and 3A, TGF-β1 augmented migration of both cell lines when compared to unstimulated controls. The 100 μM CA treatment reduced the invasion potential of C4-I cells (Figure 2B, $p < 0.05$ vs. untreated cells) and HTB-35 cells (Figure 3B, $p < 0.05$ vs. untreated cells). The exposition of cells to 10 mM Met also significantly facilitated the closure of the denuded area in C4-I cell line (Figure 2B, $p < 0.05$ vs. untreated cells) and in HTB-35 cell line (Figure 3B, $p < 0.05$ vs. untreated cells). Comparing the effect of tested compounds on scratch reduction in the two cell lines, CA inhibited the healing process in C4-I cells more effectively than Met (Figure 2B, $p < 0.05$ for CA vs. Met) while Met exerted effect greater than CA in HTB-35 cell line (Figure 3B, $p < 0.05$ for CA vs. Met). In C4-I cells treated with TGF-β1 CA/Met caused the greatest scratch reduction (up to 50%). What is more, co-treatment had greater impact on motility of the cells than each compound alone (Figure 2B, $p < 0.05$ for CA/Met vs. CA, $p < 0.05$ for CA/Met vs. Met). In HTB-35 cells Met caused a 40% reduction of cell scratch and the most effective attenuation of cell movement (Figure 3B, $p < 0.05$ for Met vs. CA, $p < 0.05$ for Met vs. CA/Met).

We examined the influence of CA and Met with and without addition of 10 ng/mL of TGF-β1 on the proliferation of C4-I and HTB-35 cells. The confluent cell cultures were exposed to compounds for 24, 48 and 72 h. The assessment of cell number revealed that CA and Met slightly reduced the growth of C4-I and HTB-35 cells after 24 and 48 h of exposure (Figure S1A). Since CA at 100 μM and Met at 10 mM significantly decreased the migratory capability of both cell lines cells, with only minor inhibition of growth (Figure S1), we subjected these concentrations to further investigation.

A

C4-I cells

B

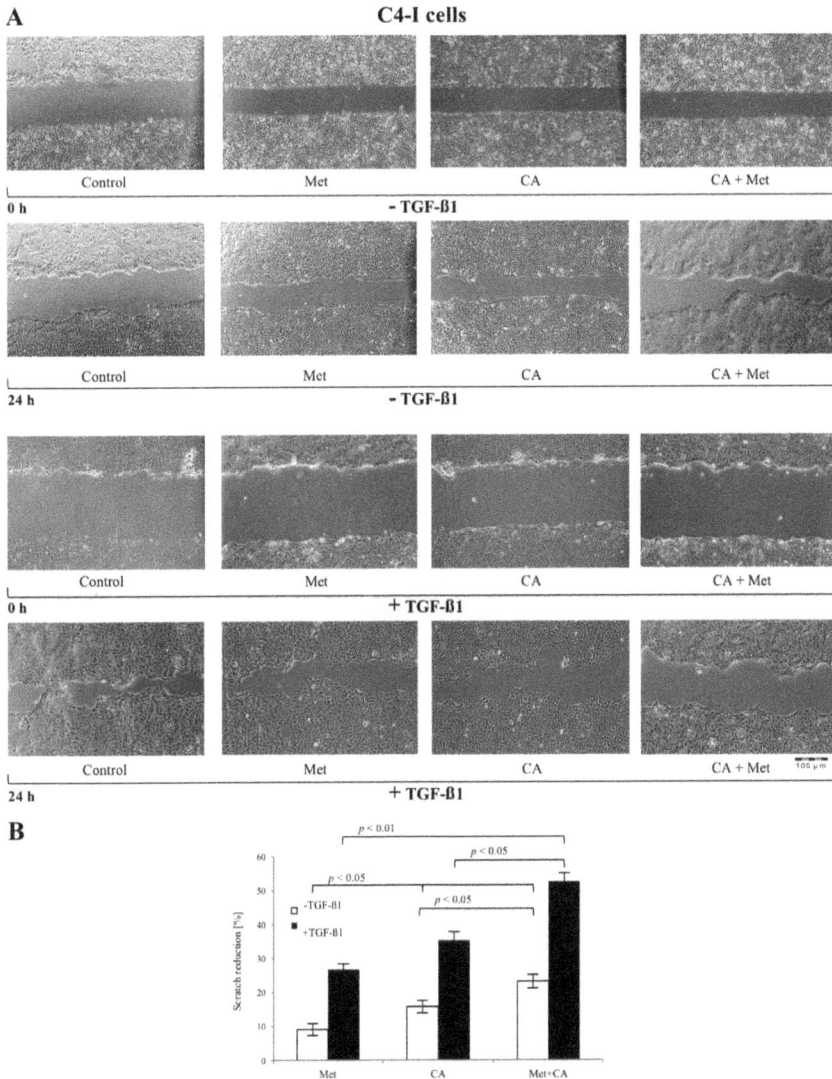

Figure 2. The effect of Caffeic Acid (CA) and Metformin (Met) on migration of C4-I cells in vitro (**A**,**B**). C4-I cells were cultured to sub-confluency and then a scratch was made on the monolayer of cells. Then the cells were incubated with addition of tested compounds (CA at 100 μM and/or Met at 10 mM) and with/without 10 ng/mL TGF-β1 for 24 h. For each scratch the images were obtained by an inverted light microscope (Olympus IX-70, Hamburg, Germany) at 0 h and 24 h. The changes of area of each wound was measured using Image J (v1.44; National Institutes of Health, Bethesda, MD, USA) and the migration rate was quantified as a change of the scratch reduction. Note that CA/Met caused the greatest scratch reduction in TGF-β1-treated cells ((**B**), $p < 0.01$ for CA/Met vs. control, $p < 0.05$ for CA/Met vs. Met, $p < 0.05$ for CA/Met vs. CA) as well as in TGF-β1-untreated cells ((**B**), $p < 0.05$ for CA/Met vs. control, $p < 0.05$ for CA/Met vs. Met, $p < 0.05$ for CA/Met vs. CA). All bars show the mean value of three experiments and error bars represent standard error of the mean (**B**).

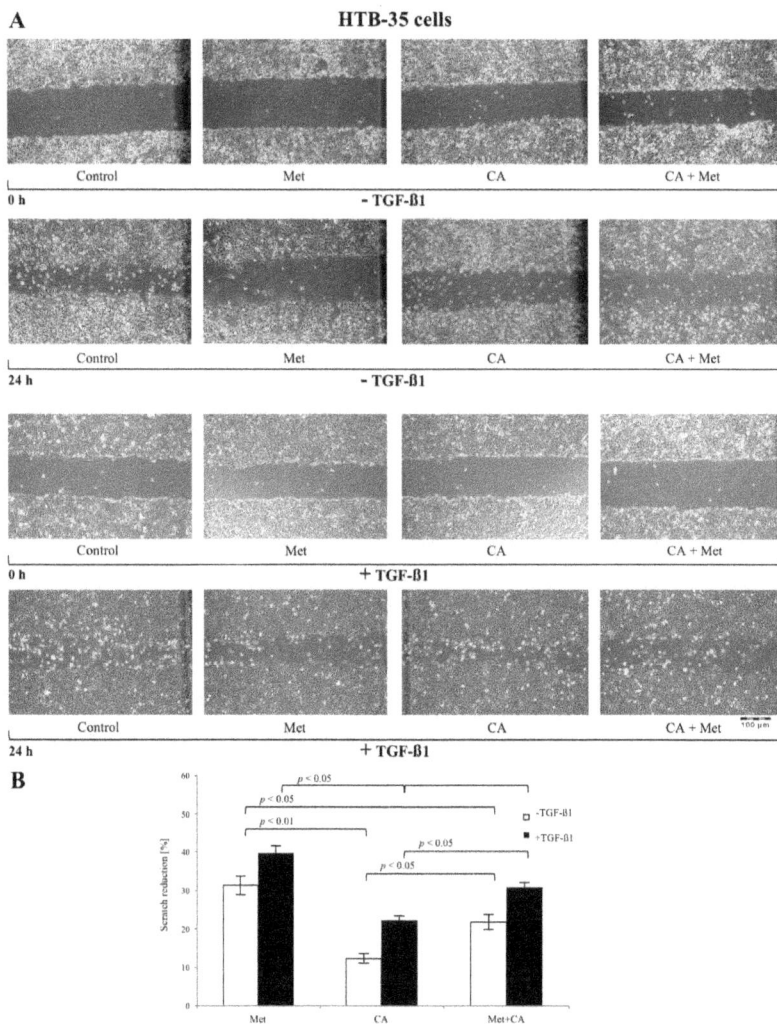

Figure 3. The effect of Caffeic Acid (CA) and Metformin (Met) on HTB-35 cells migration in vitro (**A,B**). HTB-35 cells were cultured to sub-confluency and then an injury line was made on the monolayer of cells. Then the cells were incubated with addition of tested compounds (CA at 100 µM and/or Met at 10 mM) and with/without 10 ng/mL TGF-β1 for 24 h. Representative images of the scratch assay conducted on the HTB-35 cell line following treatment of cells with CA at concentration of 100 µM, Met at 10 mM or both compounds were presented. The images were captured at 0 and 24 h after scratching (**A**). The effect of CA and/or Met on cell motility was determined by wound healing assay and presented as reduction of each scratch after 24 h of incubation ((**A**), the Image J program was used to analyze the changes in each wound area; Image J v1.44; National Institutes of Health, Bethesda, MD, USA). Note that Met exerted greater effect than CA ((**B**), $p < 0.05$ for CA vs. Met) and CA/Met in HTB-35 cell line ((**B**), $p < 0.05$ for Met vs. Met/CA). Data shown here were from a representative experiment repeated three times with similar results (**A,B**). Quantification of the scratch assay experiments are presented as mean values $\pm$ SD (**B**).

2.3. CA and Met Treatment of C4-I Cells Increases Epithelial Adhesive Markers and Decreases Mesenchymal Transcription Factors Regulating EMT

As the co-treatment with CA and Met the most effectively delayed the motility of C4-I cells, we evaluated whether the compounds may exert changes in expression of adhesion molecules representing an epithelial phenotype, such as E-cadherin, β-catenin, Occludin and Claudin. The effect of CA and Met on changes in mRNA level for the epithelial markers was measured in cultures of C4-I cells incubated for 48 h with the compounds and with addition of 10 ng/mL of TGF-β1. In parallel, for comparative reasons each culture was grown without addition of TGF-β1. CA upregulated E-cadherin, Occludin and Claudin ($p < 0.05$ vs. control) in TGF-β1-stimulated/unstimulated cells. However, as shown in Figure 4, CA/Met treatment caused the greatest increase in the expression of mRNA for *CDH1* ($p < 0.05$ vs. control), *OCLN* ($p < 0.05$ vs. control) and *CLDN1* ($p < 0.05$ vs. control) genes.

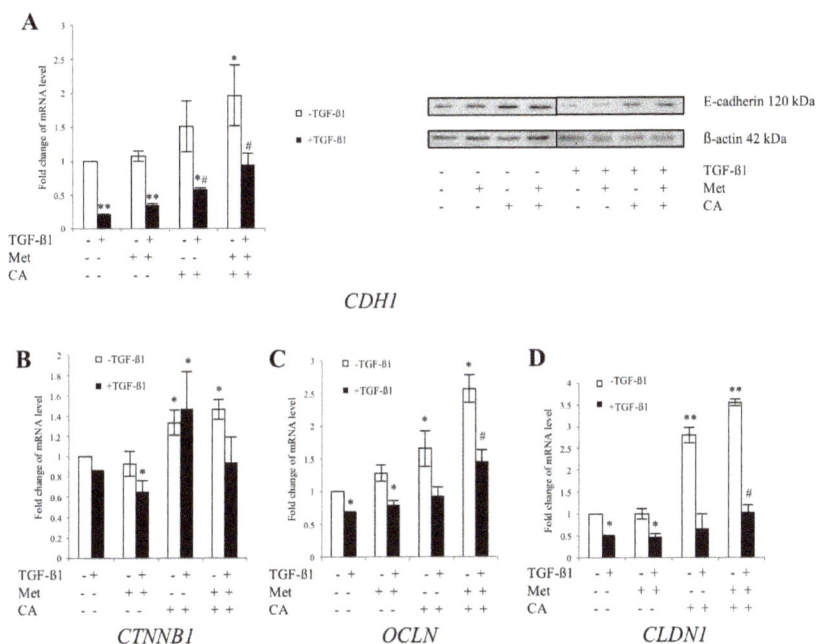

Figure 4. The effect of Caffeic Acid (CA) and Metformin (Met) on expression of epithelial markers in C4-I cells. The cells were treated for 48 h with TGF-β1 (10 ng/mL) plus CA (100 μM) and/or Met (10 mM). In parallel, cultures treated with tested compounds, but without TGF-β1 were prepared. Analysis of mRNA level for E-cadherin (*CDH1*), β-catenin (*CTNNB1*), Occludin (*OCLN*) and Claudin (*CLDN1*) was performed with Real-time PCR ((**A**) left panel, (**B–D**), respectively). Protein levels of E-cadherin were analyzed by western blot and shown in the right panel of Figure 4A (20 μg of total cell lysates were subjected to SDS-PAGE followed by immunoblotting and chemiluminescent detection, and β-actin was used as the protein loading control, the details described in Material and Methods). Note that the greatest upregulation for E-cadherin transcript was detected when CA was applied with Met ((**A**), left panel); the same effect was found for Occludin (**C**) and Claudin (**D**) (* $p < 0.05$ vs. control for E-cadherin, * $p < 0.05$ vs. control for Occludin, * $p < 0.01$ vs. control for Claudin). For qPCR the data were normalized against *GAPDH* transcript as a reference gene and levels of RNA expression were determined with the $2^{-\Delta\Delta Ct}$ method (* $p < 0.05$ and ** $p < 0.01$ vs. control, # $p < 0.05$ vs control with TGF-β1). Experiments were repeated three times with similar results and presented as mean values ± SD.

To further elucidate the mechanism of action of compounds, we examined the effect of CA and Met on the expression of mesenchymal transcriptional factors. We found that in TGF-β1-stimulated cells expression of SNAI1, a strong repressor of E-cadherin expression, was downregulated by CA ($p < 0.05$ vs. control) and CA/Met treatment ($p < 0.01$ vs. control). CA/Met inhibited *SNAI1* and upregulated *CDH1* expression after 24 of incubation of C4-I cells in TGF-β1-stimulated/unstimulated cells (Figure S2). As demonstrated in Figure 5, the concomitant action of CA and Met had also the greater impact on *SNAI1* expression following 48 h of exposition of cells to drugs. The expression of *ZEB1* was significantly downregulated by CA ($p < 0.05$ vs. control), Met ($p < 0.05$ vs. control) and CA/Met ($p < 0.05$ vs. control). Met downregulated *TWIST1* expression ($p < 0.05$ vs. control) while CA decreased mRNA for *TWIST2* ($p < 0.05$ vs. control).

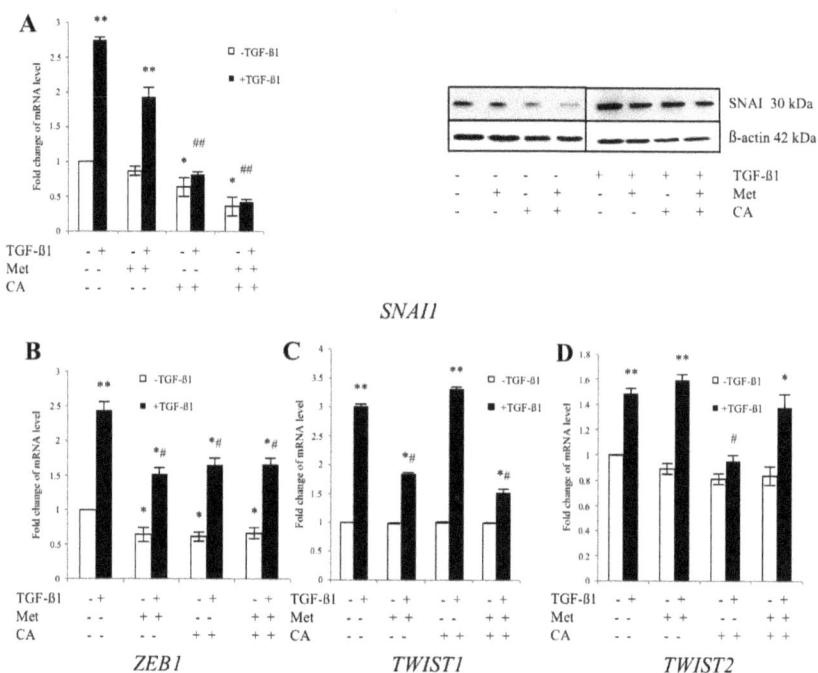

Figure 5. The effect of Caffeic Acid (CA) and Metformin (Met) on expression of transcription factors regulating EMT in C4-I cells. The cells were treated for 48 h with TGF-β1 (10 ng/mL) plus CA (100 μM) and/or Met (10 mM). In parallel, cultures of both cell lines were incubated with tested compounds, but without addition of TGF-β1. Expression of critical EMT promoters *SNAI* (**A**), *ZEB1* (**B**); *TWIST1* (**C**) and *TWIST2* (**D**) was evaluated at mRNA level by qPCR. For Snail1 protein levels were detected by western blot ((**A**), right panel). Immunoblots were prepared following SDS-PAGE separation of cell lysates as described in Materials and Methods (20 μg of total cell lysates were subjected to SDS-PAGE followed by immunoblotting and chemiluminescent detection, and β-actin was used as the protein loading control). In TGF-β1-stimulated cells the expression ((**A**), left panel) and protein amount ((**A**), right panel) of SNAI was significantly downregulated by CA and CA/Met). Met downregulated *TWIST-1* expression and CA decreased mRNA for *TWIST2*. The data were normalized against *GAPDH* transcript as a reference gene and levels of RNA expression were determined with the $2^{-\Delta\Delta Ct}$ method (* $p < 0.05$ and ** $p < 0.01$ vs. control, # $p < 0.05$ and ## $p < 0.01$ vs. control with TGF-β1). Bar graph is representative of the relative mean transcript abundance ± SD for three experiments. Results were presented from three independent experiments.

2.4. CA Downregulate the Expression of MMP-9 and Specific Tissue Inhibitor of Matrix Metalloproteinases TIMP-1 in C4-I Cells

Given that MMP-9 and MMP-2 are critical to cell invasion, we examined the expression of gelatinases at the mRNA level using real-time RT–PCR. C4-I cells were treated with CA and/or Met in the presence of 10 ng/mL of TGF-β1 in medium for 48 h. At the same time, other C4-I cell cultures were exposed only to CA and/or Met, without addition of TGF-β1. The results showed that CA reduced the expression of *MMP-9* ($p < 0.05$ vs. control) in TGF-β1-stimulated cells (Figure 6). As the activity of MMP-9 is regulated by the specific endogenous inhibitor TIMP-1, we determined the effect of tested compounds on mRNA level encoding *TIMP-1*. In fact, the mRNA level for *TIMP-1* was significantly increased in CA treated cells compared to control ($p < 0.05$ vs. control). What is more, CA/Met treatment had greatest inhibitory effect on *MMP-9* expression ($p < 0.01$ vs. control), which was in compliance with the greatest upregulation of *TIMP-1* in cells exposed to both compounds. The qPCR analysis showed that CA suppressed *MMP-2* expression ($p < 0.05$ vs. control) along with upregulation of TIMP-2 ($p < 0.05$ vs. control).

We also determined the effect of CA and Met on the mRNA level for *VEGFA* in TGF-β1-stimulated C4-I cells, since VEGFA is a potent factor facilitating tumor-induced angiogenesis. The mRNA for VEGFA was significantly decreased under exposition of cells to CA ($p < 0.05$ vs. control). Here, CA/Met had the greatest inhibitory impact on *VEGFA* gene expression (Figure 6, $p < 0.01$ for CA/Met vs. control, $p < 0.05$ for CA/Met vs. CA, $p < 0.05$ for CA/Met vs. Met).

2.5. Met Attenuates Mesenchymal Marker of Malignant HTB-35 Cells

EMT induction in tumors is associated with increased expression of molecular marker Vimentin. To evaluate the effect of CA and Met on the mRNA level for *VIM1*, HTB-35 cells were exposed to TGF-β1 and tested compounds for 48 h. The appropriate controls with addition of CA and/or Met but without TGF-β1 were made. Following incubation, qPCR and western blot analyses was performed. As presented in 2.1, unstimulated HTB-35 cells exhibited strong expression of Vimentin, while E-cadherin was weakly expressed. Treatment of cells with 10 mM metformin downregulated Vimentin at $p < 0.01$ vs. control in TGF-β1-stimulated cells, as shown in Figure 7. CA caused no distinct alteration of the expression of the mRNA level for *VIM1* ($p < 0.05$ vs. control).

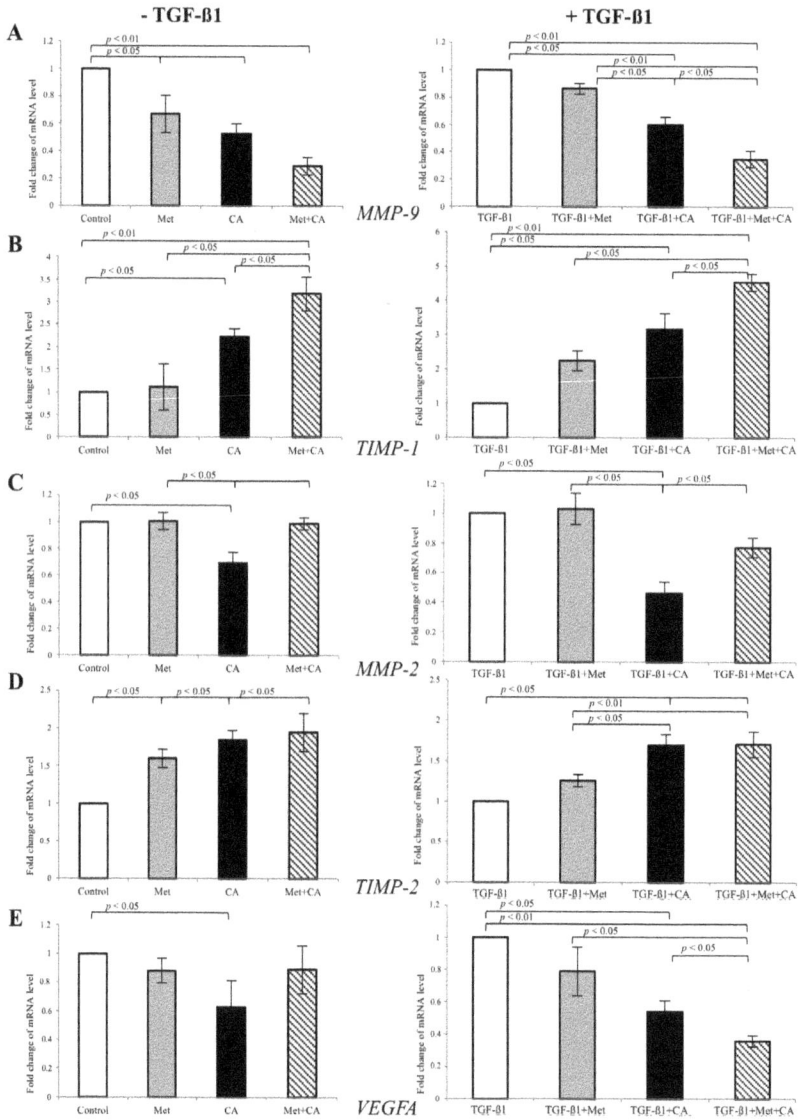

Figure 6. The effect of Caffeic Acid (CA) and Metformin (Met) on the expression of gelatinases *MMP-9* (**A**) and *MMP-2* (**C**) and tissue inhibitors *TIMP-1* (**B**) and *TIMP-2* (**D**) in C4-I cells. The cells were incubated with addition of tested compounds (CA at 100 µM and/or Met at 10 mM) and with/without 10 ng/mL TGF-β1 for 48 h. Then mRNA was isolated and purified for Real-time PCR examination. In TGF-β1-stimulated cells, CA and CA/Met caused decrease in mRNA level for *MMP-9* ($p < 0.01$ vs. control) with concomitant increase of mRNA for its specific tissue inhibitor *TIMP-1* ($p < 0.01$ vs. control). The incubation of TGF-β1-treated cells with CA/Met attenuated the expression of angiogenic molecule *VEGFA* (**E**). Data shown here were repeated three times with similar results and presented as mean values ± SD (**B**).

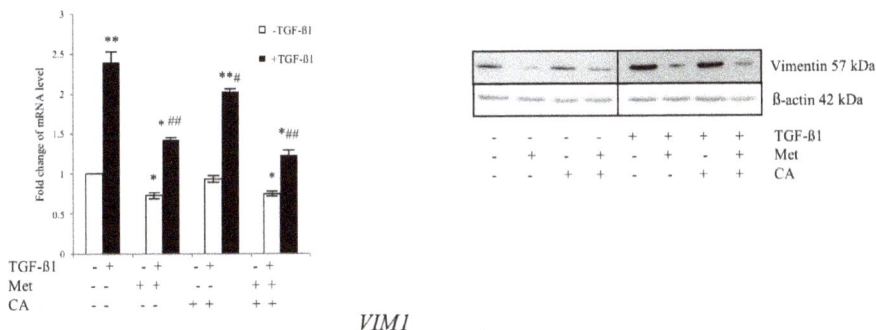

VIM1

Figure 7. Met suppress the expression of mesenchymal marker Vimentin in HTB-35 cells. The cells were treated with TGF-β1 (10 ng/mL) plus CA (100 μM) and/or Met (10 mM) for 48 h. In parallel, cell cultures were treated with tested compounds but were grown without addition of TGF-β1. The RNA expression was determined with qPCR analysis using the $2^{-\Delta\Delta Ct}$ method (*GAPDH* was a reference gene). The protein levels of Vimentin were detected by western blot (right panel). Immunoblots were prepared following SDS-PAGE separation of cell lysates as described in Materials and Methods (20 μg of total cell lysates were subjected to electrophoresis, β-actin was used as the protein loading control). In TGF-β1-stimulated cells, Met alone and applied with CA significantly decreased the transcript level for Vimentin (*VIM1*). Error bars represent the SD of the mean from triplicate results (qPCR analysis, * $p < 0.05$ and ** $p < 0.01$ vs. control, # $p < 0.05$ and ## $p < 0.01$ vs. control with TGF-β1).

2.6. Met Inhibits the Expression of CAIX in HTB-35 Cells under Hypoxic Conditions

Considering that transcription factor HIF-1α plays a primary role in mediating EMT program induced by hypoxia, we hypothesized that Met might influence the expression of HIF-1α protein. The qPCR analysis revealed that after 24 h of incubation of HTB-35 cells at 5% of O_2 level, the expression of *HIF1A* significantly increased, when compared to cells kept under normoxic conditions (21% O_2 of level), as presented in Figure 8A. Following exposition of cells to Met, the mRNA for *HIF1A* was significantly decreased (Figure 8B, $p < 0.01$ vs. control). What is more, as HIF-1α controls the expression of *CAIX*, in response to limited oxygen supply HIF-1α may cause the transcriptional activation of *CAIX* [15]. In conditions of chronic hypoxia, which develops in growing tumors, CAIX contribute to extracellular acidosis, which further promotes pro-metastatic cascade inducing cancer cell migration [16]. The results showed that following exposition of HTB-35 cells to Met in hypoxic conditions, mRNA for *CAIX* was downregulated, as measured by RT-PCR (Figure 8C).

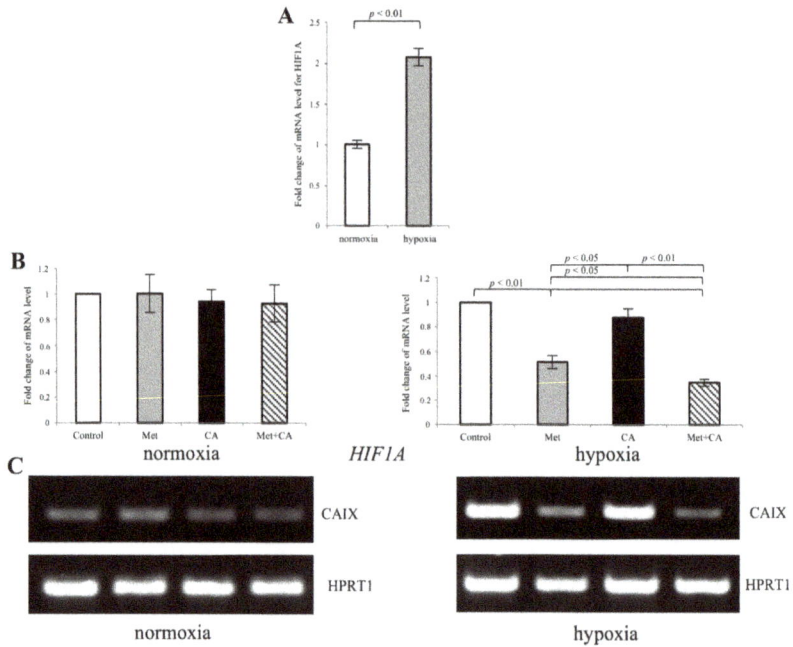

Figure 8. Met inhibits the expression of CAIX in HTB-35 cells under hypoxic conditions. The cells were incubated either in normoxia (21% O_2 level) or in hypoxia (5% O_2 level) with CA, Met or both compounds for 24 h. qPCR analysis was performed to assess the expression of HIF-1α under normoxic and hypoxic conditions (**A**). The RNA expression of *HIF1A* was determined with the $2^{-\Delta\Delta Ct}$ method. The mRNA level of *HIF1A* was significantly increased following hypoxia ((**A**), $p < 0.01$ vs. control). On the contrary, Met alone and together with CA downregulated *HIF1A* ((**B**), $p < 0.01$ vs. control for Met, $p < 0.01$ vs. control for CA/Met). RT-PCR analysis revealed that in hypoxia the expression of CAIX significantly decreased in cells exposed to Met and CA/Met ((**C**), *HPRT1* was a reference gene). Data shown here are representative of three experiments performed with similar results (**C**).

3. Discussion

EMT has recently been regarded to play a crucial role in the mechanism underlying tumor spread. Therefore, EMT regulatory pathways represent potential new targets for inhibition of progression and metastasis of malignant cells [17]. Several studies have suggested that CA and its derivatives may restrain the spread of carcinoma cells via EMT inhibition [6,7]. Met has been previously demonstrated to be beneficial in gynecologic oncology [12,13] and to inhibit EMT in human cervical cancer cell lines [3]. Therefore, these compounds were used for current studies to investigate whether they may inhibit metastatic phenotype induced by TGF-β1 in C-4I and HTB-35 cervical cancer cells. Both cell lines expressed the epithelial characteristic, but the expression of epithelial and mesenchymal markers differed depending on a particular cell line. C-4I cells, which abundantly expressed E-cadherin and weakly Vimentin, were similar to normal cervical epithelium [18]. The exposition of cells to TGF-β1 caused upregulation of Vimentin and loss of E-cadherin. HTB-35 cells had high expression of vimentin and the stimulation of cells with TGF-β1 caused greater upregulation of Vimentin. At the same time, the amount of E-cadherin was minor in TGF-β1-treated and untreated HTB-35 cells. Vessey et al. reported HTB-35 cells to be E-cadherin negative [2] and Lee et al. suggested that some cervical carcinoma cells may concomitantly express both epithelial and mesenchymal markers [1,19]. Cheng et al. have shown that unstimulated HTB-35 cells may express E-cadherin [3]. The immunofluorescence analysis of E-cadherin level and its subcellular localization in HTB-35 cell, that we had previously performed, suggested that in wild-type HTB-35 cells E-cadherin might be weakly expressed, but the protein was withdrawn from intercellular junctions [20]. Our results showed that both compounds, CA and Met, had the potential to inhibit mesenchymal phenotype induced with TGF-β1 in cervical cancer cells, but each drug acted via various proteins in the particular cell line.

The signals from primary tumor associated stroma such as TGF-β1 may trigger changes in cytoskeleton reorganization and lead to the activation of nuclear transcription factors, ZEB1, TWIST1 and TWIST2, which, once activated, implement EMT program and promote invasiveness [21]. Many cell culture experiments and in vivo studies have demonstrated that inhibition of E-cadherin expression by its transcriptional suppressor SNAI1 is a key process driving EMT [22], also in cervical cancers [1]. It has been also reported that the inhibition of E-cadherin expression is positively correlated with the tumor stage and grade in cervical cancers in humans [23]. In the current study, the exposition of C-4I cells to CA at 100 µM caused downregulation of transcript and protein for SNAI1. We may speculate that this led to the increase in the expression of E-cadherin (Figure 9). The restoration of E-cadherin expression in cervical malignant cells results in decreased motility and invasiveness [24]. The functional test showed that CA treatment alleviated movement of C-4I cells and delayed wound healing. The transcription of tight junction molecules occludins and claudins may also be suppressed by SNAI1 [25]. Besides E-cadherin, these epithelial membrane proteins have been recognized to play essential role in cytokine-induced regulation of the tight junction and loss of their expression have been correlated with increased cancer cell movement and metastasis [26]. The results demonstrated the increased expression of Occludin and Claudin following CA treatment, which suggest that SNAI1 was no longer able to repress theses downstream proteins. In the current study the effect of CA on Snail expression was detected at the level of mRNA as well as the protein, thus we may speculate that CA can act directly on the transcription factor expression. However, we expect that the exact mechanism and possible upstream proteins involved in the process will be recognized in the following study.

The results demonstrated that CA downregulated the expression of transcription factor *ZEB1*, another repressor of E-cadherin. ZEB1 induces the promotion of EMT and in combination with other factors triggers metastasis [25] and is highly expressed in aggressive cancer cell lines [1] and the overexpression of the protein may be an indicator of poor prognosis in breast, pancreatic, and lung cancer [21]. What is more, a mutual regulation between SNAI1 and ZEB1 has been recently identified [25]; SNAI1, acting as a primary EMT regulator, may enhance expression of other transcription factors, including *ZEB1* and *TWIST* [27]. In metastatic cell nucleus, ZEB1 and TWIST1 proteins prevent E-cadherin gene transcription [1]. Interestingly, the action of CA on E-cadherin expression in C-4I cells was greater when co-treated with Met. At the same time, *TWIST1* was the molecular target regulated by Met in C-4I cells. Therefore, we may speculate that inhibition of these transcription factors under combined treatment may be a possible explanation for the enhanced transcription of junctional proteins. In line with these intracellular changes leading to lower expression of mesenchymal markers and higher expression of epithelial proteins, the incubation of C-4I cells with CA/Met inhibited TGF-β1-induced migration of the cells and caused the most visible scratch reduction.

Additionally, SNAI1 is a positive regulator of Matrix metalloproteinases (MMPs), which belong to a family of zinc-dependent extracellular matrix (ECM)–degrading proteases. Matrix metalloproteinase-2 (MMP-2) and metalloproteinase-9 (MMP-9) degrade structural proteins of invaded tissues and play a crucial role in metastasis of tumor cells and angiogenesis [28]. Although the expression pattern of MMPs varies depending upon the tumor origin, type and stage, numerous studies have shown that the suppression of MMPs synthesis results in the reduction in metastatic potential of cancer cells [29]. It was reported that overexpression of MMP-9 is associated with increased invasiveness of ovarian and breast tumors, which may lead to decreased survival in patients [30]. On the other hand, it was demonstrated that CA derivatives exert strong inhibitory effect on MMP-9 activity [31]. In the present study, CA suppressed *MMP-9* in C-4I cells in two independent ways. Firstly, the suppression of SNAI1 by the action of CA might result in reduced expression of *MMP-9*. Secondly, CA treatment of C-4I cells caused upregulation of a specific tissue inhibitor *TIMP*-1 controlling the degradative activity of

MMP-9 [30]. It has been reported before that the increased synthesis of MMPs in tumor cells may be associated with enhanced angiogenic ability caused by overexpression of Vascular Endothelial Growth Factor A (VEGFA), which promotes metastatic potential of cancer cells [32]. In breast cancer cells, VEGF-A increases mRNA and protein level for SNAI1, resulting in repression of E-cadherin transcription [33]. Our findings suggest that CA, by suppressing VEGFA, may possibly contribute to the reduction of SNAI1 transcription in cervical cancer cells, but further mechanistic study is needed.

While in C-4I line CA exerted the greatest effect in inhibiting the TGF-β1-induced metastatic phenotype; Met appeared to be an effective suppressor of EMT process in HTB-35 cells. We found that in HTB-35 cells Met at the concentration of 10 mM decreased the expression of mesenchymal marker Vimentin. As discussed above, Vimentin mRNA transcript and protein was higher in more aggressive HTB-35 cells than in C-4I cell line. This finding may explain in part the divergent effect exerted by compounds in the two cervical cancer cell lines. Moreover, Met exerted the greatest effect on motility of HTB-35 cells, as measured with functional scratch test.

In cervical cancers, motility of cells may be induced by hypoxia and enhanced acidosis of surroundings tissues [15,34]. The insufficiently oxygenated environment within solid tumors may induce Hypoxia-inducible factor 1 α (HIF-1α). The activation of HIF 1α results in upregulation of several proteins involved in migration and invasion that drives EMT program and helps neoplastic cells adapt to environmental stress [16]. In order to further elucidate the molecular action of Met in HTB-35 cells, we tested the effect of the compound in the hypoxic conditions that caused increased expression of transcription factor HIF-1α. The intracellular changes following activation of HIF 1α involve the induction its downstream protein Carbonic anhydrase IX (CAIX). CAIX can act as a survival factor that protects tumor cells against enhanced acidification of tissue microenvironment and through the ability to regulate pH of cancer milieu, it facilitates the migration of malignant cells. [35]. Due to its relevant role in metastasis, CAIX was also proposed as a potential therapeutic target in cervical cancers [15,36]. In our study, Met downregulated HIF-1α, which resulted in decreased transcription of CAIX (Figure 9). We may speculate that the inhibition of CAIX in HTB-35 cells exposed to Met can impair the invasive properties of cervical malignant cells.

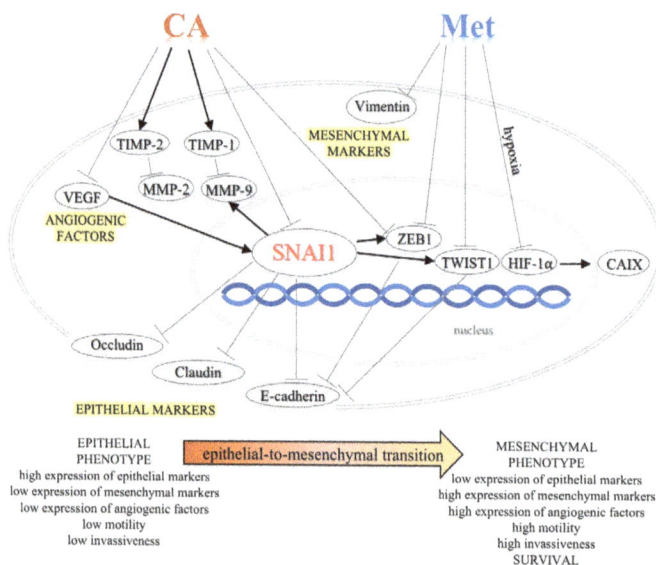

Figure 9. Schematic representation of the roles of CA and Met in inhibition of TGF-β1-induced EMT phenotype of cervical carcinoma cells. (↑—activation, ⊢—inhibition).

Among a wide range of pharmacological and biochemical effects, CA [5–11] and Met [12,13] were shown to inhibit EMT in various cancers. Given that the net balance between EMT activators and suppressors determine further progress or EMT inhibition, we can speculate that downregulation of mesenchymal transcription factors and activation of epithelial molecules by the action of drugs may reverse mesenchymal phenotype of cervical carcinoma cells and impair their motility. Emerging data suggests that such a combinatory approach targeting different molecular mechanisms may be more effective in the cancer invasiveness reduction than standard one-drug therapy [37,38]. We have recently reported that CA may expand the anti-tumor effect of Met in human epithelial cervical carcinoma cells by regulation of metabolic reprogramming [39,40].

4. Materials and Methods

4.1. Cell Culture and Treatment

The human cervical cancer cell lines C-4I (ATCC designation CRL-1594) and HTB-35 (ATCC designation HTB-35, SiHa) were purchased from the American Type Cell Culture collection (LGC Standards-ATCC, Teddington, UK). C-4I cells were kept in Waymouth's MB 752/1 medium (Life Technologies, Grand Island, NY, USA) and HTB-35 cells were grown in Dulbecco's modified Eagle's medium Lonza, Walkersville, MD, USA). Media were supplemented with 10% fetal bovine serum (BSA), (Eurex Sp z o.o., Gdansk, Poland) and 50 µg/mL of gentamicin (Sigma-Aldrich, Seelze, Germany). Cultures were grown at 37 °C in a humidified atmosphere of 5% CO_2. For experiments, C-4I cells at a density of 2.5×10^5 cells/mL and HTB-35 cells at a density of 1×10^5 cells/mL were placed in cell culture plates (Sarstedt, Numbrecht, Germany) and incubated to reach adequate confluency. Following washing with PBS solution (Lonza), the medium in each well was replaced with a fresh medium with adequate volumes of stock solution of CA (100 µM, Sigma-Aldrich), Met (10 mM, Sigma-Aldrich) or both chemicals for 48 h. Each culture was prepared with or without addition of 10 ng/mL TGF-β1 (PeproTech, Rocky Hill, NJ, USA). For cell counting, the cells were cultured to full confluency and then exposed to compounds with or without TGF-β1 for 24, 48 and 72 h. Control cells were grown in medium with addition of solvents (untreated cells), or solvents and 10 ng/mL TGF-β1 (TGF-β1-treated cells). Each experiment was conducted in triplicate. The number of cells was assessed by automatic cell counter Countess (Gibco Laboratories, Grand Island, NY, USA).

4.2. Immunoblotting

After 24 h of incubation with compounds, the cells were harvested and extracts formed by the addition of ice-cold M-PER buffer (Thermo Fisher Scientific Inc., Waltham, MA, USA) with a protease inhibitor cocktail (Merck, Darmstadt, Germany). Total protein was measured by the Bradford method. 20 µg of protein was separated on 10% SDS-polyacrylamide gel and transferred to PVDF membrane for Western blotting. The membranes were blocked with buffer contained 1% BSA in TBST (20 mM/L of Tris-hydrochloride, pH 7.5, 150 mM/L NaCl, 0.05% Tween 20; BioRad, Laboratories, Hercules, CA, USA) as reported previously [39]. Then the membranes were incubated overnight with primary antibody in Tris-buffered saline (TBS) containing 0.1% Tween 20% and 1% BSA, followed by extensive rinses and a 1h incubation with HRP-linked secondary antibody (1:4000). The following primary antibodies were used for experiments: anti-E-cadherin (Cell Signaling, Danvers, MA, USA, dilution 1: 1000), anti-Vimentin (Cell Signaling, dilution 1: 1000), anti-SNAI1 Santa Cruz Biotech (Santa Cruz, CA, USA, dilution 1: 500. β-actin (1:1000, Cell Signaling) was the loading control. The secondary HRP-conjugated antibodies were purchased in Santa Cruz Biotch. The specific immunoreactivity of each protein was measured by enhanced chemiluminescence and developed using the Super Signal West Pico Chemiluminescent Substrate Kit, Pierce Chemical, Rockford, IL, USA) using Gel Logic Imaging System 1500 (Kodak; Molecular imaging System Corestea Health Inc., Rochester, NY, USA).

4.3. Quantitative Polymerase Chain Reaction (qPCR)

The total RNA was extracted using Universal RNA purification Kit (EURx, Poland), according to vendor's protocol. The reverse polymerase transcription of mRNA was performed using MMLV reverse transcriptase (Promega, Madison, WI, USA) according to the manufacturer's protocol using ProFlex PCR System (Applied Biosystems, Foster City, CA, USA).

The real-time qPCR was performed in the QuantStudio 7 Flex (Applied Biosystems, Foster City, CA, USA) using Blank qPCR Master Mix (EURx) and the following Taq-Man human probes (Applied Biosystems): CDH1 (Hs01023894_m1), VIM (Hs00185584_m1), CTNNB1 (Hs0017025_m1), OCLN (Hs00170162_m1), CLDN1 (Hs00221623_m1), SNAI1 (Hs00195591_m1), ZEB1 (Hs00232783_m1), TWIST1 (Hs00361186_m1), TWIST2 (Hs02379973_s1), DES (Hs00157258_m1), MMP-2 (Hs00234422_m1), MMP-9 (Hs00234579_m1), TIMP-1 (Hs00171558_m1), TIMP-2 (Hs00234278_m1), VEGF (Hs00173626_m1), HIF1A (Hs00153153_m1) GAPDH (Hs99999905_m1). The data were normalized against GAPDH transcript as a reference gene and levels of RNA expression were determined with the $2^{-\Delta\Delta Ct}$ method.

4.4. Wound Healing Migration

Alteration of cell migration induced by CA, Met and CA/Met in cultures of C-4I and HTB-35 cells was estimated by means of wound healing migration (alteration of two-dimensional cellular movement). The cells were cultured to sub-confluency in 12-well culture dishes and then a scratch was made on the monolayer of cells with a sterile 10 μL plastic pipette. After washing of floating cells with PBS, the medium in each well was replaced with a new one containing addition of adequate amounts of tested compounds with/without addition of TGF-β1. Cell movement into the wound area was photographed at the initiation of the experiment (0 h) and after 24 h under an inverted light microscope (Olympus IX-70 microscope with fluorescence, Olympus, Hamburg, Germany) from the exact same location as the first picture. The area of the scratch wound was measured using Image J (v1.44; National Institutes of Health, Bethesda, MD, USA) and the migration rate was quantified as a rate of the scratch reduction (scratch area at 0 h–scratch area at 24 h) and the data was calculated as the average of 3 fields as described therein [41].

4.5. Hypoxia Conditions

HTB-35 cells were seeded into 6-well plates (Sarstedt) at a density of 1×10^5 cells/mL. After 24 h fresh medium with addition of CA, Met or both compounds were used and cells were incubated either in normoxia (21% O_2 level) or hypoxia (5% O_2 level) for 24 h. Afterwards, the cells were harvested for RNA isolation.

4.6. Reverse Transcription-Polymerase Chain Reaction (RT-PCR)

Total RNA was isolated from cells using the RNeasy Mini kit (Qiagen, Hilden, Germany) and the quantity of total RNA was measured using a NanoDrop ND-1000 Spectrophotometer (NanoDrop Technologies, Wilmington, DE, USA). cDNA synthesis was performed using oligo (dT) 15 primer and GoScript Transcriptase according to manufacturer's instructions (Promega GmbH, Mannheim, Germany). The PCR mixture contained: 1.5 uL of cDNA, 1× PCR buffer (Sigma Aldrich) 2.5 mM Magnesium Chloride (Sigma Aldrich), 0.2 mM dNTPs (Sigma Aldrich), 0.2 uM of each primer (Sigma Aldrich), and 1.25 U JumpStart™ Taq DNA Polymerase (Sigma Aldrich). The cDNA was amplified by PCR using the MJ Research PTC-200 Thermal Cycler with the following primers: CAIX (Carbonic Anhydrase 9) forward (5′-TACAGCTGAACTTCCGAGCG-3′), CAIX reverse (5′-CTAGGCTCCAGTCTCGGCTA-3′), HPRT1 (Hypoxanthine-Guanine Phosphoribosyltransferase) forward (5′-TGGCGTCGTGATTAGTGATG-3′), HPRT1 reverse (5′-TATCCAACACTTCGTGGGGT-3′). The PCR conditions for all analyzed genes were as following: denaturing at 95 °C for 5 min, followed by 30 cycles of 30 s at 95 °C, 30 s at 58 °C and 30 s at 72 °C. The reaction was completed for 10 min at 72 °C. The PCR reaction was evaluated by checking the PCR products on 1.5% w/v agarose gels. Bands were normalized by use of HPRT1 to correct for differences in loading of the cDNAs.

4.7. Statistical Analysis

The experimental data were shown as mean ±SD. Analysis was performed using one-way ANOVA followed by Duncan post-hoc test. p values < 0.05 and p < 0.01 were considered statistically significant. Calculations were carried out using the commercially available packages Statistica PL v.10 (StatSoft, Tulsa, OK, USA).

5. Conclusions

In conclusion, treatment of human squamous cancer cells with CA and with Met suppressed the motility of cells and the effect depended on a particular cell line. Both compounds regulated EMT process in C4-I and HTB-35 cells by interfering with different molecular targets. In TGF-β1-stimulated C4-I cells, CA suppressed the expression of mesenchymal transcription factor SNAI1 which resulted in enhanced expression of epithelial markers E-cadherin, Occludin and Claudin. Additionally, CA blocked *MMP-9* and upregulated *TIMP-1*, a specific inhibitor of MMP-9. In HTB-35 cells stimulated with TGF-β1, Met decreased the expression of Vimentin. By suppressing hypoxia master regulator *HIF-1α*, Met caused downregulation of *CAIX*, an enzyme involved in metastasis of aggressive malignant cells. In this study we showed that CA and Met inhibited EMT process in cancer cells via different mechanisms. However, when applied together, compounds exerted a greater effect on EMT than each compound alone.

This is the first report revealing that CA alone and CA co-treated with Met may reverse the mesenchymal phenotype of TGF-β1-treated cervical tumor cells and we believe that the use of the two small molecules may be considered a potential therapeutic approach for metastatic cervical cancer.

Supplementary Materials: Supplementary materials can be found at www.mdpi.com/1422-0067/19/1/266/s1.

Acknowledgments: This research was supported by the grant from Jagiellonian University Medical College, K/ZDS/005624.

Author Contributions: Malgorzata Tyszka-Czochara conceived and designed the experiments; Malgorzata Tyszka-Czochara and Malgorzata Lasota performed the experiments; Malgorzata Tyszka-Czochara and Marcin Majka analyzed and interpreted the data; Marcin Majka contributed reagents and materials; Malgorzata Tyszka-Czochara wrote the paper.

Conflicts of Interest: The authors declare no conflict of interest.

References

1. Lee, M.Y.; Shen, M.R. Epithelial-mesenchymal transition in cervical carcinoma. *Am. J. Transl. Res.* **2012**, *4*, 1–13. [PubMed]
2. Vessey, C.; Wilding, J.; Folarin, N.; Hirano, S.; Takeichi, M.; Soutter, P.; Stamp, G.; Pignatelli, M. Altered expression and function of E-cadherin in cervical intraepithelial neoplasia and invasive squamous cell carcinoma. *J. Pathol.* **1995**, *176*, 151–159. [CrossRef] [PubMed]
3. Cheng, K.; Hao, M. Metformin Inhibits TGF-β1-Induced Epithelial-to-Mesenchymal Transition via PKM2 Relative-mTOR/p70s6k Signaling Pathway in Cervical Carcinoma Cells. *Int. J. Mol. Sci.* **2016**, *17*, 2000. [CrossRef] [PubMed]
4. Kalluri, R.; Weinberg, R.A. The basics of epithelial-mesenchymal transition. *J. Clin. Investig.* **2009**, *119*, 1420–1428. [CrossRef] [PubMed]
5. Tseng, J.; Lin, C.; Su, L.; Fu, H.; Yang, S.; Chuu, C. CAPE suppresses migration and invasion of prostate cancer cells via activation of non-canonical Wnt signaling. *Oncotarget* **2016**, *7*, 38010–38024. [CrossRef] [PubMed]
6. Dziedzic, A.; Kubina, R.; Kabała-Dzik, A.; Wojtyczka, R.D.; Morawiec, T.; Bułdak, R.J. Caffeic acid reduces the viability and migration rate of oral carcinoma cells (SCC-25) exposed to low concentrations of ethanol. *Int. J. Mol. Sci.* **2014**, *15*, 18725–18741. [CrossRef] [PubMed]

7. Yang, Y.; Li, Y.; Wang, K.; Wang, Y.; Yin, W.; Li, L. P38/NF-κB/snail pathway is involved in caffeic acid-induced inhibition of cancer stem cells-like properties and migratory capacity in malignant human keratinocyte. *PLoS ONE* **2013**, *8*, e58915. [CrossRef] [PubMed]

8. Sirota, R.; Gibson, D.; Kohen, R. The timing of caffeic acid treatment with cisplatin determines sensitization or resistance of ovarian carcinoma cell lines. *Redox Biol.* **2017**, *11*, 170–175. [CrossRef] [PubMed]

9. Kuo, Y.Y.; Lin, H.P.; Huo, C.; Su, L.C.; Yang, J.; Hsiao, P.H.; Chiang, H.C.; Chung, C.J.; Wang, H.D.; Chang, J.Y.; et al. Caffeic acid phenethyl ester suppresses proliferation and survival of TW2.6 human oral cancer cells via inhibition of Akt signaling. *Int. J. Mol. Sci.* **2013**, *14*, 8801–8817. [CrossRef] [PubMed]

10. Cai, H.; Huang, X.; Xu, S.; Shen, H.; Zhang, P.; Huang, Y.; Jiang, J.; Sun, Y.; Jiang, B.; Wu, X.; et al. Discovery of novel hybrids of diaryl-1,2,4-triazoles and caffeic acid as dual inhibitors of cyclooxygenase-2 and 5-lipoxygenase for cancer therapy. *Eur. J. Med. Chem.* **2016**, *108*, 89–103. [CrossRef] [PubMed]

11. Kao, H.; Chang-Chien, P.; Chang, W.; Yeh, T.; Wang, J. Propolis inhibits TGF-β1-induced epithelial-mesenchymal transition in human alveolar epithelial cells via PPARγ activation. *Int. Immunopharmacol.* **2013**, *3*, 565–574. [CrossRef] [PubMed]

12. Imai, A.; Ichigo, S.; Matsunami, K.; Takagi, H.; Yasuda, K. Clinical benefits of metformin in gynecologic oncology. *Oncol. Lett.* **2015**, *10*, 577–582. [CrossRef] [PubMed]

13. Del Barco, S.; Vazquez-Martin, A.; Cufí, S.; Oliveras-Ferraros, C.; Bosch-Barrer, J.; Joven, J.; Martin-Castillo, B.; Menendez, J. Metformin: Multi-faceted protection against cancer. *Oncotarget* **2011**, *2*, 896–917. [CrossRef] [PubMed]

14. Auersperg, N. Histogenetic behavior of tumors. I. Morphologic variation in vitro and in vivo of two related human carcinoma cell lines. *J. Natl. Cancer Inst.* **1969**, *43*, 151–173. [PubMed]

15. Sedlakova, O.; Svastova, E.; Takacova, M.; Kopacek, J.; Pastorek, J.; Pastorekova, S. Carbonic anhydrase IX, a hypoxia-induced catalytic component of the pH regulating machinery in tumors. *Front. Physiol.* **2014**, *4*, e400. [CrossRef] [PubMed]

16. Svastova, E.; Pastorekova, S. Carbonic anhydrase IX: A hypoxia-controlled "catalyst" of cell migration. *Cell Adh. Migr.* **2013**, *7*, 226–231. [CrossRef] [PubMed]

17. Kalluri, R. EMT: When epithelial cells decide to become mesenchymal-like cells. *J. Clin. Investig.* **2009**, *119*, 1417–1419. [CrossRef] [PubMed]

18. Carlson, M.W.; Iyer, V.R.; Marcotte, E.M. Quantitative gene expression assessment identifies appropriate cell line models for individual cervical cancer pathways. *BMC Genom.* **2007**, *10*, 2–13. [CrossRef] [PubMed]

19. Lee, J.M.; Dedhar, S.; Kalluri, R.; Thompson, E.W. The epithelial-mesenchymal transition: New insights in signaling, development, and disease. *J. Cell Biol.* **2006**, *172*, 973–981. [CrossRef] [PubMed]

20. Miekus, K.; Pawlowska, M.; Sekuła, M.; Drabik, G.; Madeja, Z.; Adamek, D.; Majka, M. MET receptor is a potential therapeutic target in high grade cervical cancer. *Oncotarget* **2015**, *12*, 10086–10101. [CrossRef] [PubMed]

21. Katsuno, Y.; Lamouille, S.; Derynck, R. TGF-β signaling and epithelial-mesenchymal transition in cancer progression. *Curr. Opin. Oncol.* **2013**, *25*, 76–84. [CrossRef] [PubMed]

22. Zeisberg, M.; Neilson, E.G. Biomarkers for epithelial-mesenchymal transitions. *Cell* **2009**, *119*, 1429–1437. [CrossRef] [PubMed]

23. Hu, D.; Zhou, J.; Wang, F.; Shi, H.; Li, Y.; Li, B. HPV-16 E6/E7 promotes cell migration and invasion in cervical cancer via regulating cadherin switch in vitro and in vivo. *Arch. Gynecol. Obstet.* **2015**, *292*, 1345–1354. [CrossRef] [PubMed]

24. Chang, B.; Kim, J.; Jeong, D.; Jeong, Y.; Jeon, S.; Jung, S.; Yang, Y.; Kim, K.; Lim, J.; Kim, C.; et al. Klotho inhibits the capacity of cell migration and invasion in cervical cancer. *Oncol. Rep.* **2012**, *28*, 1022–1028. [CrossRef] [PubMed]

25. Meng, F.; Wu, G. The rejuvenated scenario of epithelial-mesenchymal transition (EMT) and cancer metastasis. *Cancer Metastasis Rev.* **2012**, *31*, 455–467. [CrossRef] [PubMed]

26. Wahdan-Alaswad, R.; Harrell, J.; Fan, Z.; Edgerton, S.; Liu, B.; Thor, A. Metformin attenuates transforming growth factor beta (TGF-β) mediated oncogenesis in mesenchymal stem-like/claudin-low triple negative breast cancer. *Cell Cycle* **2016**, *8*, 1046–1059. [CrossRef] [PubMed]

27. Wu, W.S.; You, R.I.; Cheng, C.C.; Lee, M.C.; Lin, T.Y.; Hu, C.T. Snail collaborates with EGR-1 and SP-1 to directly activate transcription of MMP 9 and ZEB1. *Sci. Rep.* **2017**, *7*, 17753. [CrossRef] [PubMed]

28. Zavadil, J.; Böttinger, E.P. TGF-β and epithelial-to-mesenchymal transitions. *Oncogene* **2005**, *24*, 5764–5774. [CrossRef] [PubMed]

29. Stanciu, A.; Zamfir-Chiru-Anton, A.; Stanciu, M.; Popescu, C.; Gheorghe, D. Imbalance between Matrix Metalloproteinases and Tissue Inhibitors of Metalloproteinases Promotes Invasion and Metastasis of Head and Neck Squamous Cell Carcinoma. *Clin. Lab.* **2017**, *63*, 1613–1620. [CrossRef] [PubMed]

30. Roomi, M.; Kalinovsky, T.; Rath, M.; Niedzwiecki, A. Modulation of u-PA, MMPs and their inhibitors by a novel nutrient mixture in human female cancer cell lines. *Oncol. Rep.* **2012**, *28*, 768–776. [CrossRef] [PubMed]

31. Jin, U.; Lee, J.; Kang, S.; Kim, J.; Park, W.; Kim, J.; Moon, S.; Kim, C. A phenolic compound, 5-caffeoylquinic acid (chlorogenic acid), is a new type and strong matrix metalloproteinase-9 inhibitor: Isolation and identification from methanol extract of Euonymus alatus. *Life Sci.* **2005**, *77*, 2760–2769. [CrossRef] [PubMed]

32. Mahecha, A.; Wang, H. The influence of vascular endothelial growth factor-A and matrix metalloproteinase-2 and -9 in angiogenesis, metastasis, and prognosis of endometrial cancer. *Onco Targets Ther.* **2017**, *10*, 4617–4624. [CrossRef] [PubMed]

33. Wanami, L.S.; Chen, H.Y.; Peiró, S.; García de Herreros, A.; Bachelder, R.E. Vascular endothelial growth factor-A stimulates Snail expression in breast tumor cells: Implications for tumor progression. *Exp. Cell Res.* **2008**, *314*, 2448–2453. [CrossRef] [PubMed]

34. Olive, P.; Aquino-Parsons, C.; MacPhail, S.; Liao, S.; Raleigh, J.; Lerman, M.; Stanbridge, E. Carbonic anhydrase 9 as an endogenous marker for hypoxic cells in cervical cancer. *Cancer Res.* **2001**, *61*, 8924–8929. [PubMed]

35. Andreucci, E.; Peppicelli, S.; Carta, F.; Brisotto, G.; Biscontin, E.; Ruzzolini, J.; Bianchini, F.; Biagioni, A.; Supuran, C.; Calorini, L. Carbonic anhydrase IX inhibition affects viability of cancer cells adapted to extracellular acidosis. *J. Mol. Med.* **2017**, *95*, 1341–1353. [CrossRef] [PubMed]

36. Pastorek, J.; Pastorekova, S. Hypoxia-induced carbonic anhydrase IX as a target for cancer therapy: From biology to clinical use. *Semin. Cancer Biol.* **2015**, *31*, 52–64. [CrossRef] [PubMed]

37. Karthikeyan, S.; Kanimozhi, G.; Prasad, N.R.; Mahalakshmi, R. Radiosensitizing effect of ferulic acid on human cervical carcinoma cells in vitro. *Toxicol. In Vitro* **2011**, *25*, 1366–1375. [CrossRef] [PubMed]

38. Lopez, J.S.; Banerji, U. Combine and conquer: Challenges for targeted therapy combinations in early phase trials. *Nat. Rev. Clin. Oncol.* **2017**, *14*, 57–66. [CrossRef] [PubMed]

39. Tyszka-Czochara, M.; Bukowska-Strakova, K.; Majka, M. Metformin and caffeic acid regulate metabolic reprogramming in human cervical carcinoma SiHa/HTB-35 cells and augment anticancer activity of Cisplatin via cell cycle regulation. *Food Chem. Toxicol.* **2017**, *106*, 260–272. [CrossRef] [PubMed]

40. Tyszka-Czochara, M.; Konieczny, P.; Majka, M. Caffeic Acid Expands Anti-Tumor Effect of Metformin in Human Metastatic Cervical Carcinoma HTB-34 Cells: Implications of AMPK Activation and Impairment of Fatty Acids De Novo Biosynthesis. *Int. J. Mol. Sci.* **2017**, *18*, 462. [CrossRef] [PubMed]

41. Skrzypek, K.; Kusienicka, A.; Szewczyk, B.; Adamus, T.; Lukasiewicz, E.; Miekus, K.; Majka, M. Constitutive activation of MET signaling impairs myogenic differentiation of rhabdomyosarcoma and promotes its development and progression. *Oncotarget* **2015**, *6*, 31378–31398. [CrossRef] [PubMed]

International Journal of
Molecular Sciences

MDPI

Review

Plant Secondary Metabolites as Anticancer Agents: Successes in Clinical Trials and Therapeutic Application

Ana M. L. Seca [1,2,*] and Diana C. G. A. Pinto [2]

[1] cE3c—Centre for Ecology, Evolution and Environmental Changes/Azorean Biodiversity Group & Faculty of Sciences and Technology, University of Azores, Rua Mãe de Deus, 9501-321 Ponta Delgada, Portugal
[2] Department of Chemistry & QOPNA—Organic Chemistry, Natural Products and Food Stuffs, University of Aveiro, Campus de Santiago, 3810-193 Aveiro, Portugal; diana@ua.pt
* Correspondence: ana.ml.seca@uac.pt; Tel.: +351-296-650-172

Received: 12 December 2017; Accepted: 12 January 2018; Published: 16 January 2018

Abstract: Cancer is a multistage process resulting in an uncontrolled and abrupt division of cells and is one of the leading causes of mortality. The cases reported and the predictions for the near future are unthinkable. Food and Drug Administration data showed that 40% of the approved molecules are natural compounds or inspired by them, from which, 74% are used in anticancer therapy. In fact, natural products are viewed as more biologically friendly, that is less toxic to normal cells. In this review, the most recent and successful cases of secondary metabolites, including alkaloid, diterpene, triterpene and polyphenolic type compounds, with great anticancer potential are discussed. Focusing on the ones that are in clinical trial development or already used in anticancer therapy, therefore successful cases such as paclitaxel and homoharringtonine (in clinical use), curcumin and ingenol mebutate (in clinical trials) will be addressed. Each compound's natural source, the most important steps in their discovery, their therapeutic targets, as well as the main structural modifications that can improve anticancer properties will be discussed in order to show the role of plants as a source of effective and safe anticancer drugs.

Keywords: secondary metabolites; clinical trial; anticancer therapy; vincristine; paclitaxel; homoharringtonine; ingenol mebutate; curcumin; betulinic acid

---

1. Introduction

Although cancer is the most devastating disease, causing more deaths than all coronary heart diseases or all strokes, with 14.1 million new cases and 8.2 million deaths in 2012 [1], there is a register of a continuous decline in cancer death rates that has resulted in an overall drop of 23% since 1991 [2]. Despite this progress, there is a register of 8.8 million deaths globally in 2015, and cancer is now the leading cause of death in 21 states of the United States of America [2]. The total annual economic cost of cancer in 2010 was approximately $1.16 trillion [3]. This burden is further expected to rise, with over the predicted 20 million new cancer cases expected globally by 2025 [4]. Moreover, incidence and death rates are increasing for several cancer types, for example liver and pancreas [2]. In the low- and middle-income countries, the picture is even darker, where approximately 70% of deaths are due to cancer diseases and where only one in five countries have the necessary data to drive cancer policy [3,5]. Advancing the fight against cancer requires both increased investment in cancer pathology research and in new safe, effective, inexpensive and minimal side effect anticancer agents.

For millennia, indigenous cultures around the world have used traditional herbal medicine to treat a myriad of maladies. Plants constitute a common alternative for cancer treatment in many countries, and more than 3000 plants worldwide have been reported to have anticancer properties [6,7].

Although a recent study suggests that nowadays, the traditional medicines are less used, even in populous middle-income countries [8], herbal medicine use is still common in oncology therapy worldwide [6,7,9–11]. In the last two decades, the use of herbal remedies has also been widely embraced in many developed countries as complementary and alternative medicine, but following tight legislation and under surveillance [12]. Natural products have garnered increasing attention in cancer chemotherapy because they are viewed as more biologically friendly and consequently more co-evolved with their target sites and less toxic to normal cells [13]. Moreover, there is evidence that natural product-derived anticancer drugs have alternative modes of promoting cell death [14,15]. Based on these facts, many researchers are now centering their investigations on the plants' potential to deliver natural products that can become useful to the pharmaceutical industry [16–18]. In fact, the utilization of natural products as the background to discover and develop a drug entity is still a research hot point. From small molecules approved for cancer chemotherapy between 1940 and 2014, around 49% are natural products [19].

In spite of all the beneficial potential of medicinal plants and consequently of their products, many continue without adequate monitoring to guarantee their effectiveness and safety [20,21].

The following sections offer an overview of compounds from plants that have been found to exhibit activity against different types of cancer and are now on the market as anticancer drugs or are involved in clinic trials, which means they are involved in the last stage of the development of a clinical drug. Therefore, these compounds, which constitute successful cases in cancer therapy, will be briefly discussed.

2. Secondary Metabolites from Plants as Anticancer Agents

Throughout history, plants have been a rich source of affordable natural compounds, explicitly the secondary metabolites, that possess sufficient structural complexity so that their synthesis is difficult or at this time not yet accomplished and exhibit a broad spectrum of bioactivities including antitumor activity [22,23]. Secondary metabolites are mostly small organic molecules, produced by an organism, that are not essential for its growth, development and reproduction. They can be classified based on the pathway by which they are synthesized [24]. Additionally, a simple classification includes three main groups: terpenoids (polymeric isoprene derivatives and biosynthesized from acetate via the mevalonic acid pathway), phenolics (biosynthesized from shikimate pathways, containing one or more hydroxylated aromatic rings) and the extremely diverse alkaloids (non-protein nitrogen-containing compounds, biosynthesized from amino acids such as tyrosine, with a long history in medication) [24,25]. Several new cytotoxic secondary metabolites are isolated from plants each year and constitute a source of new possibilities to explore in order to fight against cancerous diseases.

Although some natural compounds have unique anticancer effects, their use in clinical practice is not possible due to their physico-chemical properties (e.g., limited bioavailability) and/or their toxicity. On the other hand, plant occurring secondary metabolites often can be excellent leads for drug development. Thus, modifying the chemical structure of these more promising compounds is one strategic way to increase their anticancer action and selectivity, improve their absorption, distribution, metabolism and excretion properties and decrease their toxicity and side effects [26,27]. Herein we will present the most significant achievements in the area of plant secondary metabolites, some already in clinical use and others in clinical trials as anticancer agents, as well as their most efficient derivatives obtained by structural modifications.

2.1. Metabolites Used in Cancer Therapy

During the last few decades, a wide range of cytotoxic agents was discovered from plants, but very few of these managed to reach clinical use after successfully running through the entire long, selective, expensive and bureaucratic process from their chemical identification to their effectiveness in therapeutic cancer treatment. Each of these compounds has their histories of success and limitations, which has been told by many authors and which are hereinafter counted in a historical, molecular, pharmaceutical and clinical point of view.

2.1.1. Vincristine

Vincristine (**1**) has a non-symmetrical dimeric structure, composed of a two indole-type nucleus linked by a carbon–carbon bond, the vindoline portion and the catharanthine type portion (Figure 1). In 1963, the Food and Drug Administration (FDA) approved its clinical use to treat cancer. In fact, it was one of the first plant-derived anticancer agents approved by this agency [19]. It is a naturally-occurring alkaloid extracted from the leaves of *Catharanthus roseus* (L.) G.Don (formerly *Vinca rosea* L.) and has been used in chemotherapy in adult, but mainly in pediatric oncology practice against acute lymphoblastic leukemia. Its incorporation in the treatment regimen increases the survival rate to eighty percent [28]. It is also used to treat rhabdomyosarcoma, neuroblastoma, lymphomas and nephroblastoma [29,30].

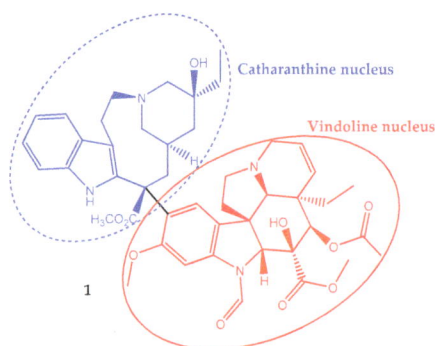

Figure 1. Chemical structure of the vinca alkaloid vincristine (**1**), an anticancer natural agent that repress cell growth by altering the microtubular dynamics.

The large interest in vincristine contrasts with its low natural occurrence, and consequently, its extraction is very expensive. This situation has stimulated an intense research effort aiming to find promising strategies to increase vincristine (and other vinca alkaloids) production. Selected enzymes' manipulation by genetic engineering to raise the metabolic flow rate toward vincristine and the use of elicitors to activate genes involved in vincristine metabolic pathways are effective strategies to increase the biotechnological production of this compound [30,31]. However, some improvements are needed before these processes become economically viable. Another possibility to obtain more vincristine is the application/optimization of high yield extraction methodologies like negative-pressure cavitation extraction [32].

Vincristine, in a concentration-dependent manner, can affect cells' division. However, the most well-known mechanism of vincristine antitumor activity involves interaction with tubulin, the basic constituent of mitotic spindle microtubules, inhibiting its polymerization and resulting in the suppression of mitosis. Therefore, it disrupts the assembly of the mitotic spindle, which in turn leads to the demise of actively-dividing cells [33]. Some authors report that at the lowest effective concentration, the anti-proliferative effect is due to a subtly change in the addition and loss of tubulins at the mitotic spindle microtubule and thus stabilizes the mitotic spindle assembly and disassembly processes that lead to metaphase arrest [30]. Once microtubule dynamics, and therefore cell division, can be perturbed by blocking the polymerization or depolymerization of tubulin in microtubules and thus impairing the mitotic spindle assembly, it seems that vincristine can act by both mechanisms depending on the concentration level. Moreover, a molecular docking study showed some evidence suggesting each part of the vincristine dimeric structure exhibits a specific role on its anticancer activity once the vindoline nucleus binds tubulin heterodimers, while the catharanthine nucleus provides a cytotoxic effect [34].

Despite the long history of vincristine clinical application in fighting cancer, there are three factors that diminish its impact in therapeutics: (i) its antitumor mechanism is cell-cycle-specific, and the duration of its exposure to tumor cells can significantly affect its antitumor activity; (ii) the pharmacokinetic behavior of vincristine in human blood is described by a bi-exponential elimination pattern with a very fast initial distribution half-life followed by a longer elimination half-life, and it has a large volume of distribution, suggesting diffuse distribution and tissue binding [35]; (iii) it may cause temporary or permanent peripheral neuropathy, which is a dose-dependent side effect influenced by several variables such as age, race, genetic profile and administration method, and older children, in particular Caucasian, seem to be more susceptible [36]. Some of these factors could be mitigated by encapsulation of vincristine into liposomes, which is intended to increase the circulation time, optimize delivery to target tissues and facilitate dose intensification without increasing toxicity [35].

In 2012, the FDA approved the use of sphingomyelin/cholesterol (SM/Chol) liposomal vincristine (Marqibo®) to treat adults with relapsed acute lymphoblastic leukemia (New Drug Application: 202497). Vincristine can be loaded into conventional liposomes like SM/Chol liposomes, but other types of liposomes, for example PEGylated liposomes, were already tested, although SM/Chol liposomal vincristine displays a relatively long circulation time, a reduced leakage rate from liposomes and an improved antitumor effect compared to PEGylated liposomal vincristine [33]. Clinical trials involving Marqibo® are underway to pediatric patients with relapsed or chemotherapy-refractory solid tumors and leukemia (ClinicalTrials.gov Identifier: NCT01222780). Moreover, other vincristine encapsulated formulations are involved in clinical studies in which they are tested against other types of cancer such as small-cell lung cancer (ClinicalTrials.gov Identifier: NCT02566993), advanced cervical cancer (ClinicalTrials.gov Identifier: NCT02471027) and liver cancer (ClinicalTrials.gov Identifier: NCT00980460).

Vincristine generally exhibits better efficacy when administered in combination with other antitumor agents. In fact, combined chemotherapy can not only enhance the destruction of tumor cells, but also decrease toxicity and drug resistance with drugs exhibiting different mechanisms of action. Therefore, open clinical trials are in progress involving combined vincristine therapy (e.g., NCT02879643; NCT01527149). Very recently, a case report was done of infantile fibrosarcoma treated by adjuvant therapy after excision, using vincristine and dactinomycin, where the duration of chemotherapy was determined according to tumor response. At the end, there was no functional impairment and no evidence of recurrence at 18 months after therapy [37].

2.1.2. Paclitaxel

The discovery of novel natural structures with significant biological relevance and with new action mechanisms have tremendous impact on the pharmaceutical industry. The discovery of (**2**) is an excellent example. Its high activity and its novel mechanism of action, tubulin-assembly promotion, is a milestone of a new era in anticancer drug discovery. Paclitaxel, isolated from the bark of *Taxus brevifolia* Nutt. (Pacific Yew) and sold under the brand name Taxol® since 1993, is a complex molecule that has become one of the most active cancer chemotherapeutic drugs known [38,39]. It is a tricyclic diterpenoid, occasionally considered as a pseudo alkaloid, that contains a complex 6,8,6-tri-cycle-fused skeleton, named the "taxane" ring system, linked to a four-member oxetane ring and having alcohol, ester, ketone and amide functions (Figure 2).

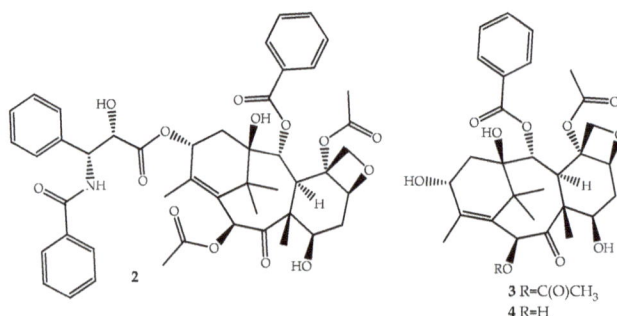

Figure 2. Chemical structure of paclitaxel (**2**), a natural microtubule inhibitor, and of its precursors baccatin III (**3**) and 10-deacetylbaccatin III (**4**).

Paclitaxel is a non-ionic molecule with high lipophilicity (log P = 3.20) that is practically insoluble in aqueous medium (aqueous solubility ~0.3–0.5 µg/mL) [40]. Due to this hydrophobicity its administration is performed in a solution containing alcohol and polyoxyethylated castor oil to enhance its delivery. The biosynthetic pathway of paclitaxel is a complex process that starts with precursor geranylgeranyl diphosphate and involves 19 steps regulated by several enzymes, and some were already characterized, but the process is not yet fully understood [41].

Although the medicinal use of paclitaxel has been achieved exclusively with purified compound from the bark of Pacific Yew, the plant's low content and the ecological impact of its harvesting have prompted extensive searches for alternative sources. The total synthesis of paclitaxel was not successful until 1994 [42], and even after several improvements [43,44], it remains a laborious work that prevents its industrial viability. More sustainable alternatives are being used: (i) the fermentation technology with microbes or plant cell culture [45]; (ii) protein engineering to elevate catalytic fitness for paclitaxel production [46]; (iii) semisynthesis from baccatin III (**3**, Figure 2) [47] or 10-deacetylbaccatin III (**4**, Figure 2) [48], two paclitaxel precursor molecules, which are non-cytotoxic and are found in much higher quantities and readily available from the needles of *Taxus baccata*, *Taxus brevifolia* and other *Taxus* species [49]. The last approach is the one employed by the pharmaceutical industry.

The introduction of paclitaxel in the last few decades has expanded the therapeutic options, mainly due to its powerful anticancer activity, and great successes in the treatment of breast, ovarian and lung cancers have been achieved [39]. Moreover, its success is also due to effectiveness on both solid and disseminated tumors and a broad spectrum of antitumor activity predicted by its unique mechanism of action, which targets the very basic elements of the cancer phenotype like cell proliferation and DNA repair [38]. In fact, paclitaxel skeleton functional groups are at special positions and ensure that β-tubulin is targeted in order to prevent the dynamic microtubule disassembly process required for proper mitotic spindle assembly and chromosome segregation during cell division. Consequently, cell death is caused in a time- and concentration-dependent manner [38].

The continuous research on the mechanism of action of paclitaxel together with the structure activity relationship (SARs) and quantitative SAR (QSAR) revealed and assigned the pharmacophores, as well as structural parts that should not be modified (Figure 3). This allowed the design of novel derivatives with the best efficacy and fewer side effects [26,50]. Based on this knowledge, two semi-synthetic derivatives were developed with great success, docetaxel (**5**) and cabazitaxel (**6**) (Figure 3). They were obtained by structural modifications restricted to the variable sections of the original structure and are now available for clinical use (Figure 3).

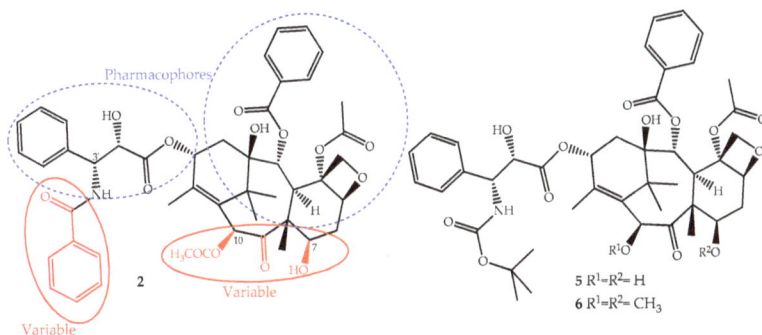

Figure 3. The parts of the paclitaxel (**2**) structure that could be modified without loss of activity and two of its derivatives, docetaxel (**5**) and cabazitaxel (**6**), available on the market for clinical use.

Although paclitaxel has been applied effectively to treat many cancer diseases, its therapeutic efficacy is starting to be limited due to multidrug resistance (MDR) development [51,52]. Although the cellular mechanisms involved in the MDR are not fully understood, it appears that the overexpression of ABCB1 (also called P-glycoprotein) and ABCC10 (also named multidrug resistance protein 7) efflux transporters, the α-/β-tubulin mutations and/or alterations in the binding regions are the main cause [51,52].

The development of new drug delivery systems and new formulations allowed paclitaxel to find its way to the tumor tissue for more direct and safe anticancer activity and to overcome paclitaxel's multidrug resistance, its poor aqueous solubility, clinical neurotoxicity and neutropenia [53–55]. For example, Lipusu®, the first paclitaxel lecithin/cholesterol liposome injectable, has been on the Chinese market since 2006 and is used in the treatment of ovarian, breast, non-SCLC, gastric and head and neck cancers [39]. This liposomal formulation Lipusu® exhibited similar antitumor effects to paclitaxel, but its toxicity is lower than that of paclitaxel under the same dosage [39,56]. Another example is Abraxane®, an injectable nanoparticle albumin-bound paclitaxel, also named *nab*-paclitaxel developed to improve the solubility of paclitaxel, which was approved in 2005 by FDA and in 2012 by European Medicines Agency (EMA) (EMA/99258/2015, EMEA/H/C/000778) [57]. Higher doses of *nab*-paclitaxel can be administered over a shorter infusion time, and consequently, there is an improvement in neuropathy side effects after the therapy discontinuation [57], although peripheral sensory neuropathy occurred more frequently with *nab*-paclitaxel compared to paclitaxel [55].

The development of paclitaxel-mimics, with a simplified structure, also allowed the discovery of docetaxel (**5**, Figure 3), on the market since 1995 under the trade name Taxotere®, a drug that has fewer side effects and improved pharmaceutical properties [58]. It is obtained by semisynthesis from 10-deacetylbaccatin-III and shares with paclitaxel the same mechanism of action and identical ABCB1 affinity, but with different pharmacokinetics and side effects [49]. It is structurally different from paclitaxel only at the C-10 (acetyl group removed) and C-3′ positions (the *N*-C(O)Ph group is replaced for an *N*-*tert*-butyl acetate group), (Figure 3) alterations that increase its water solubility and lower its lipophilicity (log *P* = 3.20). It belongs to the first generation of taxanes, used for the treatment of breast, ovarian, prostate and non-SCLCs, and exhibits a longer half-life, more rapid cellular uptake and longer intracellular retention than paclitaxel [59].

Cabazitaxel (Jevtana®) (**6**, Figure 3) was approved by the FDA in 2010 for the treatment of patients with hormone-refractory metastatic prostate cancer and tumors that are docetaxel- or paclitaxel-resistant [60]. It is also obtained by semisynthesis and is a dimethoxyl derivative of docetaxel, a structural change that increases its lipophilicity (log *P* = 3.90) and consequently its cell penetration through passive influx associated with alteration of the P-gp affinity [61]. This allows the drug to accumulate intracellularly at greater concentrations than docetaxel and explains its improved cytotoxicity and effectiveness in taxane-resistant patients [27,49].

Paclitaxel is already a blockbuster of the pharmacy industry not only due to the development of new delivery systems in cancer therapy [62] and its application in combination with other anticancer drugs (e.g., ClinicalTrials.gov Identifier: NCT02379416, NCT00584857 and NCT01288261) [63,64], but also due to its use in clinical trials for other treatments such as psoriasis [65] and botulinum neurotoxin inhibiting [66], just to mention a few examples that ensure this compound's success.

2.1.3. Homoharringtonine

Homoharringtonine (**7**) is an alkaloid with a cephalotaxine nucleus named cephalotaxine 4-methyl-2(*R*)-hydroxy-2-(4-hydroxy-4-methylpentyl)succinate (Figure 4). It was first isolated from *Cephalotaxus harringtonii* (Knight ex J.Forbes) K.Koch and *Cephalotaxus fortunei* Hook. trees, whose bark extracts were used in Chinese traditional medicine to treat cancer. Homoharringtonine and other cephalotaxine derivatives can also be found in leaves, bark and seeds of other *Cephalotaxus* species [67]. In fact, the cephalotaxine itself is very abundant in *Cephalotaxus* species leaves which can be isolated and transformed by a simple esterification into homoharringtonine, and thus, this procedure constitutes a semisynthetic methodology used for homoharringtonine industrial production [50,68].

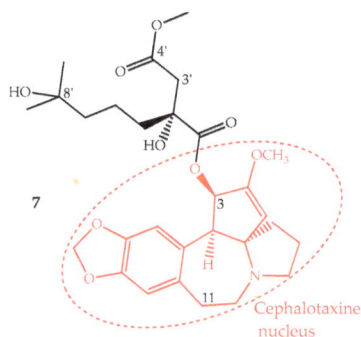

Figure 4. Chemical structure of homoharringtonine also named omacetaxine mepesuccinate (**7**) with alkaloid nucleus cephalotaxine (red).

The interest in homoharringtonine started when its potent antiproliferative activity against murine P-388 leukemia cells with IC_{50} values of 17 nM was demonstrated [69]. In fact, since the 1970s homoharringtonine or a mixture of cephalotaxine esters has been used in China to treat hematological malignancies [70]. However, only after the development of the above-mentioned semisynthetic procedure did homoharringtonine attract the attention of Western medicine.

Homoharringtonine is a first-in-class protein translation inhibitor, which means that it inhibits the elongation step of protein synthesis. In fact, homoharringtonine binds to the A-site of the large ribosomal subunit, an action that blocks the access of the charged tRNA and consequently the peptide bond formation [71]. Since this drug does not target specific proteins, its success is mainly due to the fact that it can disturb proteins with rapid turnover such as the leukemic cells' upregulated short-lived oncoproteins BCR-ABL1 and antiapoptotic proteins (Mcl-1, Myc) leading to cells apoptosis [71]. Recently, other mechanisms indicated that it could also affect signaling pathways, like the Jak-stat5 pathway, by regulating protein tyrosine kinase phosphorylation [72] and by activating the TGF-β pathway through phosphorylation of smad3 [73].

The identification of several natural cephalotaxine esters structurally similar to homoharringtonine and other derivatives obtained by semisynthesis allowed establishing some structure activity relationships, which were recently reviewed and discussed by Chang et al. [69].

The most important SAR are: (i) the cephalotaxine nucleus is much less active against the P388 cell line than its esters derivatives; (ii) an aliphatic side chain bonded to the hydroxyl group at C-3 seems to be necessary to enhance the activity; (iii) the presence of hydroxyl groups at C-11 or C-3′ decreases the activity; (iv) a free carboxylic acid at C-4′ abruptly decreases the activity; however, the methyl group can be replaced by other alkyl groups, even bulky ones, without the loss of the activity and in some cases enhancing it; (v) bulky groups bonded to 8′-OH are also tolerated; (vi) substituents bonded at 2′-OH imply a significant loss of activity (Figure 4).

There is a long track record of the clinical efficacy and safety of homoharringtonine use in the treatment of chronic myeloid leukemia. Currently, the focus is on its use in patients that experienced resistance or intolerance to multiple tyrosine kinase inhibitors (sorafenib and imatinib target) [74] and in patients carrying the T315I mutation, a variant that is unresponsive to tyrosine kinase inhibitors [74–76]. In fact, homoharringtonine was approved by the FDA in 2012 (sold under the trade name Synribo®) to be used in the treatment of chronic myeloid leukemia in patients with resistance and/or intolerance to two or more tyrosine kinase inhibitors, and it is the only natural therapeutic agent approved as a commercial drug to treat chronic myeloid leukemia.

The commercial approval of homoharringtonine and continued preclinical and clinical investigations of this compound indicate opportunities for its use in other hematological malignancies. For instance, the produced durable hematologic and cytogenetic responses regardless of mutational status [76,77] exhibit the ability to effectively kill stem/progenitor cells [77,78] and have a role in acute myeloid leukemia [79].

The homoharringtonine therapeutic efficiency continues to be evaluated, and its use is expect in the near future in other hematologic malignancies. It is being evaluated in 20 clinical trials, in mono and combined therapy, involving, for example, patients with newly-diagnosed acute myelogenous leukemia (NCT01873495), with relapsed/refractory acute myeloid leukemia carrying FLT3-ITD (NCT03170895), with myelodysplastic syndrome (NCT02159872), and in combined therapy with imatinib mesylate (NCT00114959), with quizartinib (NCT03135054) and with cytarabine and idarubicin (NCT02440568). Moreover, the subcutaneous administration of homoharringtonine does not influence its bioavailability (NCT00675350) [80] and allowed decreasing its cardiac toxicity [77]. Additionally, the FDA in 2014 approved its administration at home by the patient or a caregiver, which is indeed an improvement because patients have the opportunity to self-administer their therapy and due to homoharringtonine's stability [81].

Although homoharringtonine treatment may result in some hematologic toxicity such as myelosuppression, this should not prevent the use of this drug, once the benefits exceed the damage and the latter can be limited mainly by adequate dose adjustment and patient training for symptoms [82]. All these data show a large number of scenarios where homoharringtonine use is applied and suggest many others where it can receive approval in the near future, showing that its long history in cancer therapy is far from over.

2.2. Metabolites in Clinical Trials

In September 2007, a total of 91 plant-derived compounds was in clinical trials [83], whereas at the end of 2013, there were 100 unaltered natural products plus their derivatives involved in clinical trials, with a majority being in oncology [68].

Several semisynthetic derivatives from the plant-derived compounds camptothecin (e.g., gimatecan), combretastatin A (e.g., fosbretabulin tromethamine; combretastatin A1 diphosphate), rohitukine (e.g., alvocidib, riviciclib), triptolide (e.g., minnelide) and daidzein (e.g., phenoxodiol) [50,68,83] are in clinical trials, while the lead compounds are not involved in any clinical studies as an anticancer agent, although they exhibit relevant cytotoxic properties. Only the plant-derived lead compounds are presently in clinical trials as anticancer agents, and their derivatives are discussed below.

2.2.1. Ingenol Mebutate

The phytochemical study of *Euphorbia peplus* L. latex sap yielded several macrocyclic diterpenes [84], including ingenol mebutate (**8**, Figure 5) (also known as PEP005, ingenol-3-angelate and 3-ingenyl angelate), which was later on identified as the most active antitumor component [85]. In fact, the *Euphorbia peplus* sap has been shown, in a recent phase I/II clinical study, to be effective against human non-melanoma skin cancer [86]. This ingenene-type diterpene (Figure 5) can also be isolated from other *Euphorbia* species such as *Euphorbia paralias* L., *Euphorbia millii* Des Moul., *Euphorbia palustris* L., *Euphorbia marginata* Pursh and *Euphorbia helioscopia* L., and especially in the lower leafless stems of *Euphorbia myrsinites* L., where it is found in high quantity (547 mg/kg of dry weight) [68,87]. Ingenol mebutate has been prepared by semisynthesis using ingenol, which is isolated from the seeds of *Euphorbia lathyris* L. (yield ~100 mg/kg). The methodology involves a selective esterification of the hydroxyl group at position 3 with (Z)-2-methylbut-2-enoic acid (angelate nucleus) (Figure 5) [88]. Some efforts have been made to accomplish the ingenol total synthesis, but they are not suitable for application in the pharmaceutical industry, so the ingenol mebutate total synthesis remains undone. Ingenol mebutate is a monoester considered, in pharmacological terms, a small molecule. Its stability is pH dependent and can undergo facile acyl migration involving the hydroxyl groups, mainly the 5- and 20-OH (Figure 5). This characteristic is important from the biological activity point of view, because the free hydroxyl groups and the ester moiety at position 3 are required for the anticancer activity [89].

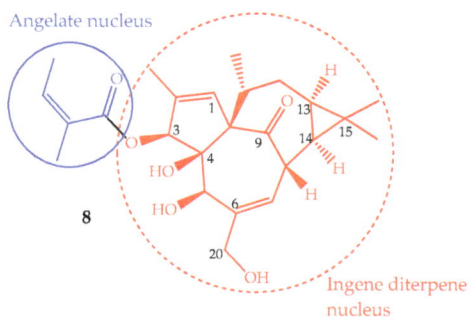

Figure 5. Chemical structure of the diterpene ingenol mebutate (**8**).

Ingenol mebutate showed potent antiproliferative effects in a dose- and time-dependent manner against several cell lines [90,91], especially against colon 205 cells line with IC_{50} = 10 nM, that means more active than staurosporine (IC_{50} = 29 nM) or doxorubicin (IC_{50} = 1.5 µM), known active compounds used as standards [90]. There is evidence that its effectiveness at damaging the tumor vasculature is related to the fact that it can be transported through the epidermis into the deep dermis via a P-glycoprotein [92]. Treatment with this compound, both in vitro (230 µM) and in vivo (42 nmol), rapidly caused swelling of mitochondria probably by loss of mitochondrial membrane potential and cell death by primary necrosis and is, therefore, unlikely to have its activity compromised by the development of apoptosis resistance in tumor cells [86]. There is evidence that this rapid action of ingenol mebutate is due to its dual action combining cytotoxic and immunomodulatory effects in which rapid lesion necrosis and antibody-dependent cellular cytotoxicity mediated by neutrophils occur [93]. The mechanism of action of ingenol mebutate is also partially related to the modulation of protein kinase C (PKC) to which it has a potent binding affinity by activating PKCδ and inhibiting PKCα [91,94]. In an in vitro assay, low isozyme selectivity was verified with a Ki ranging from 0.105–0.376 nM [95].

The above-mentioned results support the potential of ingenol mebutate for further improvements in cancer therapy; in fact, the cutaneous treatment of non-hyperkeratotic, non-hypertrophic actinic keratosis (a precancerous condition, that if untreated usually leads to a melanoma) with a gel formulation of ingenol mebutate (formerly PEP005 and marketed as Picato®) was approved by both FDA and EMA agencies in 2012 [96,97]. Unfortunately, adverse reactions associated with this application have been reported, although they are restricted to moderate "local skin responses" and included erythema, flaking/scaling, swelling, crusting, erosion/ulceration and vesiculation/postulation. However, it shows a favorable safety and tolerability profile exhibiting a lack of systemic absorption and photosensitivity [92,97].

2.2.2. Curcumin

Curcumin (9, Figure 6) or diferuloylmethane (bis-α,β-unsaturated β-diketone) is a polyphenolic compound that has been extracted from the rhizome of turmeric (*Curcuma longa* L.), a tropical Southeast Asia plant mainly used as a spice. However, the turmeric powder, which has 2–5% of curcumin, is used in Chinese and Indian traditional medicines [98]. To this ancient remedy have been attributed a wide range of beneficial properties including anti-inflammatory, antioxidant, chemopreventive, chemotherapeutic and chemo-sensitizing activity [98]. Curcumin is an orange-yellow crystalline lipophilic phenolic substance that, in solution, exists in equilibrium with its keto-enol tautomeric forms (Figure 6). It is not very soluble in water and also not very stable, although its degradation increases in basic medium [99].

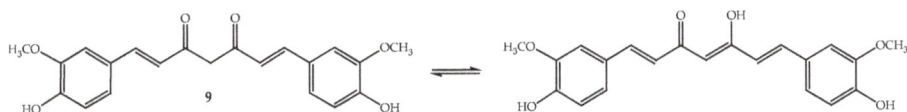

Figure 6. Chemical structure of the polyphenol curcumin (**9**).

Research interest in curcumin's anticancer properties has been developed based on the low rate occurrence of gastrointestinal mucosal cancers in Southeast Asian populations and its association with regular turmeric use in their diet [100].

A large volume of experimental data established the therapeutic efficacy of curcumin in in vitro cellular level, as well as in some ex vivo tumor-derived cancer cells/solid tumors like brain tumors, pancreatic, lung, breast, leukemia, prostate, skin cancers and hepatocellular carcinoma, including cytotoxic effects on cancer stem cells and antimetastatic activity [101–103]. This year, its possible application in colorectal, head and neck cancer chemotherapy was also reviewed [104,105]. Equally important were the assays demonstrating that curcumin was not cytotoxic to normal cells at the dosages required for therapeutic efficacy against the cancer cell lines [106,107]. The scientific interest and pharmacological potential of curcumin anticancer effects becomes also evident from the number of patents on curcumin-based therapeutics registered in the last five years [108].

Several studies have shown that curcumin can modulate a variety of cancer-related targets or pathways [102,103,109,110], which may be responsible for its effectiveness in combating cancer diseases. Recent studies demonstrate that curcumin's mechanism of action includes: (i) modulation of CYP enzymes by elevation of transcription factor Nrf2 level via the mitogen-activated protein kinase (MAPK) signaling pathway and Akt pathway [111]; (ii) mitotic catastrophe induction due to caspase activation and mitochondrial membrane polarization [14]; (iii) promotion of autophagic cell death, an important death inducer in apoptosis resistant cancer cells by beclin-1-dependent and independent pathways [14,112]; (iv) arrest of the cell cycle at the check points G1, S-phase and G2/M phase, modulating the cell cycle regulators, including upregulation of cyclin-dependent kinase inhibitors (CDKIs) [113]; (v) promotion of the inhibition of transcription factor NF-κB by preventing nuclear translocation of NF-κB and attenuating the DNA binding ability of NF-κB, contouring the problem of chemoresistance [114]; (vi) promotion of the inhibition of the crucial steps to angiogenesis by downregulation of the PGDF, VEGF and FGF expression and downregulation of MMPs via NF-κB,

ERKs, MAPKs, PKC and PI3K inhibition [115]; and (vii) inhibiting tubulin polymerization, that is curcumin binds with DNA [116,117]. Despite this knowledge about curcumin's multiple mechanisms of action, its biological properties are not fully understood. For example, does curcumin's survival and proliferative effects depend on its concentration, treatment period and cells type? On the other hand, administered doses of curcumin have been studied. Systematic in vivo doses up to 300 mg–3.5 g/kg b.w. (administered for up to 14–90 days) or clinical studies with oral intake of 1.2–12 g daily (for 6 weeks–4 months) did not demonstrate any adverse effects at the populations, animals and patients [118], although these values exceed that normally consumed (granted an acceptable daily intake level of 0.1–3 mg/kg b.w. by the Joint FAO/WHO Expert Committee on Food Additives) and also the typical intake of the Indian population (60–100 mg per day).

Moreover, curcumin has been reported to act as a chemosensitizer for some clinical anticancer drugs (e.g., gemcitabine, paclitaxel and 5-fluorouracil, doxorubicin) and exhibits a synergistic effect in combination with other natural products (e.g., resveratrol, honokiol, epigallocatechin-3-gallate, licochalcone and omega-3), aspects that could be used as an effective strategy to overcome tumor resistance and reduce recurrence [108,119,120]. These observations therefore suggest that a superior therapeutic index may be achieved with curcumin when used in combination and could be advantageous in the treatment of some tumors. Anyway, additional studies are still needed to assess the exact mechanism of curcumin's synergic effect.

Nevertheless, the clinical translation of curcumin has been significantly hampered since it is poorly absorbed, improperly metabolized and shows poor systemic bioavailability, which mandates that patients consume up to 8–10 grams of free curcumin orally each day, in order for detectable levels in the circulation [109,118]. Thus, several strategies have been proposed to counter the bioavailability issue of curcumin involving (i) the use of adjuvants like piperine, which interferes with curcumin metabolism by glucuronidation, (ii) curcumin formulations based on nanotechnology with liposomes, micelles, phospholipid, among others, and (iii) use of curcumin analogues [117,121–123]. As result of the anticancer potential of curcumin and despite its clinical therapeutic limitations, there are currently 17 open clinical studies involving curcumin, mainly studies of combined curcumin therapy with other substances for the treatment of several types of cancer.

2.2.3. Betulinic Acid

Betulinic acid (3β-hydroxy-lup-20(29)-en-28-oic acid), a lupane-type pentacyclic triterpene (**10**, Figure 7) is biosynthesized from six different isoprene units and was first identified and isolated from *Gratiola officinalis* L. and named "graciolon". It was also isolated from other species, but identified with different names (from the bark of *Platanus acerifolia* (Aiton) Willd. named "platanolic acid" and from *Cornus florida* L. named "cornolic acid"), which led to some confusion. Later on, it was confirmed that all have the same structure, and the compound was named betulinic acid. Nowadays, it is known that this triterpene is extensively spread throughout the plant kingdom (for instance *Betula* spp., *Diospyros* spp., *Syzygium* spp., *Ziziphus* spp., *Paeonia* spp., *Sarracenia flava* L., *Anemone raddeana* Regel and *Lycopodium cernuum* L., among others) and in considerable amounts (up to 2.5%) [124]. However, these sources are not sufficient to meet the growing demand for betulinic acid. Therefore, it can be obtained through a selective oxidation of betulin (lup-20(29)-en-3,28-diol) [125], far more abundant (up to 30%) in birch bark than betulinic acid [126].

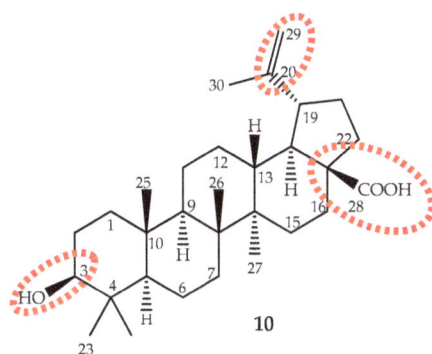

Figure 7. The chemical structure of the pentacyclic triterpene betulinic acid (**10**) and the main target to structural modifications (red).

In 1995, the first betulinic acid antitumor activity was reported by a researcher at the University of Illinois. It killed melanoma cells in mice with low IC_{50} values (IC_{50} 0.5–1.5 µg/mL) [127]. Since then, a number of researchers have conducted laboratory tests on betulinic acid to determine its antitumor properties, especially with respect to melanoma cells [128]. More recent studies suggest that betulinic acid possesses a broader spectrum of activity against other cancer cells, and consequently betulinic acid has been selected by the National Cancer Institute for addition into the Rapid Access to Intervention in Development (RAID) program.

Betulinic acid exhibits significant in vitro cytotoxicity in a variety of tumor cell lines and also inhibits the growth of solid tumors in vivo, comparable to some clinically-used drugs and showing a good selectivity index for cancer over normal cells even at doses up to 500 mg/kg b.w. [14,127,129,130]. Its anticancer proprieties have been demonstrated against colorectal lung, colon, breast, prostate, hepatocellular, bladder, head and neck, stomach, pancreatic, ovarian and cervical carcinoma, glioblastoma, chronic myeloid leukemia cells and human melanoma with IC_{50} values mainly between 1 to 13.0 µg/mL [14,124,128–132].

Betulinic acid exhibits potent anticancer activity by multiple molecular targets, the best characterized mechanism being the induction of apoptosis by direct regulation of the mitochondrial apoptotic pathway; which can be associated with mitochondrial collapse through direct opening of the permeability transition pore, decreasing mitochondrial outer membrane potential, downregulation of Bcl-2 family members, release of pro-apoptotic factors such as cytochrome c, increase of caspase activities, attenuating both constitutive and inducible STAT3 phosphorylation, nuclear translocation and its DNA binding [124,130,133]. However, there is also evidence that, in some cases, apoptosis may be induced by stabilizing p53 and downregulating NF-kB-mediated signaling [124,134].

The antimetastatic effect of betulinic acid seems to be through the prevention of the epithelial-to-mesenchymal transition in highly aggressive melanoma cells [131], while in breast cancer cells, it be by downregulation of the matrix metalloproteinases expression [133]. Betulinic acid can also induce an antiangiogenic response under hypoxia mediated by the STAT3/HIF-1α/VEGF signaling pathway [124,130], can block the cell cycle in the G1 phase through inhibition of Cyclin B1 and Hiwi in mRNA and potently induces autophagy as a survival mechanism in response to permeability transition pore opening and mitochondrial damage [14,133]. Recently, a new cell death pathway was attributed to betulinic acid in which cell death is induced through the inhibition of the stearoyl-CoA-desaturase (SCD-1), an enzyme that is overexpressed in tumor cells [135]. Proteasome inhibition assays suggest the proteasome is the main target for betulinic acid [136]. However, the regulatory effects of betulinic acid on the NF-κB pathway and on Bax or Bak expression are not well clarified [130].

Betulinic acid seems to be a very effective chemosensitizer for anticancer drug treatment in chemoresistant cell lines once it promotes the inhibition of multidrug resistance proteins in vivo and in vitro, as for example in combination with 5-fluorouracil (5-FU) and oxaliplatin [133,137]. These results clearly demonstrate that in some cases, it is possible to circumvent acquired chemoresistance by combination therapy of anticancer drugs with chemosensitizers as betulinic acid. Moreover, betulinic acid has strong synergy with mithramycin A on the inhibition of migration and invasion of pancreatic cancer cells at nontoxic concentrations by suppressing the Sp1 and uPAR level [138]. Furthermore, a synergistic effect of betulinic acid and tumor necrosis factor-related apoptosis-inducing ligand (TRAIL) combination to inhibit liver cancer progression in vitro and in vivo through targeting the p53 signaling pathway [139] revealed that betulinic acid combined with TRAIL has potential value against liver cancer.

Betulinic acid is slightly soluble in water, and therefore, its water solubility is a drawback that should be overcome to improve its absorption and bioavailability. The main targets for structure activity studies were the C-3 hydroxyl, C-20 vinyl and C-28 carboxyl groups (Figure 7). The 3-OH oxidation increased cytotoxic activity, but decreased selectivity; introduction of groups, such as amine or hydroxyl, at the C-28 position increased activity; while modifications at C-20 did not enhance cytotoxicity [14,124,140]. It can be conclude that modifications may improve the cytotoxicity and/or the water solubility, but not the selectivity. It seems that the presence of the free hydroxyl group at C-3 and the carboxyl group at C-28 are the most important features.

Recently, a clearer and more realistic method, 3D-QSAR by CoMFA and CoMSIA, shows the structure-cytotoxicity relationship of betulinic acid derivatives against human ovarian cancer cell A2780, and the main conclusions were: an electropositive group at the C-2 α-site; an electronegative and hydrogen bond acceptor group at the C-2 β-site; bulky groups at the C-3 β-site; bulky and electronegative groups at the C-3 α-site; bulky, electronegative and hydrogen bond donor or acceptor groups at the C-28 side chain; and would be beneficial to the antitumor potency (Figure 7) [130].

Due to its extraordinary potential as an antitumor agent, betulinic acid was involved in phase I/II clinical trials to evaluate its safety and effectiveness. The study involved topical applications (20% betulinic acid in ointment) to treat dysplastic nevi that can transform into melanoma. Unfortunately, at the end of 2013, the study was suspended due to funding issues (Clinical Trials database).

3. Conclusions

Cancer is becoming a high profile disease in developed and developing countries, and its treatment is a struggle with some successful cases. Nevertheless, the drugs developed by synthesis and used in chemotherapy have limitations mainly due to their toxic effects on non-targeted tissues and consequently furthering human health problems. Therefore, there is a demand for alternative treatments, and the naturally-derived anticancer agents are regarded as the best choice. As demonstrated herein, with some representative examples, secondary metabolites are themselves suitable anticancer agents leading to the development of new clinical drugs with also new anticancer mechanisms of action. Some have already become cases of success for the pharmaceutical industry. Additionally, they are excellent lead compounds, by which, through structural modifications, alternative formulations and/or using increasingly effective delivery systems, their pharmacological potential is enhanced. Recent new biotechnological solutions, using nanotech approaches, present a new hope for cancer therapy (e.g., plant drug-functionalized nanodiamonds and other nanocarriers based on anticancer drugs). Simultaneously, they provide a further step forward in the successful use of secondary metabolites for cancer therapeutic purposes [141–144]. In other cases, the success story has not yet reached its high point with its introduction in the market, but the more recent studies presented and discussed in this paper clearly show that this goal is getting closer. On the other hand, demand for plant-derived drugs is putting pressure on high-value medicinal plants and risking their biodiversity, so exploitation of these agents needs to be managed to keep up with demands and be sustainable. Fortunately, there are currently developments using new biotechnological solutions and sustainable alternative methods for the production of high-value plant metabolites.

Acknowledgments: This study was financed by Portuguese National Funds, through FCT (Fundação para a Ciência e a Tecnologia), and as applicable co-financed by the FEDER within the PT2020 Partnership Agreement by funding the Organic Chemistry Research Unit (QOPNA) (UID/QUI/00062/2013) and the cE3c Centre (UID/BIA/00329/2013).

Author Contributions: Ana M. L. Seca and Diana C. G. A. Pinto conceived of and wrote the paper.

Conflicts of Interest: The authors declare no conflict of interest.

Abbreviations

5-FU	5-fluorouracil
A2780	human ovarian carcinoma cell line
ABCB1	ATP binding cassette subfamily B member 1
ABCC10	ATP binding cassette subfamily C member 10
Akt	serine/threonine-specific protein kinase
b.w.	body weight
Bak	pro-apoptotic Bcl-2 protein
Bax	bcl-2-like protein 4
Bcl-2	B-cell lymphoma 2 protein
BCR-ABL1	breakpoint cluster region protein-Abelson murine leukemia viral oncogene homolog 1
CDKI	cyclin-dependent kinase inhibitors
colon 205	human Caucasian colon adenocarcinoma cell line
CoMFA	comparative molecular field analysis
CoMSIA	comparative molecular similarity index analysis
CYP	cytochrome P450
DNA	deoxyribonucleic acid
EMA	European Medicines Agency
ERK	extracellular signal-regulated kinases
FAO	Food And Agriculture Organization
FDA	Food And Drug Administration
FGF	fibroblast growth factor
FLT3-ITD	fms-related tyrosine kinase 3 internal tandem duplication
HIF-1α	hypoxia-inducible factor 1-alpha
HMGB1	high mobility group box 1 protein
IC$_{50}$	half maximal inhibitory concentration
MAPK	mitogen-activated protein kinase
Mcl-1	induced myeloid leukemia cell differentiation protein
MDR	multidrug resistance
miRNA	micro-ribonucleic acid
MMP	matrix metalloproteinase
mRNA	messenger ribonucleic acid
Myc	proto-oncogene
nab-paclitaxel	nanoparticle albumin-bound paclitaxel
NF-κB	nuclear factor kappa B cells
Nrf2	nuclear factor (erythroid-derived 2)-like 2
P-388	bipotential murine pre-B cell lymphoma
PEP005	ingenol mebutate
PGDF	platelet-derived growth factor
P-gp	p-glycoprotein
PI3K	phosphatidylinositol-4,5-bisphosphate 3-kinase
PICN	paclitaxel injection concentrate for nanodispersion
PKC	protein kinase C
PKCα	protein kinase C-α
PKCδ	protein kinase C-δ
QSAR	quantitative structure activity relationship

RAID	rapid access to intervention in development
SAR	structure activity relationship
SCD-1	stearoyl-CoA- desaturase 1
SCLC	small-cell lung cancer
SIRT1	NAD-dependent protein deacetylase sirtuin-1
SM/Chol	sphingomyelin/cholesterol
smad3	mothers against decapentaplegic homolog 3
Sp1	specificity protein 1
STAT3	signal transducer and activator of transcription 3
T315I	mutation resulting in an amino acid substitution at position 315 in BCR-ABL1, from a threonine (T) to an isoleucine (I).
TGF-β	transforming growth factor beta
TRAIL	tumor necrosis factor-related apoptosis-inducing ligand
tRNA	transfer ribonucleic acid
uPAR	urokinase receptor
VEGF	vascular endothelial growth factor
WHO	World Health Organization

References

1. Ferlay, J.; Soerjomataram, I.; Dikshit, R.; Eser, S.; Mathers, C.; Rebelo, M.; Parkin, D.M.; Forman, D.; Bray, F. Cancer incidence and mortality worldwide: Sources, methods and major patterns in GLOBOCAN 2012. *Int. J. Cancer* **2015**, *136*, 359–386. [CrossRef] [PubMed]
2. Siegel, R.L.; Miller, K.D.; Jemal, A. Cancer statistics, 2016. *CA Cancer J. Clin.* **2016**, *66*, 7–30. [CrossRef] [PubMed]
3. World Health Organization. *Cancer: Fact Sheets*; WHO: Geneva, Switzerland, 2017. Available online: http://www.who.int/mediacentre/factsheets/fs297/en/ (accessed on 2 October 2017).
4. Bray, F. Transitions in human development and the global cancer burden. In *World Cancer Report 2014*; Stewart, B.W., Wild, C.P., Eds.; International Agency for Research on Cancer: Lyon, France, 2014; pp. 54–68. ISBN 978-92-832-0443-5.
5. Adeloye, D.; David, R.A.; Aderemi, A.V.; Iseolorunkanmi, A.; Oyedokun, A.; Iweala, E.E.; Omoregbe, N.; Ayo, C.K. An estimate of the incidence of prostate cancer in Africa: A systematic review and meta-analysis. *PLoS ONE* **2016**, *11*, e0153496. [CrossRef] [PubMed]
6. Alves-Silva, J.M.; Romane, A.; Efferth, T.; Salgueiro, L. North African medicinal plants traditionally used in cancer therapy. *Front. Pharmacol.* **2017**, *8*, 1–24. [CrossRef] [PubMed]
7. Tariq, A.; Sadia, S.; Pan, K.; Ullah, I.; Mussarat, S.; Sun, F.; Abiodun, O.O.; Batbaatar, A.; Li, Z.; Song, D.; et al. A systematic review on ethnomedicines of anticancer plants. *Phytother. Res.* **2017**, *31*, 202–264. [CrossRef] [PubMed]
8. Oyebode, O.; Kandala, N.-B.; Chilton, P.J.; Lilford, R.J. Use of traditional medicine in middle-income countries: A WHO-SAGE study. *Health Policy Plan.* **2016**, *31*, 984–991. [CrossRef] [PubMed]
9. Diorio, C.; Salena, K.; Ladas, E.J.; Lam, C.G.; Afungcwhi, G.M.; Njuguna, F.; Marjerrison, S. Traditional and complementary medicine used with curative intent in childhood cancer: A systematic review. *Pediatr. Blood Cancer* **2017**, *64*, 1–8. [CrossRef] [PubMed]
10. Ma, L.; Wang, B.; Long, Y.; Li, H. Effect of traditional Chinese medicine combined with Western therapy on primary hepatic carcinoma: A systematic review with meta-analysis. *Front. Med.* **2017**, *11*, 191–202. [CrossRef] [PubMed]
11. Yan, Z.; Lai, Z.; Lin, J. Anticancer properties of traditional Chinese medicine. *Comb. Chem. High Throughput Screen.* **2017**, *20*, 423–429. [CrossRef] [PubMed]
12. Enioutina, E.Y.; Salis, E.R.; Job, K.M.; Gubarev, M.I.; Krepkova, L.V.; Sherwin, C.M. Herbal Medicines: Challenges in the modern world. Part 5. Status and current directions of complementary and alternative herbal medicine worldwide. *Expert Rev. Clin. Pharmacol.* **2017**, *10*, 327–338. [CrossRef] [PubMed]
13. Mishra, B.B.; Tiwari, V.K. Natural products: An evolving role in future drug discovery. *Eur. J. Med. Chem.* **2011**, *46*, 4769–4807. [CrossRef] [PubMed]

14. Gali-Muhtasib, H.; Hmadi, R.; Kareh, M.; Tohme, R.; Darwiche, N. Cell death mechanisms of plant-derived anticancer drugs: Beyond apoptosis. *Apoptosis* **2015**, *20*, 1531–1562. [CrossRef] [PubMed]
15. Khalid, E.B.; Ayman, E.E.; Rahman, H.; Abdelkarim, G.; Najda, A. Natural products against cancer angiogenesis. *Tumor Biol.* **2016**, *37*, 14513–14536. [CrossRef] [PubMed]
16. Katz, L.; Baltz, R.H. Natural product discovery: Past, present and future. *J. Ind. Microbiol. Biotechnol.* **2016**, *43*, 155–176. [CrossRef] [PubMed]
17. Kotoku, N.; Arai, M.; Kobayashi, M. Search for anti-angiogenic substances from natural sources. *Chem. Pharm. Bull.* **2016**, *64*, 128–134. [CrossRef] [PubMed]
18. Bernardini, S.; Tiezzi, A.; Laghezza Masci, V.; Ovidi, E. Natural products for human health: An historical overview of the drug discovery approaches. *Nat. Prod. Res.* **2017**. [CrossRef] [PubMed]
19. Newman, D.J.; Cragg, G.M. Natural products as sources of new drugs from 1981 to 2014. *J. Nat. Prod.* **2016**, *79*, 629–661. [CrossRef] [PubMed]
20. Ekor, M. The growing use of herbal medicines: Issues relating to adverse reactions and challenges in monitoring safety. *Front. Pharmacol.* **2014**, *4*, 1–10. [CrossRef] [PubMed]
21. Moreira, D.L.; Teixeira, S.S.; Monteiro, M.H.D.; de-Oliveira, A.C.A.X.; Paumgartten, F.J.R. Traditional use and safety of herbal medicines. *Rev. Bras. Farmacogn.* **2014**, *24*, 248–257. [CrossRef]
22. Nwodo, J.N.; Ibezim, A.; Simoben, C.V.; Ntie-Kang, F. Exploring cancer therapeutics with natural products from African medicinal plants, Part II: Alkaloids, terpenoids and flavonoids. *Anticancer Agents Med. Chem.* **2016**, *16*, 108–127. [CrossRef] [PubMed]
23. Habli, Z.; Toumieh, G.; Fatfat, M.; Rahal, O.N.; Gali-Muhtasib, H. Emerging cytotoxic alkaloids in the battle against cancer: Overview of molecular mechanisms. *Molecules* **2017**, *22*, 250. [CrossRef] [PubMed]
24. Delgoda, R.; Murray, J.E. Evolutionary perspectives on the role of plant secondary metabolites. In *Pharmacognosy: Fundamentals, Applications and Strategies*, 1st ed.; Badal, S., Delgoda, R., Eds.; Academic Press: Oxford, UK, 2017; pp. 93–100. ISBN 9780128020999.
25. Kabera, J.N.; Semana, E.; Mussa, A.R.; He, X. Plant secondary metabolites: Biosynthesis, classification, function and pharmacological properties. *J. Pharm. Pharmacol.* **2014**, *2*, 377–392.
26. Guo, Z. The modification of natural products for medical use. *Acta Pharm. Sin. B* **2017**, *7*, 119–136. [CrossRef] [PubMed]
27. Yao, H.; Liu, J.; Xu, S.; Zhu, Z.; Xu, J. The structural modification of natural products for novel drug discovery. *Expert Opin. Drug Discov.* **2017**, *12*, 121–140. [CrossRef] [PubMed]
28. Evans, A.E.; Farber, S.; Brunet, S.; Mariano, P.J. Vincristine in the treatment of acute leukaemia in children. *Cancer* **1963**, *16*, 1302–1306. [CrossRef]
29. Moore, A.; Pinkerton, R. Vincristine: Can its therapeutic index be enhanced? *Pediatr. Blood Cancer* **2009**, *53*, 1180–1187. [CrossRef] [PubMed]
30. Almagro, L.; Fernández-Pérez, F.; Pedreño, M.A. Indole alkaloids from *Catharanthus roseus*: Bioproduction and their effect on human health. *Molecules* **2015**, *20*, 2973–3000. [CrossRef] [PubMed]
31. Tang, K.; Pan, Q. Strategies for enhancing alkaloids yield in *Catharanthus roseus* via metabolic engineering approaches. In *Catharanthus Roseus: Current Research and Future Prospects*; Naeem, M., Aftab, T., Khan, M., Eds.; Springer International Publishing: Basel, Switzerland, 2017; pp. 1–16.
32. Mu, F.; Yang, L.; Wang, W.; Luo, M.; Fu, Y.; Guo, X.; Zu, Y. Negative-pressure cavitation extraction of four main vinca alkaloids from *Catharanthus roseus* leaves. *Molecules* **2012**, *17*, 8742–8752. [CrossRef] [PubMed]
33. Wang, X.; Song, Y.; Su, Y.; Tian, Q.; Li, B.; Quan, J.; Deng, Y. Are PEGylated liposomes better than conventional liposomes? A special case for vincristine. *Drug Deliv.* **2016**, *23*, 1092–1100. [CrossRef] [PubMed]
34. Sertel, S.; Fu, Y.; Zu, Y.; Rebacz, B.; Konkimalla, B.; Plinkert, P.K.; Krämer, A.; Gertsch, J.; Efferth, T. Molecular docking and pharmacogenomics of Vinca alkaloid and their monomeric precursor, vindoline and catharanthine. *Biochem. Pharmacol.* **2011**, *81*, 723–735. [CrossRef] [PubMed]
35. Douer, D. Efficacy and safety of vincristine sulfate liposome injection in the treatment of adult acute lymphocytic leukaemia. *Oncologist* **2016**, *21*, 840–847. [CrossRef] [PubMed]
36. Velde, M.E.; Kaspers, G.L.; Abbink, F.C.H.; Wilhelm, A.J.; Ket, J.C.F.; Berg, M.H. Vincristine-induced peripheral neuropathy in children with cancer: A systematic review. *Crit. Rev. Oncol. Hematol.* **2017**, *114*, 114–130. [CrossRef] [PubMed]

37. Yoshihara, H.; Yoshimoto, Y.; Hosoya, Y.; Hasegawa, D.; Kawano, T.; Sakoda, A.; Okita, H.; Manabe, A. Infantile fibrosarcoma treated with postoperative vincristine and dactinomycin. *Pediatr. Int.* **2017**, *59*, 371–374. [CrossRef] [PubMed]

38. Weaver, B.A. How Taxol/paclitaxel kills cancer cells. *Mol. Biol. Cell* **2014**, *25*, 2677–2681. [CrossRef] [PubMed]

39. Bernabeu, E.; Cagel, M.; Lagomarsino, E.; Moretton, M.; Chiappetta, D.A. Paclitaxel: What has been done and the challenges remain ahead. *Int. J. Pharm.* **2017**, *526*, 474–495. [CrossRef] [PubMed]

40. Bernabeu, E.; Gonzalez, L.; Cagel, M.; Gergic, E.P.; Moretton, M.A.; Chiappetta, D.A. Novel Soluplus1®-TPGS mixed micelles for encapsulation of paclitaxel with enhanced in vitro cytotoxicity on breast and ovarian cancer cell lines. *Colloids Surf. B Biointerfaces* **2016**, *140*, 403–411. [CrossRef] [PubMed]

41. Howat, S.; Park, B.; Oh, I.S.; Jin, Y.W.; Lee, E.K.; Loake, G.J. Paclitaxel: Biosynthesis, production and future prospects. *New Biotechnol.* **2014**, *31*, 242–245. [CrossRef] [PubMed]

42. Nicolaou, K.C.; Yang, Z.; Liu, J.J.; Ueno, H.; Nantermet, P.G.; Guy, R.K.; Claiborne, C.F.; Renaud, J.; Couladouros, E.A.; Paulvannan, K.; et al. Total synthesis of taxol. *Nature* **1994**, *367*, 630–634. [CrossRef] [PubMed]

43. Fukaya, K.; Kodama, K.; Tanaka, Y.; Yamazaki, H.; Sugai, T.; Yamaguchi, Y.; Watanabe, A.; Oishi, T.; Sato, T.; Chida, N. Synthesis of Paclitaxel. 2. Construction of the ABCD ring and formal synthesis. *Org. Lett.* **2015**, *17*, 2574–2577. [CrossRef] [PubMed]

44. Hirai, S.; Utsugi, M.; Iwamoto, M.; Nakada, M. Formal total synthesis of (−)-taxol through Pd-catalyzed eight-membered carbocyclic ring formation. *Chemistry* **2015**, *21*, 355–359. [CrossRef] [PubMed]

45. Gallego, A.; Malik, S.; Yousefzadi, M.; Makhzoum, A.; Tremouillaux-Guiller, J.; Bonfill, M. Taxol from *Corylus avellana*: Paving the way for a new source of this anti-cancer drug. *Plant Cell Tissue Organ Cult.* **2017**, *129*, 1–16. [CrossRef]

46. Li, B.-J.; Wang, H.; Gong, T.; Chen, J.-J.; Chen, T.-J.; Yang, J.-L.; Zhu, P. Improving 10-deacetylbaccatin III-10-β-*O*-acetyltransferase catalytic fitness for Taxol production. *Nat. Commun.* **2017**, *8*, 1–13. [CrossRef] [PubMed]

47. Baloglu, E.; Kingston, D.G.I. A new semisynthesis of paclitaxel from baccatin III. *J. Nat. Prod.* **1999**, *62*, 1068–1071. [CrossRef] [PubMed]

48. Mandai, T.; Kuroda, A.; Okumoto, H.; Nakanishi, K.; Mikuni, K.; Hara, K.J.; Hara, K.Z. A semisynthesis of paclitaxel via a 10-deacetylbaccatin III derivative bearing a β-keto ester appendage. *Tetrahedron Lett.* **2000**, *41*, 243–246. [CrossRef]

49. Liu, W.C.; Gonga, T.; Zhu, P. Advances in exploring alternative Taxol sources. *RSC Adv.* **2016**, *6*, 48800–48809. [CrossRef]

50. Xiao, Z.; Morris-Natschke, S.L.; Lee, K.H. Strategies for the optimization of natural leads to anticancer drugs or drug candidates. *Med. Res. Rev.* **2016**, *36*, 32–91. [CrossRef] [PubMed]

51. Wang, N.N.; Zhao, L.J.; Wu, L.N.; He, M.F.; Qu, J.W.; Zhao, Y.B.; Zhao, W.Z.; Li, J.S.; Wang, J.H. Mechanistic analysis of taxol-induced multidrug resistance in an ovarian cancer cell line. *Asian Pac. J. Cancer Prev.* **2013**, *14*, 4983–4988. [CrossRef] [PubMed]

52. Barbuti, A.M.; Chen, Z.S. Paclitaxel through the ages of anticancer therapy: Exploring its role in chemoresistance and radiation therapy. *Cancers* **2015**, *7*, 2360–2371. [CrossRef] [PubMed]

53. Nehate, C.; Jain, S.; Saneja, A.; Khare, V.; Alam, N.; Dubey, R.; Gupta, P.N. Paclitaxel formulations: Challenges and novel delivery options. *Curr. Drug Deliv.* **2014**, *11*, 666–686. [CrossRef] [PubMed]

54. Soliman, H.H. *nab*-Paclitaxel as a potential partner with checkpoint inhibitors in solid tumors. *Onco Targets Ther.* **2017**, *10*, 101–112. [CrossRef] [PubMed]

55. Zong, Y.; Wu, J.; Shen, K. Nanoparticle albumin-bound paclitaxel as neoadjuvant chemotherapy of breast cancer: A systematic review and meta-analysis. *Oncotarget* **2017**, *8*, 17360–17372. [CrossRef] [PubMed]

56. Xu, X.; Wang, L.; Xu, H.Q.; Huang, X.E.; Qian, Y.D.; Xiang, J. Clinical comparison between paclitaxel liposome (Lipusu®) and paclitaxel for treatment of patients with metastatic gastric cancer. *Asian Pac. J. Cancer Prev.* **2013**, *14*, 2591–2594. [CrossRef] [PubMed]

57. Rivera, E.; Cianfrocca, M. Overview of neuropathy associated with taxanes for the treatment of metastatic breast cancer. *Cancer Chemother. Pharmacol.* **2015**, *75*, 659–670. [CrossRef] [PubMed]

58. Wen, G.; Qu, X.X.; Wang, D.; Chen, X.X.; Tian, X.C.; Gao, F.; Zhou, X.L. Recent advances in design, synthesis and bioactivity of paclitaxel-mimics. *Fitoterapia* **2016**, *110*, 26–37. [CrossRef] [PubMed]

59. Crown, J.; O'Leary, M.; Ooi, W.S. Docetaxel and paclitaxel in the treatment of breast cancer: A review of clinical experience. *Oncologist* **2004**, *9*, 24–32. [CrossRef] [PubMed]

60. Vrignaud, P.; Semiond, D.; Benning, V.; Beys, E.; Bouchard, H.; Gupta, S. Preclinical profile of cabazitaxel. *Drug Des. Dev. Ther.* **2014**, *8*, 1851–1867. [CrossRef] [PubMed]

61. De Morree, E.; van Soest, R.; Aghai, A.; de Ridder, C.; de Bruijn, P.; Ghobadi Moghaddam-Helmantel, I.; Burger, H.; Mathijssen, R.; Wiemer, E.; de Wit, R.; et al. Understanding taxanes in prostate cancer; importance of intratumoral drug accumulation. *Prostate* **2016**, *76*, 927–936. [CrossRef] [PubMed]

62. Goyal, S.; Oak, E.; Luo, J.; Cashen, A.F.; Carson, K.; Fehniger, T.; DiPersio, J.; Bartlett, N.L.; Wagner-Johnston, N.D. Minimal activity of nanoparticle albumin-bound (nab) paclitaxel in relapsed or refractory lymphomas: Results of a phase-I study. *Leuk. Lymphoma* **2018**, *59*, 357–362. [CrossRef] [PubMed]

63. Ricci, F.; Guffanti, F.; Damia, G.; Broggini, M. Combination of paclitaxel, bevacizumab and MEK162 in second line treatment in platinum-relapsing patient derived ovarian cancer xenografts. *Mol. Cancer* **2017**, *16*, 1–13. [CrossRef] [PubMed]

64. Gill, K.K.; Kamal, M.M.; Kaddoumi, A.; Nazzal, S. EGFR targeted delivery of paclitaxel and parthenolide co-loaded in PEG-Phospholipid micelles enhance cytotoxicity and cellular uptake in non-small cell lung cancer cells. *J. Drug Deliv. Sci. Technol.* **2016**, *36*, 150–155. [CrossRef]

65. Ehrlich, A.; Booher, S.; Becerra, Y.; Borris, D.L.; Figg, W.D.; Turner, M.L.; Blauvelt, A. Micellar paclitaxel improves severe psoriasis in a prospective phase II pilot study. *J. Am. Acad. Dermatol.* **2004**, *50*, 533–540. [CrossRef] [PubMed]

66. Dadgar, S.; Ramjan, Z.; Floriano, W.B. Paclitaxel is an inhibitor and its boron dipyrromethene derivative is a fluorescent recognition agent for botulinum neurotoxin subtype A. *J. Med. Chem.* **2013**, *56*, 2791–2803. [CrossRef] [PubMed]

67. Kantarjian, H.M.; O'Brien, S.; Cortes, J. Homoharringtonine/omacetaxine mepesuccinate: The long and winding road to food and drug administration approval. *Clin. Lymphoma Myeloma Leuk.* **2013**, *13*, 530–533. [CrossRef] [PubMed]

68. Butler, M.S.; Robertson, A.A.B.; Cooper, M.A. Natural product and natural product derived drugs in clinical trials. *Nat. Prod. Rep.* **2014**, *31*, 1612–1661. [CrossRef] [PubMed]

69. Chang, Y.; Meng, F.-C.; Wang, R.; Wang, C.M.; Lu, X.-Y.; Zhang, Q.-W. Chemistry, bioactivity, and the structure-activity relationship of cephalotaxine-type alkaloids from *Cephalotaxus* sp. In *Studies in Natural Products Chemistry Bioactive Natural Products*; Atta-Ur-Rahman, F.R.S., Ed.; Elsevier Science Publishers: Amsterdam, The Netherlands, 2017; Volume 53, pp. 339–373. ISBN 978-0-444-63930-1.

70. Lü, S.; Wang, J. Homoharringtonine and omacetaxine for myeloid hematological malignancies. *J. Hematol. Oncol.* **2014**, *7*, 1–10. [CrossRef] [PubMed]

71. Gandhi, V.; Plunkett, W.; Cortes, J.E. Omacetaxine: A protein translation inhibitor for treatment of chronic myelogenous leucemia. *Clin. Cancer Res.* **2014**, *20*, 1735–1740. [CrossRef] [PubMed]

72. Li, X.; Yin, X.; Wang, H.; Huang, J.; Yu, M.; Ma, Z.; Li, C.; Zhou, Y.; Yan, X.; Huang, S.; et al. The combination effect of homoharringtonine and ibrutinib on FLT3-ITD mutant acute myeloid leukaemia. *Oncotarget* **2017**, *8*, 12764–12774. [CrossRef] [PubMed]

73. Chen, J.; Mu, Q.; Li, X.; Yin, X.; Yu, M.; Jin, J.; Li, C.; Zhou, Y.; Zhou, J.; Shanshan Suo, S.; et al. Homoharringtonine targets Smad3 and TGF-β pathway to inhibit the proliferation of acute myeloid leukaemia cells. *Oncotarget* **2017**, *8*, 40318–40326. [CrossRef] [PubMed]

74. Pasic, I.; Lipton, J.H. Current approach to the treatment of chronic myeloid leukaemia. *Leuk. Res.* **2017**, *55*, 65–78. [CrossRef] [PubMed]

75. Chung, C. Omacetaxine for treatment-resistant or treatment-intolerant adult chronic myeloid leukaemia. *Am. J. Health Syst. Pharm.* **2014**, *71*, 279–288. [CrossRef] [PubMed]

76. Cortes, J.E.; Kantarjian, H.M.; Rea, D.; Wetzler, M.; Lipton, J.H.; Akard, L.; Khoury, H.J.; Michallet, M.; Guerci-Bresler, A.; Chuah, C.; et al. Final analysis of the efficacy and safety of omacetaxine mepesuccinate in patients with chronic- or accelerated-phase chronic myeloid leukaemia: Results with 24 months of follow-up. *Cancer* **2015**, *121*, 1637–1644. [CrossRef] [PubMed]

77. Damlaj, M.; Lipton, J.H.; Assouline, S.E. A safety evaluation of omacetaxine mepesuccinate for the treatment of chronic myeloid leukaemia. *Expert Opin. Drug Saf.* **2016**, *15*, 1279–1286. [CrossRef] [PubMed]

78. Allan, E.K.; Holyoake, T.L.; Craig, A.R.; Jørgensen, H.G. Omacetaxine may have a role in chronic myeloid leukaemia eradication through downregulation of Mcl-1 and induction of apoptosis in stem/progenitor cells. *Leukaemia* **2011**, *25*, 985–994. [CrossRef] [PubMed]

79. Lam, S.S.; Ho, E.S.; He, B.L.; Wong, W.W.; Cher, C.Y.; Ng, N.K.; Man, C.H.; Gill, H.; Cheung, A.M.; Ip, H.W.; et al. Homoharringtonine (omacetaxine mepesuccinate) as an adjunct for FLT3-ITD acute myeloid leukaemia. *Sci. Transl. Med.* **2016**, *8*, 1–14. [CrossRef] [PubMed]

80. Heiblig, M.; Sobh, M.; Nicolini, F.E. Subcutaneous omacetaxine mepesuccinate in patients with chronic myeloid leukaemia in tyrosine kinase inhibitor-resistant patients: Review and perspectives. *Leuk. Res.* **2014**, *38*, 1145–1153. [CrossRef] [PubMed]

81. Shen, A.Q.; Munteanu, M.; Khoury, H.J. Updated product label allows home administration of omacetaxine mepesuccinate. *Oncologist* **2014**, *19*, 1. [CrossRef] [PubMed]

82. Akard, L.; Kantarjian, H.M.; Nicolini, F.E.; Wetzler, M.; Lipton, J.H.; Baccarani, M.; Jean Khoury, H.; Kurtin, S.; Li, E.; Munteanu, M.; et al. Incidence and management of myelosuppression in patients with chronic- and accelerated-phase chronic myeloid leukaemia treated with omacetaxine mepesuccinate. *Leuk. Lymphoma* **2016**, *57*, 654–665. [CrossRef] [PubMed]

83. Saklani, A.; Kutty, S.K. Plant-derived compounds in clinical trials. *Drug Discov. Today* **2008**, *13*, 161–171. [CrossRef] [PubMed]

84. Rizk, A.M.; Hammouda, F.M.; El-Missiry, M.M.; Radwan, H.M.; Evans, F.J. Biologically active diterpene esters from *Euphorbia peplus*. *Phytochemistry* **1985**, *24*, 1605–1606. [CrossRef]

85. Ramsay, J.R.; Suhrbier, A.; Aylward, J.H.; Ogbourne, S.; Cozzi, S.J.; Poulsen, M.G.; Baumann, K.C.; Welburn, P.; Redlich, G.L.; Parsons, P.G. The sap from *Euphorbia peplus* is effective against human nonmelanoma skin cancers. *Br. J. Dermatol.* **2011**, *164*, 633–636. [CrossRef] [PubMed]

86. Ogbourne, S.M.; Parsons, P.G. The value of nature's natural product library for the discovery of new chemical entities: The discovery of ingenol mebutate. *Fitoterapia* **2014**, *98*, 36–44. [CrossRef] [PubMed]

87. Béres, T.; Dragull, K.; Pospíšil, J.; Tarkowská, D.; Dančák, M.; Bíba, O.; Tarkowski, P.; Doležal, K.; Strnad, M. Quantitative analysis of ingenol in *Euphorbia* species via validated isotope dilution ultra-high performance liquid chromatography tandem mass spectrometry. *Phytochem. Anal.* **2018**, *29*, 23–29. [CrossRef] [PubMed]

88. Liang, X.; Grue-Sørensen, G.; Petersen, A.K.; Högberg, T. Semisynthesis of ingenol 3-angelate (PEP005): Efficient stereoconservative angeloylation of alcohols. *Synlett* **2012**, *23*, 2647–2652. [CrossRef]

89. Liang, X.; Grue-Sørensen, G.; Månsson, K.; Vedsø, P.; Soor, A.; Stahlhut, M.; Bertelsen, M.; Engell, K.M.; Högberg, T. Syntheses, biological evaluation and SAR of ingenol mebutate analogues for treatment of actinic keratosis and non-melanoma skin cancer. *Bioorg. Med. Chem. Lett.* **2013**, *23*, 5624–5629. [CrossRef] [PubMed]

90. Serova, M.; Ghoul, A.; Benhadji, K.A.; Faivre, S.; Le Tourneau, C.; Cvitkovic, E.; Lokiec, F.; Lord, J.; Ogbourne, S.M.; Calvo, F.; et al. Effects of protein kinase C modulation by PEP005, a novel ingenol angelate, on mitogen-activated protein kinase and phosphatidylinositol 3-kinase signaling in cancer cells. *Mol. Cancer Ther.* **2008**, *7*, 915–922. [CrossRef] [PubMed]

91. Benhadji, K.A.; Serova, M.; Ghoul, A.; Cvitkovic, E.; Le Tourneau, C.; Ogbourne, S.M.; Lokiec, F.; Calvo, F.; Hammel, P.; Faivre, S.; et al. Antiproliferative activity of PEP005, a novel ingenol angelate that modulates PKC functions, alone and in combination with cytotoxic agents in human colon cancer cells. *Br. J. Cancer* **2008**, *99*, 1808–1815. [CrossRef] [PubMed]

92. Collier, N.J.; Ali, F.R.; Lear, J.T. Ingenol mebutate: A novel treatment for actinic keratosis. *Clin. Pract.* **2014**, *11*, 295–306. [CrossRef]

93. Rosen, R.H.; Gupta, A.K.; Tyring, S.K. Dual mechanism of action of ingenol mebutate gel for topical treatment of actinic keratoses: Rapid lesion necrosis followed by lesion-specific immune response. *J. Am. Acad. Dermatol.* **2012**, *66*, 486–493. [CrossRef] [PubMed]

94. Matias, D.; Bessa, C.; Simões, M.F.; Reis, C.P.; Saraiva, L.; Rijo, P. Natural products as lead protein kinase c modulators for cancer therapy. In *Studies in Natural Products Chemistry Bioactive Natural Products*; Atta-Ur-Rahman, F.R.S., Ed.; Elsevier Science Publishers: Amsterdam, The Netherlands, 2017; Volume 53, pp. 45–79. ISBN 978-0-444-63930-1.

95. Kedei, N.; Lundberg, D.J.; Toth, A.; Welburn, P.; Garfield, S.H.; Blumberg, P.M. Characterization of the interaction of ingenol 3-angelate with protein kinase C. *Cancer Res.* **2004**, *64*, 3243–3255. [CrossRef] [PubMed]

96. Doan, H.Q.; Gulati, N.; Levis, W.R. Ingenol mebutate: Potential for further development of cancer immunotherapy. *J. Drugs Dermatol.* **2012**, *11*, 1156–1157. [PubMed]

97. Tzogani, K.; Nagercoil, N.; Hemmings, R.J.; Samir, B.; Gardette, J.; Demolis, P.; Salmonson, T.; Pignatti, F. The European Medicines Agency approval of ingenol mebutate (Picato) for the cutaneous treatment of non-hyperkeratotic, non-hypertrophic actinic keratosis in adults: Summary of the scientific assessment of the Committee for Medicinal Products for Human Use (CHMP). *Eur. J. Dermatol.* **2014**, *24*, 457–463. [CrossRef] [PubMed]

98. Kocaadam, B.; Şanlier, N. Curcumin, an active component of turmeric (*Curcuma longa*), and its effects on health. *Crit. Rev. Food Sci. Nutr.* **2017**, *57*, 2889–2895. [CrossRef] [PubMed]

99. Wang, Y.J.; Pan, M.H.; Cheng, A.L.; Lin, L.I.; Ho, Y.S.; Hsieh, C.Y.; Lin, J.K. Stability of curcumin in buffer solutions and characterization of its degradation products. *J. Pharm. Biomed. Anal.* **1997**, *15*, 1867–1876. [CrossRef]

100. Sinha, R.; Anderson, D.E.; McDonald, S.S.; Greenwald, P. Cancer risk and diet in India. *J. Postgrad. Med.* **2003**, *49*, 222–228. [PubMed]

101. Perrone, D.; Ardito, F.; Giannatempo, G.; Dioguardi, M.; Troiano, G.; Lo Russo, L.; De Lillo, A.; Laino, L.; Lo Muzio, L. Biological and therapeutic activities and anticancer properties of curcumin. *Exp. Ther. Med.* **2015**, *10*, 1615–1623. [CrossRef] [PubMed]

102. Pavan, A.R.; Silva, G.D.; Jornada, D.H.; Chiba, D.E.; Fernandes, G.F.; Man Chin, C.; Dos Santos, J.L. Unraveling the anticancer effect of curcumin and resveratrol. *Nutrients* **2016**, *8*, 628. [CrossRef] [PubMed]

103. Imran, M.; Saeed, F.; Nadeem, M.; Arshad, U.M.; Ullah, A.; Suleria, H.A. Cucrmin, anticancer and antitumor perspectives—A comprehensive review. *Crit. Rev. Food Sci. Nutr.* **2016**, *22*, 1–23. [CrossRef] [PubMed]

104. Redondo-Blanco, S.; Fernández, J.; Gutiérrez-del-Río, I.; Villar, C.J.; Lombó, F. New insights toward colorectal cancer chemotherapy using natural bioactive compounds. *Front. Pharmacol.* **2017**, *8*, 1–22. [CrossRef] [PubMed]

105. Borges, G.Á.; Rêgo, D.F.; Assad, D.X.; Coletta, R.D.; De Luca Canto, G.; Guerra, E.N. In vivo and in vitro effects of curcumin on head and neck carcinoma: A systematic review. *J. Oral Pathol. Med.* **2017**, *46*, 3–20. [CrossRef] [PubMed]

106. Sordillo, P.P.; Helson, L. Curcumin and cancer stem cells: Curcumin has asymmetrical effects on cancer and normal stem cells. *Anticancer Res.* **2015**, *35*, 599–614. [PubMed]

107. Yu, H.J.; Ma, L.; Jiang, J.; Sun, S.Q. Protective effect of curcumin on neural myelin sheaths by attenuating interactions between the endoplasmic reticulum and mitochondria after compressed spinal cord. *J. Spine* **2016**, *5*, 1–6. [CrossRef]

108. Di Martino, R.M.C.; Luppi, B.; Bisi, A.; Gobbi, S.; Rampa, A.; Abruzzo, A.; Bellutia, F. Recent progress on curcumin-based therapeutics: A patent review (2012–2016). Part I: Curcumin. *Expert Opin. Ther. Pat.* **2017**, *27*, 579–590. [CrossRef] [PubMed]

109. Kumar, G.; Mittal, S.; Sak, K.; Tuli, H.S. Molecular mechanisms underlying chemopreventive potential of curcumin: Current challenges and future perspectives. *Life Sci.* **2016**, *148*, 313–328. [CrossRef] [PubMed]

110. Kunnumakkara, A.B.; Bordoloi, D.; Harsha, C.; Banik, K.; Gupta, S.C.; Aggarwal, B.B. Curcumin mediates anticancer effects by modulating multiple cell signaling pathways. *Clin. Sci.* **2017**, *131*, 1781–1799. [CrossRef] [PubMed]

111. Schwertheim, S.; Wein, F.; Lennartz, K.; Worm, K.; Schmid, K.W.; Sheu-Grabellus, S.Y. Curcumin induces G2/M arrest, apoptosis, NF-κB inhibition, and expression of differentiation genes in thyroid carcinoma cells. *J. Cancer Res. Clin. Oncol.* **2017**, *143*, 1143–1154. [CrossRef] [PubMed]

112. Yang, C.; Ma, X.; Wang, Z.; Zeng, X.; Hu, Z.; Ye, Z.; Shen, G. Curcumin induces apoptosis and protective autophagy in castration-resistant prostate cancer cells through iron chelation. *Drug Des. Dev. Ther.* **2017**, *11*, 431–439. [CrossRef] [PubMed]

113. Dasiram, J.D.; Ganesan, R.; Kannan, J.; Kotteeswaran, V.; Sivalingam, N. Curcumin inhibits growth potential by G1 cell cycle arrest and induces apoptosis in p53-mutated COLO 320DM human colon adenocarcinoma cells. *Biomed. Pharmacother.* **2017**, *86*, 373–380. [CrossRef] [PubMed]

114. Uwagawa, T.; Yanaga, K. Effect of NF-κB inhibition on chemoresistance in biliary–pancreatic cancer. *Surg. Today* **2015**, *45*, 1481–1488. [CrossRef] [PubMed]

115. Fu, Z.; Chen, X.; Guan, S.; Yan, Y.; Lin, H.; Hua, Z.C. Curcumin inhibits angiogenesis and improves defective hematopoiesis induced by tumor-derived VEGF in tumor model through modulating VEGF-VEGFR2 signaling pathway. *Oncotarget* **2015**, *6*, 19469–19482. [CrossRef] [PubMed]

116. Haris, P.; Mary, V.; Aparna, P.; Dileep, K.V.; Sudarsanakumar, C. A comprehensive approach to ascertain the binding mode of curcumin with DNA. *Spectrochim. Acta A Mol. Biomol. Spectrosc.* **2017**, *175*, 155–163. [CrossRef] [PubMed]

117. Ramya, P.V.; Angapelly, S.; Guntuku, L.; Singh Digwal, C.; Nagendra Babu, B.; Naidu, V.G.; Kamal, A. Synthesis and biological evaluation of curcumin inspired indole analogues as tubulin polymerization inhibitors. *Eur. J. Med. Chem.* **2017**, *127*, 100–114. [CrossRef] [PubMed]

118. Gupta, S.C.; Patchva, S.; Aggarwal, B.B. Therapeutic roles of curcumin: Lessons learned from clinical trials. *AAPS J.* **2013**, *15*, 195–218. [CrossRef] [PubMed]

119. Klippstein, R.; Bansal, S.S.; Al-Jamal, K.T. Doxorubicin enhances curcumin's cytotoxicity in human prostate cancer cells in vitro by enhancing its cellular uptake. *Int. J. Pharm.* **2016**, *514*, 169–175. [CrossRef] [PubMed]

120. Pimentel-Gutiérrez, H.J.; Bobadilla-Morales, L.; Barba-Barba, C.C.; Ortega-De-La-Torre, C.; Sánchez-Zubieta, F.A.; Corona-Rivera, J.R.; González-Quezada, B.A.; Armendáriz-Borunda, J.S.; Silva-Cruz, R.; Corona-Rivera, A. Curcumin potentiates the effect of chemotherapy against acute lymphoblastic leukaemia cells via downregulation of NF-κB. *Oncol. Lett.* **2016**, *12*, 4117–4124. [CrossRef] [PubMed]

121. Rahimi, H.R.; Nedaeinia, R.; Shamloo, A.S.; Nikdoust, S.; Oskuee, R.K. Novel delivery system for natural products: Nano-curcumin formulations. *Avicenna J. Phytomed.* **2016**, *6*, 383–398. [PubMed]

122. Liu, W.; Zhai, Y.; Heng, X.; Che, F.Y.; Chen, W.; Sun, D.; Zhai, G. Oral bioavailability of curcumin: Problems and advancements. *J. Drug Target* **2016**, *24*, 694–702. [CrossRef] [PubMed]

123. Puneeth, H.R.; Ananda, H.; Kumar, K.S.S.; Rangappa, K.S.; Sharada, A.C. Synthesis and antiproliferative studies of curcumin pyrazole derivatives. *Med. Chem. Res.* **2016**, *25*, 1842–1851. [CrossRef]

124. Ali-Seyed, M.; Jantan, I.; Vijayaraghavan, K.; Bukhari, S.N. Betulinic acid: Recent advances in chemical modifications, effective delivery, and molecular mechanisms of a promising anticancer therapy. *Chem. Biol. Drug Des.* **2016**, *87*, 517–536. [CrossRef] [PubMed]

125. Pichette, A.; Liu, H.; Roy, C.; Tanguay, S.; Simard, F.; Lavoie, S. Selective oxidation of betulin for the preparation of betulinic acid, an antitumoral compound. *Synth. Commun.* **2004**, *34*, 3925–3937. [CrossRef]

126. Holonec, L.; Ranga, F.; Crainic, D.; Truţa, A.; Socaciu, C. Evaluation of betulin and betulinic acid content in birch bark from different forestry areas of western carpathians. *Not. Bot. Horti Agrobot.* **2012**, *40*, 99–105. [CrossRef]

127. Pisha, E.; Chai, H.; Lee, I.S.; Chagwedera, T.E.; Farnsworth, N.R.; Cordell, G.A.; Beecher, C.W.; Fong, H.H.; Kinghorn, A.D.; Brown, D.H. Discovery of betulinic acid a selective inhibitor of human-melanoma that functions by induction of apoptosis. *Nat. Med.* **1995**, *10*, 1046–1051. [CrossRef]

128. Singh, S.; Sharma, B.; Kanwar, S.S.; Kumar, A. Lead phytochemicals for anticancer drug development. *Front. Plant Sci.* **2016**, *7*, 1–13. [CrossRef] [PubMed]

129. Lee, S.Y.; Kim, H.H.; Park, S.U. Recent studies on betulinic acid and its biological and pharmacological activity. *EXCLI J.* **2015**, *14*, 199–203. [CrossRef] [PubMed]

130. Zhang, D.M.; Xu, H.G.; Wang, L.; Li, Y.J.; Sun, P.H.; Wu, X.M.; Wang, G.J.; Chen, W.M.; Ye, W.C. Betulinic acid and its derivatives as potential antitumor agents. *Med. Res. Rev.* **2015**, *35*, 1127–1155. [CrossRef] [PubMed]

131. Gheorgheosu, D.; Duicu, O.; Dehelean, C.; Soica, C.; Muntean, D. Betulinic acid as a potent and complex antitumor phytochemical: A minireview. *Anticancer Agents Med. Chem.* **2014**, *14*, 936–945. [CrossRef] [PubMed]

132. Ali, M.T.M.; Zahari, H.; Aliasak, A.; Lim, S.M.; Ramasamy, K.; Macabeoacabeo, A.P.G. Synthesis, characterization and cytotoxic activity of betulinic acid and sec-betulinic acid derivatives against human colorectal carcinoma. *Orient. J. Chem.* **2017**, *33*, 242–248. [CrossRef]

133. Luo, R.; Fang, D.; Chu, P.; Wu, H.; Zhang, Z.; Tang, Z. Multiple molecular targets in breast cancer therapy by betulinic acid. *Biomed. Pharmacother.* **2016**, *84*, 1321–1330. [CrossRef] [PubMed]

134. Shankar, E.; Zhang, A.; Franco, D.; Gupta, S. Betulinic acid-mediated apoptosis in human prostate cancer cells involves p53 and nuclear factor-kappa b (NF-κB) pathways. *Molecules* **2017**, *22*, 264. [CrossRef] [PubMed]

135. Potze, L.; Di Franco, S.; Grandela, C.; Pras-Raves, M.L.; Picavet, D.I.; van Veen, H.A.; van Lenthe, H.; Mullauer, F.B.; van der Wel, N.N.; Luyf, A.; et al. Betulinic acid induces a novel cell death pathway that depends on cardiolipin modification. *Oncogene* **2016**, *35*, 427–437. [CrossRef] [PubMed]

136. Waechter, F.; Silva, G.S.; Willig, J.; de Oliveira, C.B.; Vieira, B.D.; Trivella, D.B.; Zimmer, A.R.; Buffon, A.; Pilger, D.A.; Gnoatto, S. Design, synthesis, and biological evaluation of betulinic acid derivatives as new antitumor agents for leukaemia. *Anticancer Agents Med. Chem.* **2017**. [CrossRef] [PubMed]

137. Jung, G.R.; Kim, K.J.; Choi, C.H.; Lee, T.B.; Han, S.I.; Han, H.K.; Lim, S.C. Effect of betulinic acid on anticancer drug-resistant colon cancer cells. *Basic Clin. Pharmacol. Toxicol.* **2007**, *101*, 277–285. [CrossRef] [PubMed]

138. Gao, Y.; Jia, Z.L.; Kong, X.Y.; Li, Q.; Chang, D.Z.; Wei, D.Y.; Le, X.D.; Huang, S.D.; Huang, S.Y.; Wang, L.W.; Xie, K.P. Combining betulinic acid and mithramycin a effectively suppresses pancreatic cancer by inhibiting proliferation, invasion, and angiogenesis. *Cancer Res.* **2011**, *71*, 5182–5193. [CrossRef] [PubMed]

139. Xu, Y.; Li, J.; Li, Q.J.; Feng, Y.L.; Pan, F. Betulinic acid promotes TRAIL function on liver cancer progression inhibition through p53/Caspase-3 signaling activation. *Biomed. Pharmacother.* **2017**, *88*, 349–358. [CrossRef] [PubMed]

140. Yogeeswari, P.; Sriram, D. Betulinic acid and its derivatives: A review on their biological properties. *Curr. Med. Chem.* **2005**, *12*, 657–666. [CrossRef] [PubMed]

141. Gismondi, A.; Reina, G.; Orlanducci, S.; Mizzoni, F.; Gay, S.; Terranova, M.L.; Canini, A. Nanodiamonds coupled with plant bioactive metabolites: A nanotech approach for cancer therapy. *Biomaterials* **2015**, *38*, 22–35. [CrossRef] [PubMed]

142. Gismondi, A.; Nanni, V.; Reina, G.; Orlanducci, S.; Terranova, M.L.; Canini, A. Nanodiamonds coupled with 5,7-dimethoxycoumarin, a plant bioactive metabolite, interfere with the mitotic process in B16F10 cells altering the actin organization. *Int. J. Nanomed.* **2016**, *11*, 557–574. [CrossRef] [PubMed]

143. Gupta, C.; Prakash, D.; Gupta, S. Cancer treatment with nano-diamonds. *Front. Biosci. (Sch. Ed.)* **2017**, *9*, 62–70. [CrossRef]

144. Raj, R.; Mongia, P.; Sahu, S.K.; Ram, A. Nanocarriers based anticancer drugs: Current scenario and future perceptions. *Curr. Drug Targets* **2016**, *17*, 206–228. [CrossRef] [PubMed]

International Journal of
Molecular Sciences

MDPI

Article

Comparison of the Hepatoprotective Effects of Four Endemic Cirsium Species Extracts from Taiwan on CCl$_4$-Induced Acute Liver Damage in C57BL/6 Mice

Zi-Wei Zhao [1], Jen-Chih Chang [2], Li-Wei Lin [3], Fan-Hsuan Tsai [3], Hung-Chi Chang [4],* and Chi-Rei Wu [1],*

[1] Department of Chinese Pharmaceutical Sciences and Chinese Medicine Resources, College of Pharmacy, China Medical University, Taichung 40402, Taiwan; wei111783@gmail.com
[2] Taichung Armed Forces General Hospital, Taichung 404, Taiwan; chang.jenchih@gmail.com
[3] The School of Chinese Medicines for Post-Baccalaureate, I-Shou University, Kaohsiung County 82445, Taiwan; lwlin@isu.edu.tw (L.-W.L.); asura0734@isu.edu.tw (F.-H.T.)
[4] Department of Golden-Ager Industry Management, College of Management, Chaoyang University of Technology, Taichung 41394, Taiwan
* Correspondence: changhungchi@cyut.edu.tw (H.-C.C.); crw@mail.cmu.edu.tw (C.-R.W.); Tel.: +886-4-2332-3000 (ext. 5212) (H.-C.C.); +886-4-2205-3366 (ext. 5506) (C.-R.W.)

Received: 6 April 2018; Accepted: 27 April 2018; Published: 30 April 2018

Abstract: Species of Cirsium (Asteraceae family) have been used in folk hepatoprotective medicine in Taiwan. We collected four Cirsium species—including the aerial part of *Cirsium arisanense* (CAH), the aerial part of *Cirsium kawakamii* (CKH), the flower part of *Cirsium japonicum* DC. var. *australe* (CJF), and Cirsii Herba (CH)—and then made extractions from them with 70% methanol. We compared the antioxidant contents and activities of these four Cirsium species extracts by a spectrophotometric method and high-performance liquid chromatography–photodiode array detector (HPLC-DAD). We further evaluated the hepatoprotective effects of these extracts on CCl$_4$-induced acute liver damage in C57BL/6 mice. The present study found CAH possesses the highest antioxidant activity among the four Cirsium species, and these antioxidant activities are closely related to phenylpropanoid glycoside (PPG) contents. The extracts decreased serum ALT and AST levels elevated by injection with 0.2% CCl$_4$. However, only CJF and CH decreased hepatic necrosis. Silibinin decreased serum alanine aminotransferase (ALT) and aspartate aminotransferase (AST) levels and hepatic necrosis caused by CCl$_4$. CJF and CH restored the activities of hepatic antioxidant enzymes and decreased hepatic malondialdehyde (MDA) levels. CJF further restored the expression of hepatic antioxidant enzymes including Cu/Zn-superoxide dismutase (Cu/Zn-SOD), Mn-superoxide dismutase (Mn-SOD), and glutathione S-transferase (GST) proteins. HPLC chromatogram indicated that CKH, CJF, and CH contained silibinin diastereomers (α and β). Only CJF contained diosmetin. Hence, the hepatoprotective mechanism of CJF against CCl$_4$-induced acute liver damage might be involved in restoring the activities and protein expression of the hepatic antioxidant defense system and inhibiting hepatic inflammation, and these hepatoprotective effects are related to the contents of silibinin diastereomers and diosmetin.

Keywords: Cirsium species; phenylpropanoid glycosides; CCl$_4$; liver damage; antioxidant; inflammation

1. Introduction

The liver is an accessory digestive gland and plays a key role in the metabolism, detoxification, and secretory functions of the vertebrate body. A wide variety of viruses, drugs, and toxic chemicals can cause liver damage through their direct toxicity and/or metabolic toxic products. Drug-induced liver damage is responsible for 5% of all hospital admissions and 50% of patients suffered with acute liver failure [1]. It is known that carbon tetrachloride (CCl$_4$) causes liver damage and hepatocyte apoptosis/necrosis in vitro and in vivo [2–5]. Liver damage from CCl$_4$ is a common model used to measure the efficiency of many hepatoprotective drugs [6]. CCl$_4$ is converted by hepatic cytochrome P450 2E1 (CYP2E1) into highly reactive radicals such as trichloromethyl (CCl3$^\bullet$) free radical and trichloromethylperoxy radical (CCl3OO$^\bullet$) [7]. Then, these radicals attack cellular macromolecules and cause lipid peroxidation, protein degradation, and DNA damage. The process is followed by the release of hepatic inflammatory cytokines such as interleukin-1β (IL-1β) and tumor necrosis factor-α (TNF-α), which leads to eventual damage, including hepatocellular necrosis [2,4,7].

Cirsii Herba (CH), listed in the Chinese Pharmacopoeia, is often mixed with *Cirsium japonicum* DC and *Cirsium setosum* (Willd.) MB in the traditional Chinese material market because they are similar in morphology and taxonomy. Chinese physicians usually use CH to treat hemorrhage, hypertension, and hepatitis. Recent pharmacological reports indicated that CH exhibits various pharmacological activities such as hemostatic, hepatoprotective, antidiabetic, anti-inflammatory, antibacterial, and sedative activities [8–16]. Due to its taxonomy (Asteraceae family) and morphology being similar to *Silybum marianum* (L.) Gaertn. (milk thistle), the Cirsium genus is commonly known as thistles and has been used to treat inflammation, diabetes, and hepatitis. *S. marianum* (L.) Gaertn has been used to protect liver injury caused by chemical and environmental toxins for a long time. Silymarin, the major active ingredient of *S. marianum*, includes flavonolignan isomers: 50–70% silibinin, isosilibinin, silydianin, and silychristin [17,18]. Silymarin also exerts hepatoprotective effects against alcoholic-induced liver disease, non-alcoholic fatty liver disease, and CCl$_4$-induced liver damage [11,19,20]. Ten species of the Cirsium genus are described in the Taiwan flora [21], and *Cirsium arisanense* Kitam. (CA), *Cirsium kawakamii* Hayata (CK), and *Cirsium japonicum* DC. var. *australe* Kitam. (CJ) are known as "Formosan thistles" and commonly used as folk medicines in Taiwan. However, there are few scientific studies reporting the phytoconstituents and hepatoprotective activities of these three endemic Cirsium species from Taiwan. Therefore, this study compared the antioxidant activities of four Cirsium species against the ABTS (2,2′-azino-bis(3-ethylbenzothiazoline-6-sulphonic acid)) radical in vitro and their hepatoprotective effects against CCl$_4$-induced acute liver damage in C57BL/6 mice. Silymarin and silibinin, the major ingredients of *Silybum marianum* (L.) Gaertn. and the most commonly known hepatoprotective agents, were used as positive controls. In addition, the mechanism of the hepatoprotective effect was studied regarding their potential antioxidant and anti-inflammatory properties. Moreover, previous phytochemical studies have revealed that flavonoids, lignans, phenylpropanoids, triterpenes, volatile oils, and steroids exist in CH, especially flavonoids [11,12,22–25]. Therefore, we also compared the phytoconstituents of four Cirsium species by spectrophotometric assay and high-performance liquid chromatography–photodiode array detector (HPLC-DAD).

2. Results

2.1. CCl$_4$-Induced Acute Liver Damage in C57BL/6 Mice

2.1.1. Serum Biochemical Levels of CCl$_4$-Induced Acute Liver Damage in C57BL/6 Mice

Serum alanine aminotransferase (ALT) and aspartate aminotransferase (AST) levels were used as biochemical markers for acute liver damage. One single-dose administration with 0.2% CCl$_4$ resulted in a significant rise in the levels of serum ALT and AST compared with the control group in C57BL/6 mice ($p < 0.001$) (Figure 1). Pre-treatment with all doses of the four Cirsium species extracts for 7 days significantly prevented serum ALT and AST levels from being elevated by 0.2% CCl$_4$ ($p < 0.001$) (Figure 1). Silibinin (positive control) at a dose of 25 mg/kg also prevented the elevation of serum ALT and AST levels, but silymarin (positive control) at a dose of 100 mg/kg only prevented the elevation of serum ALT levels ($p < 0.001$) (Figure 1).

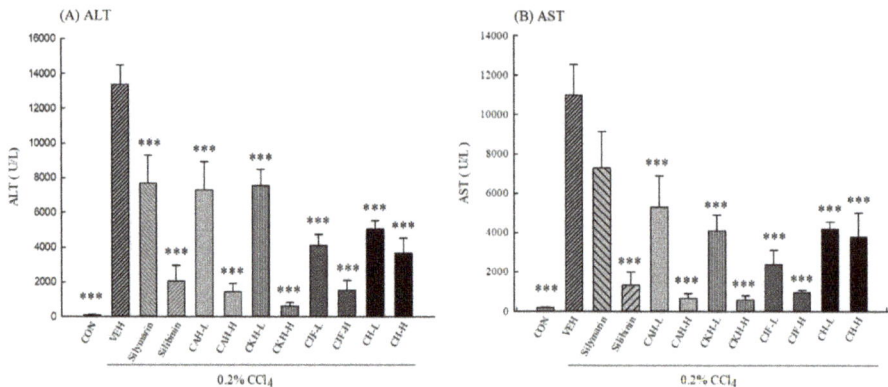

Figure 1. Effects of methanolic extracts of four Cirsium species (0.5, 5, 50 mg/kg), silymarin (100 mg/kg), and silibinin (25 mg/kg) on serum alanine aminotransferase (ALT) and aspartate aminotransferase (AST) levels of CCl$_4$-indcued acute liver damage in C57BL/6 mice. (**A**) ALT; (**B**) AST. CAH: The aerial part of *C. arisanense* Kitam. CKH: The aerial part of *C. kawakamii* Hayata. CJF: The flower part of *C. japonicum* DC. var. *australe* Kitam. CH: Cirsii Herba. CAH-L, CAH-H, CKH-, CKH-H, CJF-L, CJF-H, CH-L, and CH-H were continuously administered for 7 days. One hour after last treatment, acute liver damage was induced by injection with 0.2% CCl$_4$. Columns indicate mean ± SEM (*n* = 10). *** *p* < 0.001 compared with CCl$_4$-indcued acute liver damage mice.

2.1.2. Hepatic Histopathology of CCl$_4$-Induced Acute Liver Damage in C57BL/6 Mice

Hepatic histopathology provided supporting evidence for the serum biochemical analysis. The hepatic histopathology in control groups showed normal hepatocytes (Figure 2A). One single-dose administration with 0.2% CCl$_4$ caused a loss of hepatic architecture including vacuole formation, inflammatory infiltration, and necrosis in C57BL/6 mice (Figure 2B). Silibinin, CJF, and CH had different degrees of improving effects on CCl$_4$-induced histological changes (Figure 2C–H). There were higher hepatocyte necrosis levels in one single-dose administration with 0.2% CCl$_4$ (*p* < 0.001) (Figure 2I). Only CJF-L, CJF-H, CH-L, and CH-H inhibited the liver-damage phenomenon induced by CCl$_4$ (*p* < 0.05, 0.001, 0.01, 0.001, respectively) (Figure 2I). Silymarin, CAH, and CKH did not decrease hepatocyte necrosis levels induced by CCl$_4$ (*p* > 0.05) (Figure 2I).

2.1.3. Hepatic Antioxidant Activities and MDA Levels of CCl$_4$-Induced Acute Liver Damage in C57BL/6 Mice

To clarify the role of the antioxidative defense system on the hepatoprotective effects of four Cirsium species extracts against CCl$_4$-induced acute liver damage in C57BL/6 mice, we measured the activities of the hepatic antioxidant defense system, including the level of glutathione (GSH), the activities of related antioxidant enzymes, and the level of oxidative damage markers such as malondialdehyde (MDA). One single-dose intraperitoneal injection with 0.2% CCl$_4$ decreased the activities of hepatic antioxidant enzymes including superoxide dismutase (SOD), glutathione peroxidase (GPx), glutathione reductase (GR), and catalase in rats (*p* < 0.001) (Figure 3). We further found one single-dose intraperitoneal injection with 0.2% CCl$_4$ also decreased hepatic GSH levels but increased hepatic MDA levels in C57BL/6 mice (*p* < 0.001) (Figure 4). CAH-L and CAH-H could restore hepatic SOD and catalase activities and decrease hepatic MDA levels (*p* < 0.01, 0.001), but only CAH-H restored hepatic GPx and GR activities (*p* < 0.05) (Figures 3 and 4). Only CKH-H could restore hepatic SOD and GR activities and GSH levels (*p* < 0.05, 0.001), but CKH-L and CKH-H decreased hepatic MDA levels (*p* < 0.001) (Figures 3 and 4). CJF-L and CJF-H could restore hepatic GPx, GR, and SOD activities and decrease hepatic MDA levels, but only CJF-H restored hepatic catalase activity and GSH levels ($p < 0.001$) (Figures 3 and 4). Only CH-H could restore the activities of hepatic antioxidant enzymes ($p < 0.05$, 0.001), but CKH-L and CKH-H decreased hepatic MDA levels ($p < 0.001$) (Figures 3 and 4). Silymarin (100 mg/kg) and silibinin (25 mg/kg) restored the activities of all hepatic antioxidant enzymes and decreased hepatic MDA levels ($p < 0.01$, 0.001) (Figures 3 and 4). However, only silibinin could restore hepatic GSH level decreased by CCl_4 ($p < 0.05$) (Figure 4).

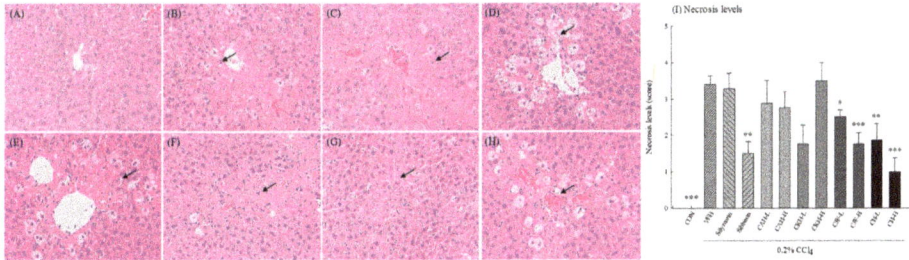

Figure 2. Effects of methanolic extracts of four Cirsium species (0.5, 5, 50 mg/kg), silymarin (100 mg/kg), and silibinin (25 mg/kg) on hepatic histopathology of CCl_4-indcued acute liver damage in C57BL/6 mice. (**A**) Hematoxylin and eosin (H&E) stain of control group; (**B**) H&E stain of CCl_4 group; (**C**) H&E stain of silymarin group; (**D**) H&E stain of silibinin group; (**E**) H&E stain of CAH-H group; (**F**) H&E stain of CKH-H group; (**G**) H&E stain of CJF-H; (**H**) H&E stain of CH-H; (**I**) Necrosis levels. CAH: The aerial part of *C. arisanense* Kitam. CKH: The aerial part of *C. kawakamii* Hayata. CJF: The flower part of *C. japonicum* DC. var. *australe* Kitam. CH: Cirsii Herba. CAH-L, CAH-H, CKH-, CKH-H, CJF-L, CJF-H, CH-L, and CH-H were continuously administered for 7 days. One hour after last treatment, acute liver damage was induced by injection with 0.2% CCl_4. Representative images (**A–H**) of hematoxylin and eosin (H&E) with 400× amplification. The black arrows indicate hepatocyte necrosis. Columns (**I**) indicate mean ± SEM ($n = 10$). * $p < 0.05$, ** $p < 0.01$, *** $p < 0.001$ compared with CCl_4-indcued acute liver damage mice.

2.1.4. Hepatic Cytokine Levels of CCl_4-Induced Acute Liver Damage in C57BL/6 Mice

To clarify the anti-inflammatory mechanism on the hepatoprotective effects of four Cirsium species extracts against CCl_4-induced acute liver damage in C57BL/6J mice, we measured the levels of hepatic cytokines, including IL-1β and TNF-α. One single-dose intraperitoneal injection with 0.2% CCl_4 elevated the levels of hepatic cytokines in C57BL/6 mice ($p < 0.001$) (Figure 5). CAH, CJF, and CH-H decreased hepatic IL-1β levels ($p < 0.001$) (Figure 5). Only CJF-H decreased hepatic TNF-α levels ($p < 0.001$) (Figure 5). Silibinin (25 mg/kg) decreased hepatic IL-1β levels ($p < 0.001$) (Figure 5). However, silymarin (100 mg/kg) and CKH did not decrease the levels of hepatic cytokines elevated by CCl_4 ($p > 0.05$) (Figure 5).

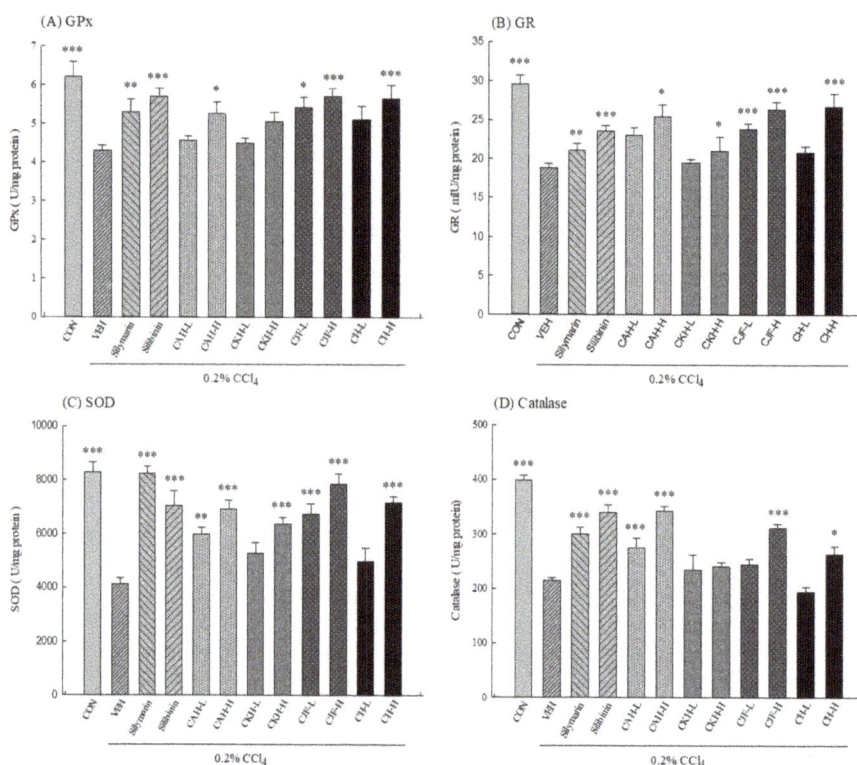

Figure 3. Effects of methanolic extracts of four Cirsium species (0.5, 5, 50 mg/kg), silymarin (100 mg/kg) and silibinin (25 mg/kg) on hepatic antioxidant enzyme activities of CCl$_4$-induced acute liver damage in C57BL/6 mice. (**A**) glutathione peroxidase (GPx); (**B**) glutathione reductase (GR); (**C**) superoxide dismutase (SOD); (**D**) catalase. CAH: The aerial part of *C. arisanense* Kitam. CKH: The aerial part of *C. kawakamii* Hayata. CJF: The flower part of *C. japonicum* DC. var. *australe* Kitam. CH: Cirsii Herba. CAH-L, CAH-H, CKH-, CKH-H, CJF-L, CJF-H, CH-L, and CH-H were continuously administered for 7 days. One hour after last treatment, acute liver damage was induced by injection with 0.2% CCl$_4$. Columns indicate mean ± SEM (n = 10). * $p < 0.05$, ** $p < 0.01$, *** $p < 0.001$ compared with CCl$_4$-indcued acute liver damage mice.

2.1.5. Antioxidant Protein Expression of CCl$_4$-Induced Acute Liver Damage in C57BL/6 Mice

The protein immunoblot assay is shown in Figure 6A. One single-dose intraperitoneal injection with 0.2% CCl$_4$ decreased hepatic GST, Cu/Zn-SOD, and Mn-SOD expression levels in C57BL/6 mice ($p < 0.05$, 0.01) (Figure 6B–D). Only CJF-H restored hepatic GST, Cu/Zn-SOD, and Mn-SOD expression levels downregulated by CCl$_4$ ($p < 0.05$, 0.01, respectively) (Figure 6B–D).

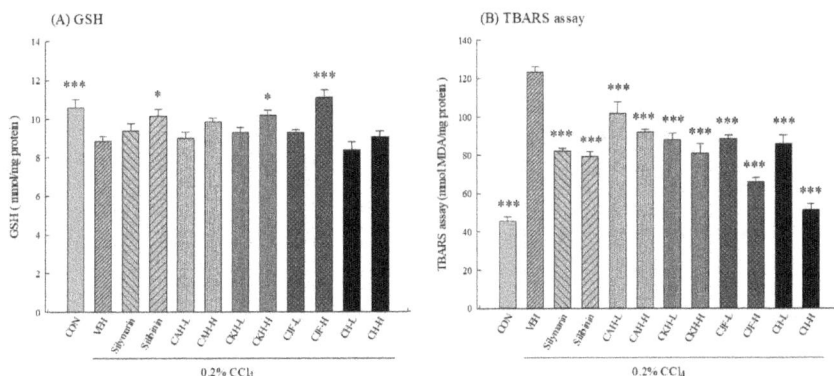

Figure 4. Effects of methanolic extracts of four Cirsium species (0.5, 5, 50 mg/kg), silymarin (100 mg/kg), and silibinin (25 mg/kg) on hepatic glutathione (GSH) and malondialdehyde (MDA) levels of CCl$_4$-indcued acute liver damage in C57BL/6 mice. (**A**) GSH; (**B**) MDA. TBARS: thiobarbituric acid reactive substances. CAH: The aerial part of *C. arisanense* Kitam. CKH: The aerial part of *C. kawakamii* Hayata. CJF: The flower part of *C. japonicum* DC. var. *australe* Kitam. CH: Cirsii Herba. CAH-L, CAH-H, CKH-, CKH-H, CJF-L, CJF-H, CH-L, and CH-H were continuously administered for 7 days. One hour after last treatment, acute liver damage was induced by injection with 0.2% CCl$_4$. Columns indicate mean ± SEM (n = 10). * $p < 0.05$, *** $p < 0.001$ compared with CCl$_4$-indcued acute liver damage mice.

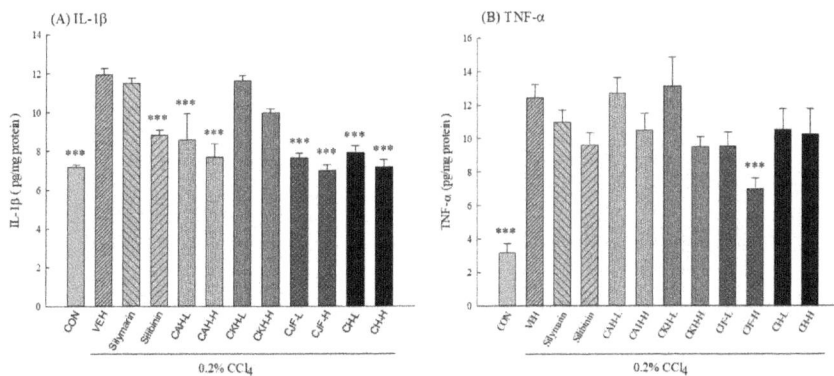

Figure 5. Effects of methanolic extracts of four Cirsium species (0.5, 5, 50 mg/kg), silymarin (100 mg/kg), and silibinin (25 mg/kg) on hepatic cytokine levels of CCl$_4$-indcued acute liver damage in C57BL/6 mice. (**A**) interleukin-1β (IL-1β); (**B**) tumor necrosis factor-α (TNF-α). CAH: The aerial part of *C. arisanense* Kitam. CKH: The aerial part of *C. kawakamii* Hayata. CJF: The flower part of *C. japonicum* DC. var. *australe* Kitam. CH: Cirsii Herba. CAH-L, CAH-H, CKH-L, CKH-H, CJF-L, CJF-H, CH-L, and CH-H were continuously administered for 7 days. One hour after last treatment, acute liver damage was induced by injection with 0.2% CCl$_4$. Columns indicate mean ± SEM (n = 10). *** $p < 0.001$ compared with CCl$_4$-indcued acute liver damage mice.

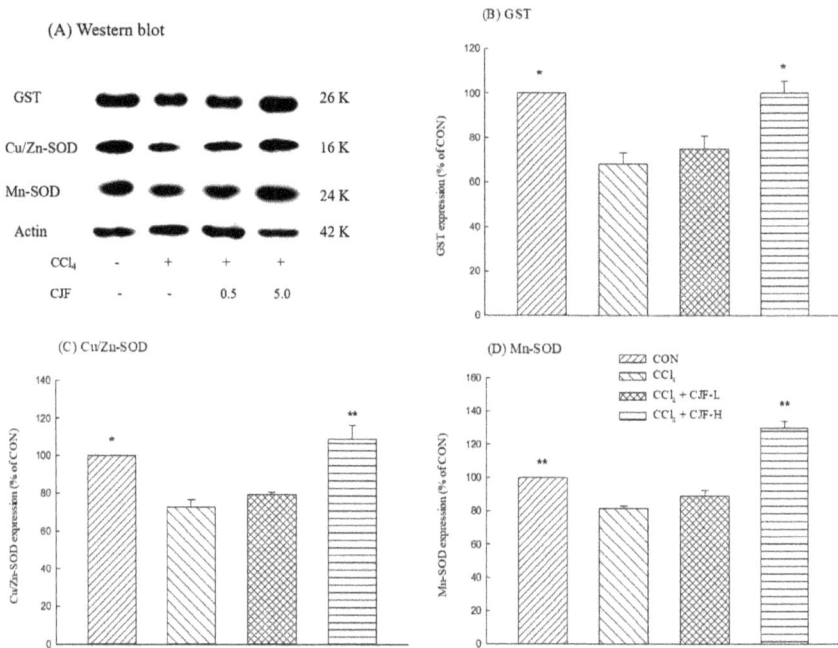

Figure 6. Effects of methanolic extracts of the flower part of *C. japonicum* DC. var. *australe* Kitam. (CJF; 0.5, 5 mg/kg) on the GST, Cu/Zn-SOD, and Mn-SOD expression of CCl_4-indcued acute liver damage in C57BL/6 mice. (**A**) Western blot; (**B**) GST; (**C**) Cu/Zn-SDO; (**D**) Mn-SOD. CJF-L and CJF-H were continuously administered for 7 days. One hour after last treatment, acute liver damage was induced by injection with 0.2% CCl_4. Columns indicate mean $\pm$ SEM ($n = 4$). * $p < 0.05$, ** $p < 0.01$ compared with CCl_4-indcued acute liver damage mice.

2.2. Antioxidant Ingredient Contents and Activities of Four Cirsium Species Extracts

2.2.1. Total Phenolics (TPs) Contents, Total Phenylpropanoid Glycosides (PPGs) Contents, and HPLC Analysis of Four Cirsium Species Extracts

The contents of TPs and PPGs in four Cirsium species extracts are shown in Table 1. There are the higher contents of TPs and PPGs in CAH compared with the other three Cirsium species extracts. The order of the TPs contents of the other three Cirsium species extracts is CKH > CJF > CH. However, the order of the PPGs contents for the other three Cirsium species extracts is CJF > CH > CKH. Furthermore, the phytoconstituents of the four Cirsium species extracts were assayed using HPLC. Their chromatographs are shown in Figure 7. The certain phytoconstituent peak zones differ among the four Cirsium species extracts. The calibration curves for apigenin, diosmetin, silibinin α, silibinin β, silydianin, silychristin, isosilibinin α, and isosilibinin β were drawn in the concentration range of 2.5–20 μg/mL. The correlation coefficients of the calibration plots were equal to 0.992–0.994, indicating good linearity in four cases. The methanolic extracts of the four Cirsium species contained silibinin α, and CAH had the highest content among the four Cirsium species. CKH, CJF, and CH contained silibinin β. Only CAH and CJF contained silydianin. Diosmetin only existed in CJF (Table 1).

Table 1. The phytoconstituents of methanolic extracts of four Cirsium species.

Samples	Yields (%)	TPs (mg of Catechin/g)	PPGs (mg of Verbascoside/g)	Silydianin	Silibinin α	Silibinin β	Diosmetin
CAH	23.9	109.65 ± 0.57	175.20 ± 5.45	1.23 ± 0.06	2.53 ± 0.01	-	-
CKH	4.5	93.91 ± 1.07	2.30 ± 0.00$_0$	-	1.11 ± 0.02	25.14 ± 0.23	-
CJF	10	49.52 ± 2.11	28.64 ± 1.01	3.28 ± 0.04	0.88 ± 0.01	3.74 ± 0.01	8.66 ± 0.06
CH	15.1	44.86 ± 1.77	25.33 ± 1.24	-	0.38 ± 0.01	0.34 ± 0.01	-

TPs: Total phenolics. PPGs: Total phenylpropanoid glycosides. CAH: The aerial part of *C. arisanense* Kitam. CKH: The aerial part of *C. kawakamii* Hayata. CJF: The flower part of *C. japonicum* DC. var. *australe* Kitam. CH: Cirsii Herba. Columns indicate mean ± SD ($n = 3$).

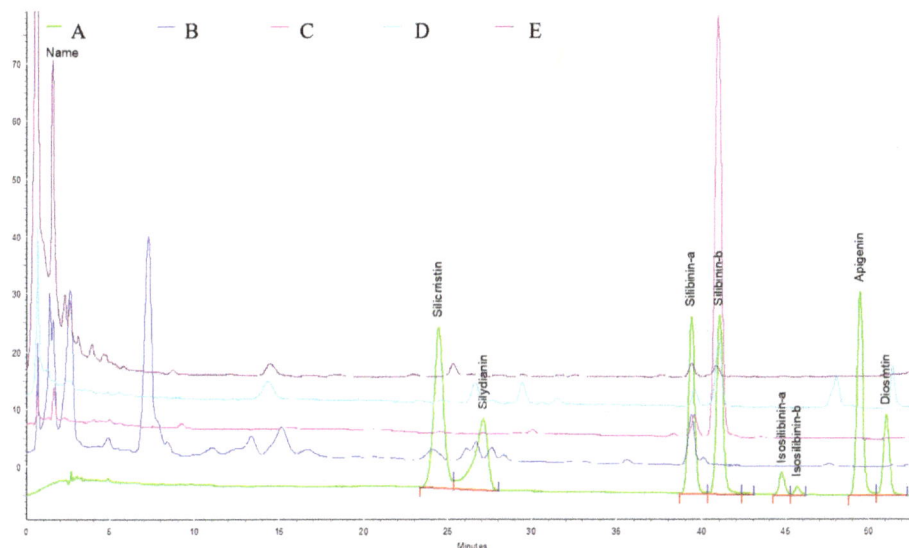

Figure 7. High-performance liquid chromatography (HPLC) chromatograms of methanolic extracts of four Cirsium species at 280 nm. Trace: (**A**) Standard, (**B**) CAH: The aerial part of *C. arisanense* Kitam., (**C**) CKH: The aerial part of *C. kawakamii* Hayata. (**D**) CJF: The flower part of *C. japonicum* DC. var. *australe* Kitam. (**E**) CH: Cirsii Herba.

2.2.2. Antioxidant Activities of Four Cirsium Species Extracts

The Trolox Equivalent Antioxidant Capacity (TEAC) assay is a simple and rapid method that is commonly used to assess total antioxidant capacity related to the scavenging 2,2′-azino-bis(3-ethylbenzothiazoline-6-sulphonic acid) (ABTS) radical and reactive oxygen species (ROS) in vitro. The total antioxidant capacities of the four Cirsium species extracts in the TEAC assay are shown in Figure 8A and expressed as a Trolox equivalent (mmol trolox/g sample). CAH had the highest antioxidant capacity against ABTS radical compared with the other three Cirsium species. Then, we further examined the relationship between TEAC and the contents of antioxidant phytoconstituents by Pearson correlation analysis. Only the PPGs content was positively and highly correlated with TEAC ($r = 0.99$) (Figure 8B).

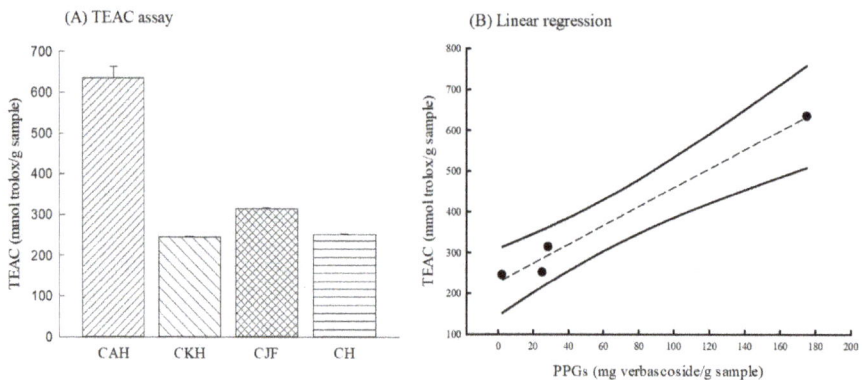

Figure 8. Antioxidant activity of methanolic extracts of four Cirsium species. (**A**) Trolox equivalent antioxidant capacity (TEAC); (**B**) the relationship between TEAC and PPGs content. Columns indicate mean ± SD (*n* = 3). CAH: The aerial part of *C. arisanense* Kitam. CKH: The aerial part of *C. kawakamii* Hayata. CJF: The flower part of *C. japonicum* DC. var. *australe* Kitam. CH: Cirsii Herba.

3. Discussion

The Cirsium genus is similar to *Silybum marianum* (L.) Gaertn based on the taxonomy, morphology, and folk medical use. They are commonly known as thistles and have been used to treat inflammation and hepatitis. *Silybum marianum* and its ingredient silymarin also exert hepatoprotective effects against alcohol-induced liver disease, non-alcoholic fatty liver disease, and CCl₄-induced liver damage [11,19,20]. CCl₄ is the oldest and most widely used hepatic toxin, and one single-dose injection with CCl₄ in rodents is often used as an acute liver damage model for screening hepatoprotective drugs [6]. CCl₄-induced liver damage is mainly due to the lipid peroxidation of hepatocyte membranes by free radicals derived from CCl₄ metabolites. Membrane destruction of hepatocytes leads to the release of hepatic enzymes such as ALT and AST into blood circulation. Consequently, serum ALT and AST levels are used as biochemical markers of liver damage [26]. In this study, an acute CCl₄ liver-damage model was used to compare the hepatoprotective efficacy of four Cirsium species extracts, with silymarin and silibinin being used as positive controls. We found that one single-dose injection with CCl₄ drastically elevated serum ALT and AST levels in C57BL/6 mice, which indicates CCl₄-induced acute liver damage. The present results demonstrated that all four Cirsium species extracts at any dose could decrease serum ALT and AST levels elevated by CCl₄. Further, from the serum biochemical analysis by hepatic histopathological stain, we found CCl₄ caused dysfunction of hepatic cellular structures, including vacuole formation, inflammatory infiltration, and necrosis in C57BL/6 mice. However, there was minimal disruption of the hepatic cellular structure and less necrosis against CCl₄-induced acute liver injury when pretreatment with only CJF and CH (at any dose) and silibinin for 7 days, compared with the CCl₄ group. Previous reports indicated that CH (*C. japonicum* DC or *C. setosum*) exhibited hepatoprotective effects against CCl₄-induced hepatotoxicity in L02 cells or HL-7702 cells [11,12]. Ku et al. reported that the root, but not aerial part, of CA possessed hepatoprotective properties against tacrine-induced hepatotoxicity in Hep 3B cells and C57BL/6 mice [27]. Therefore, our present results from serum biochemical and histopathological data demonstrate that only CJF and CH, among the four Cirsium species, possessed convincing hepatoprotective potency against CCl₄-induced acute liver damage, although all four endemic Cirsium species decreased serum ALT and AST levels. Two positive controls, silymarin and silibinin, could decrease serum ALT and AST levels elevated by CCl₄, but only silibinin prevented dysfunction of the hepatic cellular structure. Wu et al. indicated that silymarin did not decrease serum ALT and AST levels and histopathological alteration caused by CCl₄ in mice [28], although many reports have found that silymarin at 200–800 mg/kg decreased liver damage caused by acetaminophen, ethanol, and CCl$_4$ [29–31]. Hence, our results confirmed other reports [32] that silibinin is a major active constituent of silymarin, and at 20–50 mg/kg prevented liver damage caused by acetaminophen, ethanol, and CCl$_4$.

Experimental and clinical reports indicate that oxidative stress plays a critical role in the development of drug-induced liver damage [33]. CCl$_4$-induced liver damage is mostly due to reactive free-radical generation from the dehalogenation of CCl$_4$ by CYP2E1. These reactive free radicals, including trichloromethyl radical (CCl3•) and trichloromethylperoxy radical (CCl3OO•), cause the depletion of hepatic antioxidant status and the exacerbation of hepatic lipid peroxidation [7]. The major hepatic antioxidant defense system against free radicals includes SOD, catalase, and GSH redox cycle. SOD, the first enzyme involved in the antioxidant defense system, scavenges superoxide anions, which are produced from the mitochondria electron transfer chain. When superoxide anions are transformed into H$_2$O$_2$ by SOD, catalase then continuously detoxifies it to H$_2$O. The GSH redox cycle, mainly including GSH, GPx, and GR, modulates the redox-mediated responses of hepatic cells induced by external or intracellular stimuli. GSH, a cytosolic tripeptide, is the major non-enzymatic regulator of intracellular redox homeostasis. GSH scavenges hydroxyl radicals directly and is as a cofactor in detoxifying hydrogen peroxide, lipid peroxides, and alkyl peroxides by catalytic action of the detoxifying enzyme GPx. GPx, a selenocysteine-containing enzyme, reduces lipid hydroperoxides to their corresponding alcohols and hydrogen peroxide to water in the liver. GR is essential for the glutathione redox cycle and maintains adequate levels of reduced cellular GSH, which catalyzes the reduction of oxidized GSH to reduce GSH. In this regard, the enhancement of the hepatic antioxidant system capacity may be an effective therapeutic strategy for the alleviation and treatment of liver damage [33,34]. Our present data showed that one single-dose injection with CCl$_4$ induced significant depletion of GSH levels, dysfunction of SOD, catalase, GR, and GPX, and enhanced lipid peroxidation in liver homogenates. Hence, the results confirmed that CCl$_4$ decreases the function of antioxidant enzymes and the GSH redox cycle to alter the intracellular redox status, causing oxidative stress and enhancing lipid peroxidation in the liver. Here, our results showed that pretreatment with CJF or silibinin for 7 days significantly restored GSH levels and the activities of antioxidant enzymes, including SOD, catalase, GPx, and GR, and reduced MDA levels in mice liver homogenate. Pretreatment with CH also restored the activities of antioxidant enzymes but not the GSH redox cycle. However, CH still reduced hepatic MDA levels. Previous studies have shown that *C. japonicum* DC can decrease hepatic lipid peroxidation along with an increase in hepatic content of reduced glutathione in ethanol-treated rats. CJF, similar to silibinin, might play a role in restoring hepatic redox capacity and antioxidant enzyme activities, resulting in a protective effect against CCl$_4$-induced acute liver damage. Other Cirsium species are not so comprehensive in restoring hepatic redox capacity and antioxidant enzyme activities against CCl$_4$-induced acute liver damage.

Moreover, GST is a phase II metabolic enzyme. It increases cellular GSH levels and protects cells against oxidative stress through conjugating free radicals with GSH [35]. SOD consists of three isoforms in mammals: the cytoplasmic Cu/Zn-SOD (SOD1), the mitochondrial Mn-SOD (SOD2), and the extracellular Cu/Zn-SOD (SOD3). Cu/Zn-SOD (SOD1) is mainly localized in the cytosol, with a smaller fraction in the intermembrane space of mitochondria. Cu/Zn-SOD (SOD1) comprises 90% of the total SOD and has great physiological significance and therapeutic potential. Mn-SOD is a manganese (Mn) containing enzyme and is localized in the mitochondrial matrix. The essential role of Mn-SOD is in maintaining mitochondrial function [36]. Hence, we further explored the role of GST, Cu/Zn-SOD, and Mn-SOD proteins on the hepatoprotective effects of CJF against CCl$_4$-induced acute liver damage in C57BL/6 mice. CCl$_4$ decreased the expression of GST, Cu/Zn-SOD, and Mn-SOD proteins in liver homogenates. CJF, at a high dose, can restore the expression of GST, Cu/Zn-SOD, and Mn-SOD proteins downregulated by CCl$_4$. Therefore, the hepatoprotective mechanism of CJF is related to restoring the activities of the hepatic antioxidant defense system through reversing the GSH redox cycle and the activities and expression of antioxidant enzymes.

During the pathological events of acute liver damage with CCl$_4$, inflammatory processes play a crucial role. Resident macrophages in the liver, known as Kupffer cells, are activated and rapidly release pro-inflammatory cytokines such as TNF-α and IL-1β. These pro-inflammatory cytokines play an important role in a complex network involved in the regulation of inflammatory responses. The increase in TNF-α and IL-1β levels is correlated with the histopathological evidence of hepatic necrosis and the increase in serum ALT and AST levels. Therefore, inhibiting the activity of macrophages and the release of pro-inflammatory cytokines provides an excellent therapeutic strategy to alleviate liver inflammation and damage [37]. Furthermore, we also investigated the effects of four Cirsium species on the levels of hepatic pro-inflammatory cytokines, such as TNF-α and IL-1β, in CCl$_4$-induced acute liver damage. This study confirmed that CCl$_4$ administration caused the increase in hepatic TNF-α and IL-1β levels in C57BL/6 mice. Only CJF markedly reduced the increased TNF-α and IL-1β levels caused by CCl$_4$ in mice liver homogenates. CH and silibinin only decreased the high IL-1β levels caused by CCl$_4$. Thus, the present results demonstrate that CJF has hepatoprotective potency against CCl$_4$-induced acute liver damage by its antioxidative and anti-inflammatory features via reversal of the hepatic antioxidant defense system and suppression of pro-inflammatory cytokine expressions.

Finally, we compared antioxidant activities and the contents of four Cirsium species in vitro because their hepatoprotective effects are partially related to their antioxidant activities. In this study, CAH possessed the highest scavenging activity against the ABTS radical. CAH also had the highest TPs and PPGs content. According to the Pearson's analysis, there is a close correlation between antioxidant (scavenging ABTS radical) activities and PPGs contents. However, this result is inconsistent with the hepatoprotective result. We further assayed the phytoconstituents of the four Cirsium species by HPLC-DAD. There are different HPLC fingerprints among the four Cirsium species to distinguish them from this result. CKH, CJF, and CH contained silibinin diastereomers (α and β), with their order being CKH > CJF > CH. CJF still contained silydianin and diosmetin. CAH only contained silydianin and silibinin α, inconsistent with other reports stating that no related compounds existed in CAH [27]. This study provides the first reported results concerning the phytoconstituents of CAH, CKH, and CJF. The results regarding the phytoconstituents of CH are consistent with other reports [11]. However, we found that silibinin diastereomers might not be the active ingredients of CKH, CJF, and CH associated with their hepatoprotective effects against CCl$_4$ because there is no correlation between the contents of silibinin diastereomers and hepatoprotective activity. The hepatoprotective phytoconstituents of CH might include flavonoids, such as silibinin diastereomers and phenylethanoid glycosides, because the hepatoprotective ingredients of *C. japonicum* are flavonoids and the hepatoprotective ingredients of *C. setosum* are phenylethanoid glycosides [11,12]. The efficacy of CJF, which showed the best hepatoprotective effects among the four Cirsium species, might be due to silibinin diastereomers and diosmetin because diosmetin could inhibit CYP 2E1 [38] and it possesses antioxidant and anti-inflammatory activities against endotoxin-induced acute liver injury [39]. Therefore, the hepatoprotective phytoconstituents of CH against CCl$_4$-induced acute liver damage must be clarified. The synergistic effects of silibinin diastereomers and diosmetin against CCl$_4$-induced acute liver damage deserve to be investigated in the future. The hepatoprotective effects of CJF against CCl$_4$-induced hepatic cirrhosis must also be investigated.

4. Materials and Methods

4.1. Collection and Preparation of Plant Materials

Three Cirsium species materials including the aerial part of CA (CAH), the aerial part of CK (CKH), and the flower part of CJ (CJF), were identified and provided by Hung-Chi Chang. CH was obtained from the Herb Garden of China Medical University, Taichung City. Four Cirsium species materials were extracted with 10 times volume of 70% methanol, and the resulting extracts were concentrated under reduced pressure to obtain the four Cirsium species methanolic extracts. To assess phytoconstituents and ROS-scavenging activities, the extracts were dissolved in distilled water. To compare the hepatoprotective effects on CCl_4-induced acute liver damage, the four Cirsium species methanolic extracts were prepared using 0.5% carboxymethylcellulose (CMC).

4.2. CCl₄-Induced Acute Liver Damage in C57BL/6 Mice

4.2.1. Animals

Male C57BL/6 mice (20–25 g) were obtained from BioLASCO Taiwan Co., Ltd. They were housed in groups of six, chosen at random, in wire-mesh cages (39 cm × 26 cm × 21 cm) in a temperature (23 ± 1 °C) and humidity (60%) regulated environment with a 12 h–12 h light/dark cycle (light phase: 8:00 a.m. to 8:00 p.m.). The Institutional Animal Care and Use Committee of China Medical University approved the experimental protocol (No. CMUIACUC-2017392), and mice were cared for according to the Guiding Principles for the Care and Use of Laboratory Animals. After one week of acclimatization, the mice were used in CCl_4-induced acute liver damage.

4.2.2. CCl₄-Induced Acute Liver Damage in C57BL/6 Mice

The mice were randomly divided into 12 groups of 10 mice each. In the control group and CCl_4 group, mice were orally given 0.5% CMC (0.1 mL/10 g body weight) daily for 7 days. In the treatment groups, other mice were separated into 10 groups: (1) CCl_4 + silymarin: this group served as a positive control. Mice were orally given silymarin (100 mg/kg) daily for 7 days; (2) CCl_4 + silibinin: this group also served as a positive control. Mice were orally given silibinin (25 mg/kg) daily for 7 days; (3) CCl_4 + CAH-L: CAH-L (5 mg/kg) was orally given daily for 7 days; (4) CCl_4 + CAH-H: CAH-H (50 mg/kg) was orally given daily for 7 days; (5) CCl_4 + CKH-L: CKH-L (5 mg/kg) was orally given daily for 7 days; (6) CCl_4 + CKH-H: CKH-H (50 mg/kg) was orally given daily for 7 days; (7) CCl_4 + CJF-L: CJF-L (0.5 mg/kg) was orally given daily for 7 days; (8) CCl_4 + CJF-H: CJF-H (5 mg/kg) was orally given daily for 7 days; (9) CCl_4 + CH-L: CH-L (5 mg/kg) was orally given daily for 7 days; (10) CCl_4 + CH-H: CH-H (50 mg/kg) was orally given daily for 7 days. On the seventh day, all mice except those in the control groups were intraperitoneally given 0.2% CCl_4 (dissolved in olive oil, 0.1 mL/10 g body weight) 1 h after the last administration, while the control group received olive oil [40]. All mice were fasted for 24 h and subsequently blood was collected via retroorbital sinus plexus under isoflurane anesthesia. Then, all mice were sacrificed, and their livers were dissected and used for histopathological (formalin fixed) and biochemical (frozen −80 °C) studies.

4.2.3. Assessment of Liver Function

After blood collection, serum was separated by centrifugation at 3000 rpm at room temperature (RT) for 30 min and was kept at −20 °C for further biochemical analysis. Serum ALT and AST values were measured with commercially available Roche diagnostic kits.

4.2.4. Hepatic Antioxidant Defense System and MDA Levels

All hepatic tissues were homogenized in 9 vol ice-cold phosphate buffered saline. Homogenates were centrifuged at 12,000 rpm for 15 min at 4 °C, and the aliquots of supernatants were separated and stored at −80 °C until use. The aliquots of supernatants were used to determine antioxidant enzyme activities for catalase, GPx, GR, and SOD, and the levels of MDA and GSH. The activities of antioxidant enzymes and the levels of MDA and GSH were measured with a spectrophotometric microplate reader, as we previously reported [41]. First, catalase activity was determined with the decrease in the absorbance of amplex red at 560 nm. SOD activity was measured kinetically with the production of nitroblue tetrazolium, the absorbance of which is at 560 nm. The activities of GPx and GR were measured by a Cayman assay kit. SOD and catalase was expressed as U/mg of protein. GPx activities were expressed as U/mg of protein. GR activities were expressed as mIU/mg of protein. GSH standard solution or the supernatant solution (20 µg/50 µL) was pipetted into each well of a

96-well plate. The reaction solution, including 660 µM DTNB (5,5′-dithio-bis-(2-nitrobenzoic acid)), 900 µM NADPH (β-nicotinamide adenine dinucleotide phosphate), and 4.5 U/mL GR, was added to each well and then recorded at 405 nm for 5 min in a microplate reader. GSH levels were expressed as mmol/mg of protein. The thiobarbituric acid reactive substances (TBARS) assay was used to measure MDA levels. MDA standard solution or the supernatant solution was pipetted into 1.5 mL tubes, and a thiobarbituric acid (TBA) test was performed. Next, the absorbance of the above reaction solution was determined at 532 nm. TBARS assays were expressed as MDA equivalents (mmol MDA/mg of protein).

4.2.5. Hepatic Cytokines Levels

Hepatic tissues were homogenized with protease inhibitor solution (0.4 M NaCl, 0.05% Tween 20, 0.5% bovine serum albumin, 0.1 mM phenylmethylsulfonylfluoride, 0.1 mM benzethonium chloride, 10 mM EDTA, 10 µg/mL aprotinin). Homogenates were centrifuged at 12,000 rpm at 4 °C, aliquoted, and stored at −80 °C until analysis. IL-1β and TNF-α protein levels were assessed using ELISA kits (R&D Systems, Abingdon, UK) according to the manufacturer protocol. The levels of IL-1β and TNF-α were expressed as pg/mg of protein.

4.2.6. Histopathological Stain

A portion of the left lobe of the liver was preserved in 10% neutral formalin solution, processed, and paraffin embedded, as per the standard protocol. Sections of 4 µm in thickness were cut, deparaffinized, dehydrated, and stained with hematoxylin and eosin (H&E) for the estimation of hepatocyte necrosis and vacuolization. Morphological changes were observed including cell gross necrosis, ballooning degeneration, and inflammatory infiltration.

4.2.7. Western Blot

Hepatic tissues were subjected to Western blot analyses to determinate the protein levels of Cu/Zn-SOD, Mn-SOD, and GST. Briefly, the hepatic tissues were cut into small pieces and homogenized in 9× cold lysis buffer (20 mM HEPES pH 7.0, 10 mM KCl, 0.5% NP-40) with a tissue grinder. The homogenate was incubated for 10 min and centrifuged at 12,000 rpm for 20 min to obtain the cytoplasmic supernatant. The aliquots of cytoplasmic supernatants were stored at −80 °C until use. The protein concentration was quantified using a Bradford protein assay kit (Bio-Rad Ltd. Inc., Hercules, CA, USA) and followed by electrophoretic separation through SDS-PAGE. After transferring the protein samples to PVDF membranes, the samples were blocked with 5% non-fat dry milk and 0.1% tween-20 in tris-buffered saline at room temperature for 1 h. Then, the membranes were incubated with primary antibodies against Cu/Zn-SOD, Mn-SOD, and GST (Santa Cruz Biotechnology, Dallas, TX, USA) overnight at 4 °C and subsequently incubated with horseradish peroxidase-conjugated goat anti-rabbit or goat anti-mouse Immunoglobulin G (IgG). The images were scanned using an LAS-4000 mini imaging system (Fujifilm, Kanagawa, Japan), and the optical density data was analyzed using MultiGauge v3.0 software (Fujifilm). For the Western blot analyses, β-actin (Proteintech, Rosemont, IL, USA) served as an internal control.

4.3. Measurement of Antioxidant Phytoconstituent Contents and Activities

4.3.1. Measurement of Phytoconstituents Using a Spectrophotometric Reader

The contents of all phytochemicals, including TPs and PPGs, were assayed using a 96-well BioTek PowerWave™ 340 microtiter spectrophotometric reader (BioTek Inc., Winooski, VT, USA). The measured method for TPs contents is based on forming blue-colored products through a redox reaction with Folin-Ciocalteu's reagent and measuring its absorbance at 725 nm. The TPs contents of the four Cirsium species methanolic extracts were expressed as milligrams of catechin equivalents per gram of the four extracts [28]. The measured method for PPGs contents is based on forming colored products through PPGs with Arnow reagent (containing 5% (*w/v*) sodium nitrate and 5% sodium molybdate) and measuring its absorbance at 525 nm. The PPGs contents of the four extracts were expressed as milligrams of verbascoside equivalents per gram of the extracts [28].

4.3.2. Determination of Phytoconstituents Using HPLC-DAD

The extracts were dissolved in methanol and then filtered using a 0.22 μm filter. Stock solutions of the standards including apigenin, diosmetin, silibinin α, silibinin β, silydianin, silychristin, isosilibinin α, and isosilibinin β were prepared in methanol to the final concentration 10 mg/mL. Working solutions at 2.5–20 μg/mL were prepared freshly every day by dilution of the standard solutions with methanol. All standard and sample solutions were injected into 10 μL in triplicate. The Shimadzu VP series HPLC and Shimadzu Class-VPTM chromatography data systems were used. All chromatographic operations were performed at 25 °C. The chromatographic peaks of all standards were confirmed by comparing their retention times and ultraviolet (UV) spectra. A LiChrospher® RP-18e (250 × 4 mm, 5 μm) column (Merck KGaA, Darmstadt, Germany) was used. The separating conditions included mobile phases and gradient program conditions, modified from the description by Saleh et al. [42]. The mobile phase, same as the description by Saleh et al., was 0.5:35:65 phosphoric acid:methanol:water (solvent A) and 0.5:70:30 phosphoric acid:methanol:water (solvent B). The flow rate was modified with 0.8 mL/min. The gradient program conditions were modified as shown in Table 2.

Table 2. HPLC gradient program conditions of methanolic extracts of the four Cirsium species.

Time (min)	Solvent A (%)	Solvent B (%)
0–15	90	10
15–25	70	30
25–35	55	45
35–45	35	65
45–50	0	100
50–55	0	100
55–65	90	10

Solvent A: phosphoric acid:methanol:water = 0.5:35:65. Solvent B: phosphoric acid:methanol:water = 0.5:70:30.

4.3.3. Trolox Equivalent Antioxidant Capacity (TEAC) Assay

TEAC was measured by the ABTS radical scavenging assay. Briefly, the ABTS radical was prepared from 8 mM ABTS solution and 8.4 mM potassium persulfate solution at a ratio of 2:1. After storage in the dark at room temperature (RT) for 12–16 h, the radical solution was further diluted with ethanol to reach an initial absorbance value (0.70 ± 0.05) at 734 nm. One hundred and seventy-five microliters of the diluted ABTS solution were mixed with 25 μL of the four Cirsium species methanolic extract solutions or Trolox standard. The inhibition percentage (I%) of the radical-scavenging capacity was calculated using the following equation: $I\% = ((A_{ABTS} - A_{blank}) - (A_{s\text{-}ABTS} - A_{s\text{-}blank})) / (A_{ABTS} - A_{blank}) \times 100$, where A_{ABTS} is the absorbance of the ABTS solution, A_{blank} is the absorbance of ethanol instead of ABTS, $A_{s\text{-}ABTS}$ is the absorbance of the ABTS solution in the presence of the sample, and $A_{s\text{-}blank}$ is the absorbance of ethanol in the presence of the sample. The TEAC values are expressed as Trolox equivalents (mmol trolox/g sample) [28].

4.4. Statistical Analysis

A one-way analysis of variance (ANOVA) and then Dunnett's test were applied to data concerning serum biochemical levels, the activities of hepatic antioxidant enzymes, hepatic GSH and MDA levels, hepatic cytokines levels, and the levels of Cu/Zn-SOD, Mn-SOD, and GST expression. Significant differences in all statistical evaluations were calculated using SPSS software (version 22, IBM, Armonk, NY, USA) and p values <0.05 were considered to indicate significance.

5. Conclusions

From our present results, we suggest that four Cirsium species in Taiwan possess antioxidant and hepatoprotective effects. *C. arisanense* Kitam. possesses the highest antioxidant activity among the four Cirsium species, and its antioxidant activities are closely related to its PPG contents. *C. japonicum* DC. var. *australe* Kitam. possesses the highest hepatoprotective potency against CCl_4-induced acute liver damage among the four Cirsium species, and these hepatoprotective effects are related to its silibinin diastereomers (α and β) and diosmetin contents. The hepatoprotective mechanism of *C. japonicum* DC. var. *australe* Kitam. against CCl_4-induced acute liver damage might be involved in restoring the activities and protein expression of the hepatic antioxidant defense system and inhibiting hepatic inflammation to decrease hepatocyte necrosis. In the future, the antioxidative and anti-inflammatory cellular signaling pathway of the hepatoprotective effects against CCl_4-induced acute liver damage and the hepatoprotective effects of *C. japonicum* DC. var. *australe* Kitam. against chronic liver fibrosis should be conducted.

Author Contributions: Z.-W.Z., J.-C.C., and C.-R.W. performed the in vivo experiments and analyzed the data; L.-W.L. and F.-H.T. performed the in vitro and HPLC experiments; H.-C.C. collected and identified four Cirsium species materials; C.-R.W. conceived, designed, and supervised the study, and drafted and revised the manuscript.

Acknowledgments: We would like to thank the financial support of Ministry of Science and Technology MOST 104-2320-B-039-027 MY2, MOST 104-2622-B-039-005-CC2, and MOST 105-2622-B-039-003-CC2, and of Taichung Armed Forces General Hospital.

Conflicts of Interest: The authors declare no conflict of interest.

Abbreviations

ABTS	2,2′-Azino-bis(3-ethylbenzothiazoline-6-sulphonic acid)
ALT	Alanine aminotransferase
AST	Aspartate aminotransferase
ANOVA	One-way analysis of variance
CAH	The aerial part of *C. arisanense* Kitam.
CAH-L	Low dose (5 mg/kg) of CAH
CAH-H	High dose (50 mg/kg) of CAH
CH	Cirsii Herba
CH-L	Low dose (5 mg/kg) of CH
CH-H	High dose (50 mg/kg) of CH
CJF	The flower part of *C. japonicum* DC. var. *australe* Kitam.
CJF-L	Low dose (0.5 mg/kg) of CJF
CJF-H	High dose (5 mg/kg) of CJF
CKH	The aerial part of *C. kawakamii* Hayata
CKH-L	Low dose (5 mg/kg) of CKH
CKH-H	High dose (50 mg/kg) of CKH
CMC	Carboxymethylcellulose
Cu/Zn-SOD	Cu/Zn-superoxide dismutase
DAD	Photodiode array detector
DTNB	5,5′-Dithio-bis-(2-nitrobenzoic acid)
GPx	Glutathione peroxidase
GR	Glutathione reductase
GSH	Glutathione
GST	Glutathione *S*-transferase
H&E	Hematoxylin and eosin
IL-1β	Interleukin-1β
MDA	Malondialdehyde
Mn-SOD	Mn-superoxide dismutase

NADPH	β-Nicotinamide adenine dinucleotide phosphate
PPGs	Phenylpropanoid glycosides
ROS	Reactive oxygen species
RT	Room temperature
SOD	Superoxide dismutase
TBA	Thiobarbituric acid
TBARS	Thiobarbituric acid reactive substances
TEAC	Trolox equivalent antioxidant capacity
TNF-α	Tumor necrosis factor-α
TPs	Total phenolics

References

1. Ostapowicz, G.; Fontana, R.J.; Schiodt, F.V.; Larson, A.; Davern, T.J.; Han, S.H.; McCashland, T.M.; Shakil, A.O.; Hay, J.E.; Hynan, L.; et al. Results of a prospective study of acute liver failure at 17 tertiary care centers in the United States. *Ann. Intern. Med.* **2002**, *137*, 947–954. [CrossRef] [PubMed]
2. Zhang, W.; Dong, Z.; Chang, X.; Zhang, C.; Rong, G.; Gao, X.; Zeng, Z.; Wang, C.; Chen, Y.; Rong, Y.; et al. Protective effect of the total flavonoids from *Apocynum venetum* L. on carbon tetrachloride-induced hepatotoxicity in vitro and in vivo. *J. Physiol. Biochem.* **2018**. [CrossRef] [PubMed]
3. Yang, C.; Li, L.; Ma, Z.; Zhong, Y.; Pang, W.; Xiong, M.; Fang, S.; Li, Y. Hepatoprotective effect of methyl ferulic acid against carbon tetrachloride-induced acute liver injury in rats. *Exp. Ther. Med.* **2018**, *15*, 2228–2238. [CrossRef] [PubMed]
4. Sadek, K.M.; Saleh, E.A.; Nasr, S.M. Molecular hepatoprotective effects of lipoic acid against carbon tetrachloride-induced liver fibrosis in rats: Hepatoprotection at molecular level. *Hum. Exp. Toxicol.* **2018**, *37*, 142–154. [CrossRef] [PubMed]
5. Peng, X.; Dai, C.; Liu, Q.; Li, J.; Qiu, J. Curcumin attenuates on carbon tetrachloride-induced acute liver injury in mice via modulation of the Nrf2/HO-1 and TGF-β1/Smad3 pathway. *Molecules* **2018**, *23*. [CrossRef] [PubMed]
6. Recknagel, R.O.; Glende, E.A., Jr.; Dolak, J.A.; Waller, R.L. Mechanisms of carbon tetrachloride toxicity. *Pharmacol. Ther.* **1989**, *43*, 139–154. [CrossRef]
7. Weber, L.W.; Boll, M.; Stampfl, A. Hepatotoxicity and mechanism of action of haloalkanes: Carbon tetrachloride as a toxicological model. *Crit. Rev. Toxicol.* **2003**, *33*, 105–136. [CrossRef] [PubMed]
8. Yang, X.; Shao, H.; Chen, Y.; Ding, N.; Yang, A.; Tian, J.; Jiang, Y.; Li, G.; Jiang, Y. In renal hypertension, *Cirsium japonicum* strengthens cardiac function via the intermedin/nitric oxide pathway. *Biomed. Pharmacother.* **2018**, *101*, 787–791. [CrossRef] [PubMed]
9. Chang, N.; Li, Y.; Zhou, M.; Gao, J.; Hou, Y.; Jiang, M.; Bai, G. The hemostatic effect study of Cirsium setosum on regulating α1-ARs via mediating norepinephrine synthesis by enzyme catalysis. *Biomed. Pharmacother.* **2017**, *87*, 698–704. [CrossRef] [PubMed]
10. Yoo, O.K.; Choi, B.Y.; Park, J.O.; Lee, J.W.; Park, B.K.; Joo, C.G.; Heo, H.J.; Keum, Y.S. Ethanol extract of *Cirsium japonicum* var. *ussuriense* Kitamura exhibits the activation of nuclear factor erythroid 2-related factor 2-dependent antioxidant response element and protects human keratinocyte HaCaT cells against oxidative DNA damage. *J. Cancer Prev.* **2016**, *21*, 66–72. [CrossRef] [PubMed]
11. Ma, Q.; Wang, L.H.; Jiang, J.G. Hepatoprotective effect of flavonoids from *Cirsium japonicum* DC on hepatotoxicity in comparison with silymarin. *Food Funct.* **2016**, *7*, 2179–2184. [CrossRef] [PubMed]
12. Ma, Q.; Guo, Y.; Luo, B.; Liu, W.; Wei, R.; Yang, C.; Ding, C.; Xu, X.; He, M. Hepatoprotective phenylethanoid glycosides from Cirsium setosum. *Nat. Prod. Res.* **2016**, *30*, 1824–1829. [CrossRef] [PubMed]
13. Wan, Y.; Liu, L.Y.; Hong, Z.F.; Peng, J. Ethanol extract of *Cirsium japonicum* attenuates hepatic lipid accumulation via AMPK activation in human HepG2 cells. *Exp. Ther. Med.* **2014**, *8*, 79–84. [CrossRef] [PubMed]
14. De la Pena, J.B.; Kim, C.A.; Lee, H.L.; Yoon, S.Y.; Kim, H.J.; Hong, E.Y.; Kim, G.H.; Ryu, J.H.; Lee, Y.S.; Kim, K.M.; et al. Luteolin mediates the antidepressant-like effects of *Cirsium japonicum* in mice, possibly through modulation of the GABA$_A$ receptor. *Arch. Pharm. Res.* **2014**, *37*, 263–269. [CrossRef] [PubMed]

15. Dela Pena, I.J.; Lee, H.L.; Yoon, S.Y.; Dela Pena, J.B.; Kim, H.K.; Hong, E.Y.; Cheong, J.H. The ethanol extract of Cirsium japonicum increased chloride ion influx through stimulating GABA$_A$ receptor in human neuroblastoma cells and exhibited anxiolytic-like effects in mice. *Drug Discov. Ther.* **2013**, *7*, 18–23. [CrossRef] [PubMed]

16. Yin, J.; Heo, S.I.; Wang, M.H. Antioxidant and antidiabetic activities of extracts from *Cirsium japonicum* roots. *Nutr. Res. Pract.* **2008**, *2*, 247–251. [CrossRef] [PubMed]

17. Wellington, K.; Jarvis, B. Silymarin: A review of its clinical properties in the management of hepatic disorders. *BioDrugs* **2001**, *15*, 465–489. [CrossRef] [PubMed]

18. Saller, R.; Meier, R.; Brignoli, R. The use of silymarin in the treatment of liver diseases. *Drugs* **2001**, *61*, 2035–2063. [CrossRef] [PubMed]

19. Abenavoli, L.; Milic, N.; Capasso, F. Anti-oxidant therapy in non-alcoholic fatty liver disease: The role of silymarin. *Endocrine* **2012**, *42*, 754–755. [CrossRef] [PubMed]

20. Lieber, C.S.; Leo, M.A.; Cao, Q.; Ren, C.; DeCarli, L.M. Silymarin retards the progression of alcohol-induced hepatic fibrosis in baboons. *J. Clin. Gastroenterol.* **2003**, *37*, 336–339. [CrossRef] [PubMed]

21. Yang, Y.P.; Lu, S.Y. Cirsium. In *Flora of Taiwan*, 2nd ed.; Huang, T.C., Ed.; Editorial Committee of the Flora of Taiwan, Department of Botany, National Taiwan University: Taipei, Taiwan, 2000; Volume 4, pp. 903–911.

22. Li, L.; Sun, Z.; Shang, X.; Li, J.; Wang, R.; Zhu, J. Triterpene compounds from *Cirsium setosum*. *Zhongguo Zhong Yao Za Zhi* **2012**, *37*, 951–955. [PubMed]

23. Ganzera, M.; Pocher, A.; Stuppner, H. Differentiation of *Cirsium japonicum* and *C. setosum* by TLC and HPLC-MS. *Phytochem. Anal.* **2005**, *16*, 205–209. [CrossRef] [PubMed]

24. Shang, D.L.; Ma, Q.G.; Wei, R.R. Cytotoxic phenylpropanoid glycosides from *Cirsium japonicum*. *J. Asian Nat. Prod. Res.* **2016**, *18*, 1122–1130. [CrossRef] [PubMed]

25. Lai, W.C.; Wu, Y.C.; Danko, B.; Cheng, Y.B.; Hsieh, T.J.; Hsieh, C.T.; Tsai, Y.C.; El-Shazly, M.; Martins, A.; Hohmann, J.; et al. Bioactive constituents of *Cirsium japonicum* var. *australe*. *J. Nat. Prod.* **2014**, *77*, 1624–1631. [CrossRef] [PubMed]

26. Hu, Z.; Lausted, C.; Yoo, H.; Yan, X.; Brightman, A.; Chen, J.; Wang, W.; Bu, X.; Hood, L. Quantitative liver-specific protein fingerprint in blood: A signature for hepatotoxicity. *Theranostics* **2014**, *4*, 215–228. [CrossRef] [PubMed]

27. Ku, K.L.; Tsai, C.T.; Chang, W.M.; Shen, M.L.; Wu, C.T.; Liao, H.F. Hepatoprotective effect of *Cirsium arisanense* Kitamura in tacrine-treated hepatoma Hep 3B cells and C57BL mice. *Am. J. Chin. Med.* **2008**, *36*, 355–368. [CrossRef] [PubMed]

28. Wu, C.R.; Lin, W.H.; Hseu, Y.C.; Lien, J.C.; Lin, Y.T.; Kuo, T.P.; Ching, H. Evaluation of the antioxidant activity of five endemic *Ligustrum* species leaves from Taiwan flora in vitro. *Food Chem.* **2011**, *127*, 564–571. [CrossRef] [PubMed]

29. Song, Z.; Deaciuc, I.; Song, M.; Lee, D.Y.; Liu, Y.; Ji, X.; McClain, C. Silymarin protects against acute ethanol-induced hepatotoxicity in mice. *Alcohol Clin. Exp. Res.* **2006**, *30*, 407–413. [CrossRef] [PubMed]

30. Papackova, Z.; Heczkova, M.; Dankova, H.; Sticova, E.; Lodererova, A.; Bartonova, L.; Poruba, M.; Cahova, M. Silymarin prevents acetaminophen-induced hepatotoxicity in mice. *PLoS ONE* **2018**, *13*, e0191353. [CrossRef] [PubMed]

31. Letteron, P.; Labbe, G.; Degott, C.; Berson, A.; Fromenty, B.; Delaforge, M.; Larrey, D.; Pessayre, D. Mechanism for the protective effects of silymarin against carbon tetrachloride-induced lipid peroxidation and hepatotoxicity in mice. Evidence that silymarin acts both as an inhibitor of metabolic activation and as a chain-breaking antioxidant. *Biochem. Pharmacol.* **1990**, *39*, 2027–2034. [CrossRef]

32. Kidd, P.; Head, K. A review of the bioavailability and clinical efficacy of milk thistle phytosome: A silybin-phosphatidylcholine complex (Siliphos). *Altern. Med. Rev.* **2005**, *10*, 193–203. [PubMed]

33. Arauz, J.; Ramos-Tovar, E.; Muriel, P. Redox state and methods to evaluate oxidative stress in liver damage: From bench to bedside. *Ann. Hepatol.* **2016**, *15*, 160–173. [PubMed]

34. Feher, J.; Lang, I.; Deak, G.; Cornides, A.; Nekam, K.; Gergely, P. Free radicals in tissue damage in liver diseases and therapeutic approach. *Tokai J. Exp. Clin. Med.* **1986**, *11 Suppl*, 121–134. [PubMed]

35. Townsend, D.M.; Tew, K.D. The role of glutathione-*S*-transferase in anti-cancer drug resistance. *Oncogene* **2003**, *22*, 7369–7375. [CrossRef] [PubMed]

36. Fukai, T.; Ushio-Fukai, M. Superoxide dismutases: Role in redox signaling, vascular function, and diseases. *Antioxid. Redox. Signal.* **2011**, *15*, 1583–1606. [CrossRef] [PubMed]

37. Malhi, H.; Guicciardi, M.E.; Gores, G.J. Hepatocyte death: A clear and present danger. *Physiol. Rev.* **2010**, *90*, 1165–1194. [CrossRef] [PubMed]
38. Chen, J.J.; Zhang, J.X.; Zhang, X.Q.; Qi, M.J.; Shi, M.Z.; Yang, J.; Zhang, K.Z.; Guo, C.; Han, Y.L. Effects of diosmetin on nine cytochrome P450 isoforms, UGTs and three drug transporters in vitro. *Toxicol. Appl. Pharmacol.* **2017**, *334*, 1–7. [CrossRef] [PubMed]
39. Yang, Y.; Gong, X.B.; Huang, L.G.; Wang, Z.X.; Wan, R.Z.; Zhang, P.; Zhang, Q.Y.; Chen, Z.; Zhang, B.S. Diosmetin exerts anti-oxidative, anti-inflammatory and anti-apoptotic effects to protect against endotoxin-induced acute hepatic failure in mice. *Oncotarget* **2017**, *8*, 30723–30733. [CrossRef] [PubMed]
40. Jiang, W.; Gao, M.; Sun, S.; Bi, A.; Xin, Y.; Han, X.; Wang, L.; Yin, Z.; Luo, L. Protective effect of L-theanine on carbon tetrachloride-induced acute liver injury in mice. *Biochem. Biophys. Res. Commun.* **2012**, *422*, 344–350. [CrossRef] [PubMed]
41. Kuo, H.C.; Chang, H.C.; Lan, W.C.; Tsai, F.H.; Liao, J.C.; Wu, C.R. Protective effects of *Drynaria fortunei* against 6-hydroxydopamine-induced oxidative damage in B35 cells via the PI3K/AKT pathway. *Food Funct.* **2014**, *5*, 1956–1965. [CrossRef] [PubMed]
42. Saleh, I.A.; Vinatoru, M.; Mason, T.J.; Abdel-Azim, N.S.; Shams, K.A.; Aboutabl, E.; Hammouda, F.M. Extraction of silymarin from milk thistle (*Silybum marianum*) seeds—A comparison of conventional and microwave-assisted extraction methods. *J. Microw. Power Electromagn. Energy* **2017**, *51*, 124–133. [CrossRef]

International Journal of
Molecular Sciences

MDPI

Article

Metabolic Reprogramming by 3-Iodothyronamine (T1AM): A New Perspective to Reverse Obesity through Co-Regulation of Sirtuin 4 and 6 Expression

Fariba M. Assadi-Porter [1,2,*], Hannah Reiland [1], Martina Sabatini [3], Leonardo Lorenzini [3], Vittoria Carnicelli [3], Micheal Rogowski [4], Ebru S. Selen Alpergin [1,5], Marco Tonelli [2], Sandra Ghelardoni [3], Alessandro Saba [3], Riccardo Zucchi [3] and Grazia Chiellini [3,6,***]**

[1] Department of Integrative Biology, University of Wisconsin-Madison, Madison, WI 53706, USA; hreiland5@gmail.com (H.R.); eselinselen@gmail.com (E.S.S.A.)
[2] National Magnetic Resonance Facility at Madison, Madison, WI 53706, USA; tonelli@nmrfam.wisc.edu
[3] Department of Surgical Pathology, Medicine, Molecular and Critical Area, University of Pisa, 56126 Pisa, Italy; marti.saba88@gmail.com (M.S.); lorenzini.leo@gmail.com (L.L.); vittoria.carnicelli@discau.unipi.it (V.C.); sandra.ghelardoni@med.unipi.it (S.G.); alessandro.saba@med.unipi.it (A.S.); riccardo.zucchi@med.unipi.it (R.Z.)
[4] School of Medicine, Division of Cardiovascular Disease, University of Alabama at Birmingham, Birmingham, AL 35233, USA; rogowskimp@gmail.com
[5] Department of Biological Chemistry, Johns Hopkins University, Baltimore, MD 21205, USA
[6] Department of Biochemistry, University of Wisconsin-Madison, 433 Babcock Drive, Madison, WI 53706-1544, USA
[*] Correspondence: fariba@nmrfam.wisc.edu (F.M.A.-P.); g.chiellini@bm.med.unipi.it (G.C.)

Received: 22 April 2018; Accepted: 15 May 2018; Published: 22 May 2018

Abstract: Obesity is a complex disease associated with environmental and genetic factors. 3-Iodothyronamine (T1AM) has revealed great potential as an effective weight loss drug. We used metabolomics and associated transcriptional gene and protein expression analysis to investigate the tissue specific metabolic reprogramming effects of subchronic T1AM treatment at two pharmacological daily doses (10 and 25 mg/kg) on targeted metabolic pathways. Multi-analytical results indicated that T1AM at 25 mg/kg can act as a novel master regulator of both glucose and lipid metabolism in mice through sirtuin-mediated pathways. In liver, we observed an increased gene and protein expression of *Sirt6* (a master gene regulator of glucose) and *Gck* (glucose kinase) and a decreased expression of *Sirt4* (a negative regulator of fatty acids oxidation (FAO)), whereas in white adipose tissue only *Sirt6* was increased. Metabolomics analysis supported physiological changes at both doses with most increases in FAO, glycolysis indicators and the mitochondrial substrate, at the highest dose of T1AM. Together our results suggest that T1AM acts through sirtuin-mediated pathways to metabolically reprogram fatty acid and glucose metabolism possibly through small molecules signaling. Our novel mechanistic findings indicate that T1AM has a great potential as a drug for the treatment of obesity and possibly diabetes.

Keywords: 3-iodothyronamine; metabolomics; obesity; glucose and lipid metabolism; sirtuins

1. Introduction

3-Iodothyronamine (T1AM) (Figure 1) was identified in 2004 as a novel chemical messenger [1], putatively derived from thyroid hormone de–iodination and decarboxylation. T1AM interacts directly with a specific G-protein coupled receptor known as trace amine-associated receptor 1 (TAAR1), which is expressed in brain and in many other tissues [2,3]. Subsequent investigations have confirmed that T1AM can be detected in blood and in most tissues, with concentrations in the nanomolar range [4–6].

The number of potential T1AM targets has expanded, in addition to TAAR1, T1AM has been reported to interact with TAAR5 [7], adrenergic receptors [8] and several monoamine transporters [9], suggesting that it should be considered as a multi-targeted ligand.

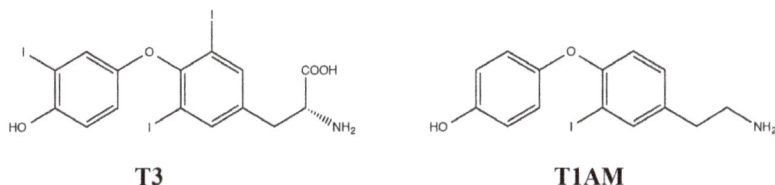

Figure 1. Structures of Thyroid hormone (T3) and 3-iodothyronamine (T1AM).

The physiological functional responses that were initially observed after administration of exogenous T1AM in the micromolar range included reduced body temperature and cardiac contractility [1,10,11]. However, in subsequent investigations neurologic, endocrine and metabolic responses were elicited at much lower dosages in the submicromolar range that are closer to the physiological range [12,13].

In the hamster and mouse, administration of exogenous T1AM produced persistent weight loss [14,15], raising interest for the potential role of this compound in the treatment of obesity. The use of cavity ring-down spectroscopy (CRDS) and ^{13}C-glucose dosing to measure lipid oxidation by real-time breath $^{13}CO_2/^{12}CO_2$ (δ^{13}C)), showed that an intraperitoneal injection of T1AM (10 mg/kg daily for up to one-week) resulted in a switch from carbohydrate to lipid metabolism lasting for a maximum of four days [15]. The metabolome profiles identified an unusually rapid action of T1AM on both glucose and lipid metabolism associated with weight loss regime, followed by a slower action on lipolysis. However, after day 5 of treatment with T1AM, protein catabolism was stimulated indicating a potential adverse effect of T1AM occurring on a longer time scale.

On the basis of these findings, the purpose of the present study was to provide a comprehensive insight into the metabolic responses to T1AM chronic treatment. We hypothesized that mice given injections of T1AM would experience a switch from carbohydrate to lipid metabolism, leading to observable levels of small molecule lipid intermediates that may impact associated gene expression levels. To test this hypothesis, mice were exposed to two different T1AM dosages (10 and 25 mg/kg) and two different treatment times (4 and 7 days). Since in our previous pilot study we observed that the metabolic effects of T1AM were similar under 4 h fasting and non-fasting conditions [15], in the present study we also introduced a short-term fasting period (4 h), that is adequate to clear immediate diet [16]. Since the effects of T1AM on glucose and lipid metabolism appear to outlast all the other effects, we used nuclear magnetic resonance (NMR) spectroscopy to obtain unbiased metabolomics profiles of tissues and analyzed changes in the expression of genes known to have a significant role in molecular mechanisms relevant to altered glucose and lipid metabolism profiles. In addition, we also employed high performance liquid chromatography coupled with tandem mass spectrometry (LC/MS-MS) to determine the final concentrations of non-metabolized T1AM in different tissues at the end of T1AM treatments to link to its tissue specific physiological effects.

2. Results

2.1. Food Consumption, Activity Levels, Weight Loss and Lipid Profiles

We administered T1AM to spontaneously overweight female Institute for Cancer Research; Caesarean Derived-1 (CD-1®) mice. We used two different dosages (10 and 25 mg/kg/day) for 7 days to examine the effects of T1AM on food intake, activity levels, body weight and lipid profiles. No significant differences in food consumption between T1AM-treated and control animals were observed during the study period. Video monitoring did not reveal any noticeable difference in activity level between the three groups of animals.

Chronic administration of 10 mg/kg/day T1AM showed a 10% body weight loss by day 7 of treatment (Table 1). Body weight loss reached 18% after day 7 of treatment with 25 mg/kg/day T1AM.

Table 1. Effect of T1AM treatment on weight, serum cholesterol, triglyceride and glucose in spontaneously overweight female CD-1 mice.

Assay	Saline	10 mg/kg/day	25 mg/kg/day
Weight loss (g)	−1.6 ± 2.2	−3.8 ± 0.7 [b]	−8.1 ± 3.2 [a]
Serum Cholesterol (mg/dL)	90.8 ± 8.57	86.7 ± 7.84	77.7 ± 3.61 [a]
Serum TG (mg/dL)	49.0 ± 3.46	52.11 ± 10.61	67.3 ± 4.23 [a]
Serum glucose (mg/dL)	151 ± 21.60	165 ± 35.74	161.52 ± 24.02

Data are shown as mean ± SEM. Significance assigned by [a] $p < 0.05$ and [b] $p < 0.07$ versus control.

During the seven days of treatment control mice only lost 4% of their initial body weight, thus the net body weight loss was 6% after treatment with 10 mg/kg/day T1AM and 14% for 25 mg/kg/day T1AM. Blood samples taken on day 7 revealed that the high dose T1AM treatment produced a significant increase in total plasma triglycerides ($67.3 ± 4.23$ vs. $49.0 ± 3.46$ mg/dL, $p < 0.05$) and a significant decrease in plasma cholesterol ($77.7 ± 3.61$ vs. $90.8 ± 8.57$ mg/dL, $p < 0.05$), without any significant change in glycaemia (Table 1).

2.2. Nuclear Magnetic Resonance (NMR) Metabolome Analysis

2.2.1. Plasma Metabolome

Metabolic profiling of plasma using ^{1}H-NMR analysis identified 22 metabolites that belong to major metabolic pathways. Partial least squares discriminant analysis (PLSDA) of plasma metabolome profiles showed that treatments with T1AM (10 or 25 mg/kg/day) resulted in clear separations between T1AM treated and control mice at days 4 and 7, indicating marked effects of T1AM treatment on primary metabolism (Figure 2A). Score plots are shown in Figure 2B indicating major contributing metabolites involving carbohydrate, lipid, amino acids, nucleotide and antioxidant metabolism pathways.

Heat map representations were used to visualize the magnitude and directionality of change in metabolite levels at two doses and two times points in plasma (Figure 3A) under short-term experimental fasting conditions.

Collectively, higher dose T1AM treatment appeared to elicit extensive changes in metabolism by increasing or decreasing the level of metabolites involved in major metabolic pathways, including antioxidant metabolic intermediates (Asc and Bet). However, statistically significant changes were only observed for a smaller subset of metabolites belonging to glycolysis (Ala and lactate), lipolysis (acetate and 3-HB) and amino acid metabolism (the branched chain amino acid (Ile) and (Ala)) (Figure 3B). Changes appeared to follow the dose level and the period of treatment with T1AM. Specifically, plasma samples taken at Day 4 and 7 of treatment revealed increasing lactate (Lac) concentrations for both T1AM doses, however, the concentration level of lactate in the low dose T1AM only reached significance on Day 7 of treatment, indicating increased glycolysis. On the other hand, the glucogenic amino acid, Ala showed an increased trend at 10 mg/kg at Day 4 and 7 while reached significance ($p < 0.05$) at 25 mg/kg dose and at Day 7 of treatment as compared to the saline treated group. In addition, the plasma level of the ketone body, 3-hydroxybutyrate (3-HB) and that of an intermediate in lipid metabolism, acetate, dramatically decreased from Day 4 to Day 7 at both T1AM dosages. These results suggest a high demand for these metabolites as carbon sources for energy metabolism under our short-term fasting condition. As expected, 3-HB and acetate levels in the control (saline) groups were not changed from Day 4 to 7, indicating the observed changes are due to T1AM treatment.

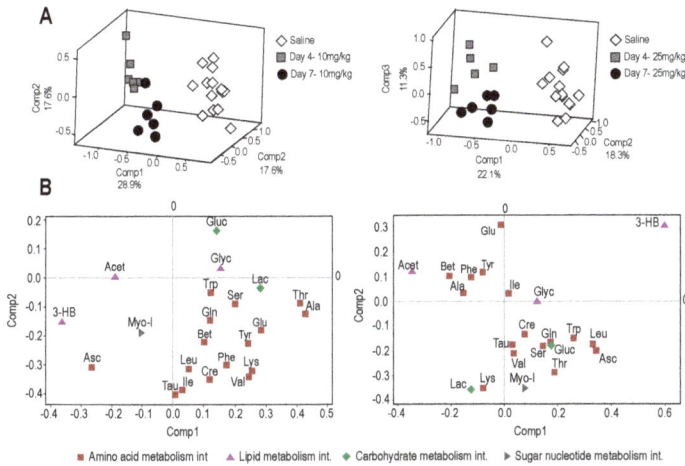

Figure 2. Partial least square discriminant analysis (PLSDA) plots of plasma with and without T1AM treatments show separation of metabolic profiles in obese groups. (**A,B**) show PLSDA and score plots for T1AM treatment with 10 mg/kg body weight/day and 25mg/kg body weight/day at two time points (Day 4 and Day 7). Dark gray squares and black circles represent T1AM treatment groups for 4 and 7 days and white diamonds show saline (control) groups. Abbrev: Fatty acid metabolism (black triangles): Glyc, glycerol; Acet, acetate; 3-HB, 3-hydroxy butyrate; antioxidant (Asc, ascorbate) and amino acid metabolism (red rectangular): Tau, taurine; Leu, leucine; Ile, isoleucine; Cre, creatine; Bet, betaine; Gln, glutamine; Trp, tryptophan; Tyr, tyrosine; Ser, serine; Glu, glutamate; Lys, lysine; Val, valine; Thr, threonine; Ala, alanine. Carbohydrate metabolism (green diamonds): Gluc, glucose; Lac, lactate and Myo-I, myoinositol.

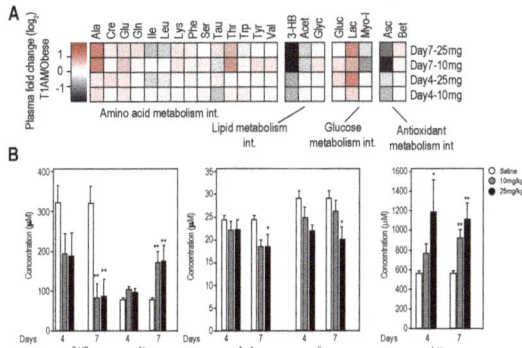

Figure 3. Plasma metabolome profiles. (**A**) Heat map of plasma metabolite fold changes indicates increases or decreases in metabolite concentrations in response to different doses and time periods of T1AM treatments. Differential changes are shown for each identified and quantified metabolite at 10 and 25 mg/kg/day at days 4 and 7. Increased or decreased metabolome levels are color coded for each dose and time point. (**B**) shows bar plots of significantly changed metabolites levels in the T1AM-treated groups (10 mg/kg/day or 25 mg/kg/day) vs. saline group. The X-axis shows metabolites ID for Days 4 and 7; and shaded codes are defined as: control mice (white bar), 10 mg/kg/day T1AM-treated mice (gray bar) and 25 mg/kg/day.T1AM-treated mice (black bar). Y-axis shows relative concentration levels to formate (internal standard). In all analyses, statistical significance is shown as: * $p < 0.05$, saline vs. treatment, ** $p < 0.01$ saline vs. treatment, respectively.

2.2.2. Liver and Muscle Metabolome Profiles

^{1}H-NMR analysis of liver extracts after seven days of T1AM treatment revealed more pronounced dose dependent effects. Indeed, most of the primary hepatic metabolites reached significance at the high T1AM dosage (25 mg/kg). With the exception of 2-hydroxy butyrate (2-HB) all changes in metabolite concentrations showed dose-dependent increases in liver. The observed effects were on lipid catabolism, mitochondrial energy metabolism, amino acid and nucleotide metabolism. In particular, evidence for increased fatty acid oxidation in the liver, the primary site of fatty acid metabolism, comes from increased levels of ketone bodies (3-HB and acetone), carnitine (carn, a fatty acyl carrier that mediates mitochondrial fatty acid oxidation) and succinate (a mitochondrial energy metabolism substrate), as shown in Figure 4A.

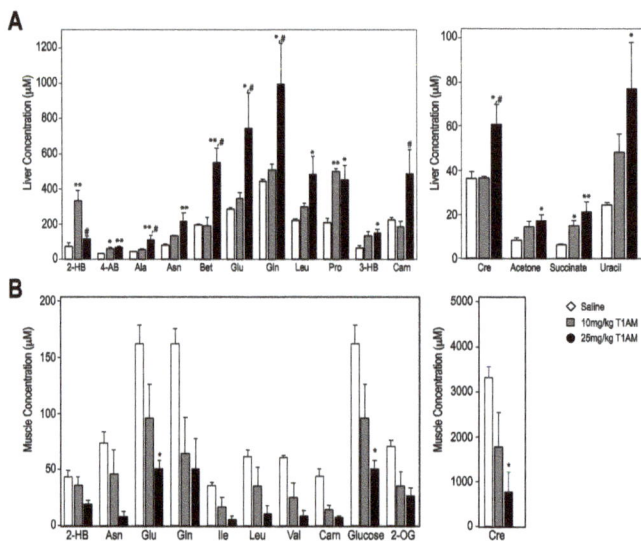

Figure 4. Selected metabolites from liver (**A**) and muscle (**B**) from ^{1}H-NMR-based metabolomics profiles at two T1AM doses for seven days treatment. The X-axis shows metabolites IDs and the Y-axis shows relative concentration levels (M) to formate. Bars are shaded according to each treatment group as: control mice (white), treatment with 10 mg/kg/day (gray) and 25 mg/kg/day (black). The Y-axis shows relative concentration levels to formate (internal standard). All data are expressed as the mean $\pm$ SEM (standard error of mean). In all analyses, statistical significance is shown as: * $p < 0.05$ vs. saline; ** $p < 0.01$ vs. saline; # $p < 0.05$ 10 mg vs. 25 mg.

Other primary hepatic metabolites such as amino acids (alanine (Ala), asparagine (Asn), proline (Pro) and leucine (Leu)) and their nitrogen metabolic byproducts (creatine (Cren)) and sugar nucleotide uracil levels were all increased in the T1AM group in a dose-dependent manner. However, changes in liver metabolome profiles for low dose T1AM treatment were less pronounced as compared to the higher T1AM dosage. Only 2-hydroxybutyrate (2-HB), 4-amino-butyrate (4-AB), Pro and succinate reached significance ($p < 0.05$) at 10 mg/kg T1AM dosage.

^{1}H-NMR metabolome analysis of skeletal muscle extracts (Figure 4B) revealed decreasing trends in amino acid and glucose concentrations relative to the saline group. The levels of glucose and amino acids including branched chain amino acids (BCAA, that is, Leu, Ile and Val) were all decreased in a dose dependent manner. In addition, glutamate (Glu), glutamine (Gln), asparagine (Asn) and creatine (Cre) levels were also decreased. Although only Glu and Cre reached statistical significance ($p < 0.05$),

these data suggested that T1AM shifts metabolism in muscles towards favoring glucose and amino acids to support its energy demand.

2.3. RT-qPCR Results

We recently provided the first evidence that T1AM at a subchronic, lower pharmacological dose administration, induced a rapid shift in metabolic pathways from carbohydrate to lipid oxidation [15]. The molecular mechanism by which T1AM favors lipid over glucose catabolism is not known. However, considering the lasting effects of T1AM on fatty acid metabolism [14,15], we speculated that the shift in energy utilization observed in mice after T1AM treatment arises from changes in gene expression. Therefore, targeted gene expression profiles were analyzed in selected metabolically active tissues, including liver, adipose tissue, skeletal muscle and heart, of mice treated with two different T1AM dosages (10 and 25 mg/kg daily up to seven days) as compared with saline treated mice. Our results revealed that at least in tissues principally involved in energy metabolism, such as liver, adipose tissue and skeletal muscle, T1AM administration uniquely impacted the expression of a large set of genes linked to carbohydrate and lipid metabolism (Figure 5A).

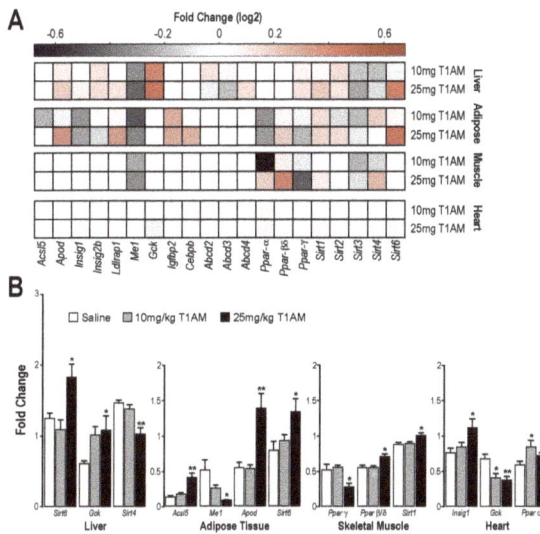

Figure 5. Changes in gene expression evidenced by RT-qPCR in metabolically active tissues. (**A**) Heat map representation of gene expression in liver, adipose tissue, skeletal muscle and heart, respectively. Differential changes in gene expression (Saline treated—T1AM treated) are color coded. Dark and light red shades show > 0; black and gray shades show < 0 in $\log_2$ base. (**B**) Fold changes in gene expression over the appropriate control are plotted on a log2 scale ($n = 5$). All data are expressed as the mean ± SEM. * $p < 0.05$ vs. saline treated; ** $p < 0.01$ vs. saline treated.

In liver (Figure 5B), T1AM (25 mg/kg/day) significantly up-regulated the expression of genes related to glucose homeostasis and fat metabolism, namely *Sirt6* (sirtuin (silent mating type information regulation 2 homolog) 6) and *Gck* (glucokinase). *Sirt6* regulates glycolysis, triglyceride synthesis and fat metabolism by deacetylating histone H3 lysine 9 in the promoter of many genes involved in these processes [17]. Liver specific deletion of *Sirt6* in mice results in fatty liver formation due to enhanced glycolysis and triglyceride synthesis [18]. *Gck* plays a central role as a glucose sensor in the regulation of glucose homeostasis. In addition, a decreased expression of *Sirt4* (sirtuin (silent mating type information regulation 2 homolog) 4), a gene known to repress fatty acid (FA) oxidation while promoting lipid anabolism [19], was observed in liver after treatment with T1AM (25 mg/kg/day).

In white adipose tissue (Figure 5B), the highest T1AM treatment regime, specifically impacted the expression of genes targeting lipid metabolism and lipoprotein function. *Acsl5* (acyl-CoA synthetase long-chain family member 5), up-regulated by T1AM, is the only ACSL isoform localized on the mitochondrial outer membrane and it is believed to play an important role in the beta-oxidation of fatty acids [20]. *Me1* (malic enzyme 1, NADP+-dependent, cytosolic), a lipogenic enzyme that generates NADPH required for fatty acid synthesis, appeared to be down regulated by T1AM. *Apod* (Apolipoprotein D) and *Sirt6* were both up-regulated. APOD is an apolipoprotein structurally related to the lipocalin family proteins that is involved in diverse aspects of metabolism, including lipid transport. Aberrant APOD expression is associated with abnormal lipid metabolism [21]. SIRT6 overexpression has been shown to repress the expression of selected peroxisome proliferator-activated receptor gamma (PPARγ) target genes and key lipid metabolism genes [22] important for triggering lipolysis. A recent report highlighted that SIRT6 deficiency in fat tissue predisposes mice to obesity, insulin resistance and hepatosteatosis [23].

In skeletal muscle tissue (Figure 5B), treatment with T1AM at the highest dose significantly impacted the expression of peroxisome proliferator-activated receptors (*Ppars*), which act as lipid sensors, namely *Pparγ* (related to lipid anabolism, down-regulated) and *Pparβ/δ* (related to fatty acid oxidation, up-regulated). Both of these receptors are also linked to SIRT1 protein activity, with *Pparγ* inhibiting *Sirt1* expression, while *Pparβ/δ* increasing *Sirt1* mRNA levels, which in turn results in increased lipid oxidation, mitochondrial biogenesis and increased insulin sensitivity [24]. An increased expression level of *Sirt1* was observed consistently in skeletal muscle of the T1AM treated group.

Previous studies showed that T1AM treatment has a protective effect in the heart due to a reduction in oxidative stress [25]. Our results indicate that in the heart, T1AM administration impacted the expression of several genes involved in cardiac energy metabolism. High dose T1AM treatment increased *Insig-1* (insulin-induced gene 1), that is a gene known to have anti-lipogenic action. The glucose level sensor *Gck*, was down regulated at both T1AM dosages, whereas only the lower dosage of T1AM up-regulated *Pparα* gene known to play a major role in the control of cardiac energy metabolism (Figure 5B).

2.4. Western Blot of Tissues

Our transcriptional analysis results indicated that mice treated with the highest dose of T1AM undergo significant tissue specific changes in gene expression. Protein expression studies by Western blotting confirmed over-expression of sirtuin 6 (SIRT6) and glucokinase (GCK) proteins in liver (Figure 6).

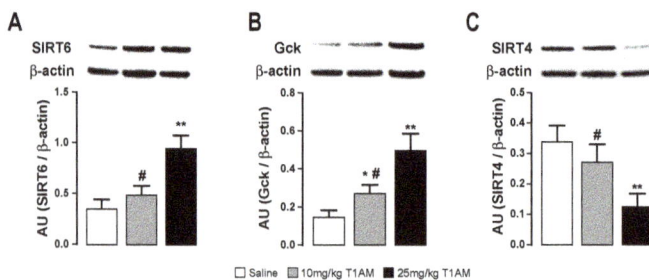

Figure 6. Effects of T1AM treatment on protein expression in liver. Immunodetection for (**A**) Sirtuin (silent mating type information regulation 2 homolog) 6 (SIRT6), (**B**) Glucokinase (GCK) and (**C**) SIRT4 was carried out on protein lysates separated on SDS-PAGE gels. A representative experiment is shown. Results are the mean ± SEM of the densitometry of three different gels. * $p < 0.05$ vs. saline treated; ** $p < 0.01$ vs. saline treated; # $p < 0.05$ T1AM 10 mg/kg vs. T1AM 25 mg/kg.

In addition, after treatment with 25 mg/kg/day T1AM, a decreased expression of SIRT4 was also observed (Figure 6). In agreement with the transcriptional findings, the expression of SIRT6 in adipose tissue showed an increasing trend when T1AM was used at the highest dose, however, this trend did not reach statistical significance and it is not reported in Figure 6.

2.5. Tissue T1AM Concentration

The LC/MS-MS assay of T1AM tissue distribution showed that T1AM is distributed in a tissue specific manner (Table 2).

Table 2. T1AM distribution in tissues by liquid chromatography mass spectrometry (LC/MS-MS).

Tissue	Saline	Low T1AM (10 mg/kg/day)	High T1AM (25 mg/kg/day)
Liver [+]	7.68 ± 0.85	318.3 ± 35.09 [a]	767.0 ± 165.40 [b]
White Adipose [+]	0.493 ± 0.17	1.71 ± 0.30 [b]	16.96 ± 3.96 [b]
Muscle [+]	19.84 ± 3.57	56.52 ± 12.64	412.86 ± 109.14 [b]
Heart [+]	18.15 ± 4.38	37.62 ± 0.76	68.19 ± 14.10 [b]

[+] Concentration of T1AM (pmol/g) determined in tissues of saline and T1AM treated mice. [a] $p < 0.01$ vs. control; [b] $p < 0.05$ vs. control.

After 7 days of treatment with 10 mg/kg/day, the remaining concentration of T1AM in adipose tissue, muscle and liver, were 3- to 40-fold higher than controls, respectively. In mice treated for 7 days with 25 mg/kg/day T1AM, a 35- to 100-fold increase was observed in adipose and liver over the control baselines, respectively, whereas in muscle there was only a 20-fold increase as compared to controls. On the other hand, in the heart, the relative amount of T1AM was only about 4-fold higher than control after 7 days of treatment with the highest dose of the drug.

3. Discussion

Recently, we provided the first evidence that sub-chronic treatment with the lowest pharmacological dose of T1AM (10 mg/kg/day) in normal obese mice increased lipolysis associated with significant weight loss but independent of food consumption [15]. In order to get a thorough evaluation of the metabolic response in mice to chronic treatment with T1AM, in the present study we used metabolomics and associated transcriptional gene and protein expression analyses to investigate the tissue specific metabolic reprogramming effects of sub-chronic T1AM treatment (7 days) at two pharmacological doses (10 and 25 mg/kg/day).

Our current observation shows that the net weight loss after treatment with 10 mg/kg/day T1AM for 7 days was 6%, while 7 days of treatment with 25 mg/kg/day T1AM resulted in a 14% net weight loss without affecting any apparent food intake and animal behavior consistent with previous findings [14,15,26]. As expected, the higher dose was more effective at inducing a larger weight loss. Consistent with weight loss and T1AM distribution in tissues, biochemical assays, metabolome profiling and gene and protein expression studies demonstrated significant changes predominantly when the highest T1AM dosage was used and after seven days of treatment.

Multivariate statistical analysis of the ^{1}H-NMR based metabolomics data sets reveals marked effects of T1AM treatment on primary metabolism, including amino acid, lipid, carbohydrate and sugar nucleotide metabolism. Increased ketone bodies (3-HB and acetone) in liver were indicative of increased lipid oxidation. Even though no significant differences were observed in plasma glucose levels at both doses, measurement of the plasma metabolome indicated that under the short-term fasting condition, treatment with T1AM induced a high demand for energy metabolism through consumption of glycolytic and lipolytic metabolites. At higher dose (25 mg/kg) of T1AM treatment we observed 3-HB, acetate and Ile were reduced, while lactate was increased when compared to the control group injected with saline. In our previous pilot study [15] the level of 3-HB was increased in plasma under fed condition and injection of ^{13}C-glucose at 10 mg/kg T1AM dosage, indicating increased lipid oxidation in plasma despite abundance of glucose for immediate energy metabolism. In both studies, animals did not show increased food consumption, however, under a short-term fasting condition, we postulate that increased level of T1AM treatment elicited an increase in energy demand by other tissues. The observed decrease in 3-HB and acetate plasma levels might indeed reflect increased uptake/consumption by other tissues to meet the higher energy demand. A caveat of the present study is that we cannot completely exclude that the observed effects of T1AM may be partially be induced by the three morning-fasting periods to which the mice were subjected.

Measurement of plasma lipid levels revealed that sub-chronic treatment with T1AM at the highest dose induced a reduction of plasma total cholesterol accompanied by increased triglycerides (TG) levels. The latter increase might be related to an increasing mobilization of triglycerides from the adipose tissue, consistent with increased plasma glycerol levels and in agreement with the metabolic changes related to a higher energy demand after administering T1AM at the highest dose.

Changes in the levels of liver small molecular size metabolic markers (e.g., lactate, acetate, ketone bodies and amino acids), confirm the efficacy of the treatment to induce weight loss. Administration of T1AM at the highest dose (25 mg/kg/day) affected the expression of numerous genes, especially in metabolically active tissues, such as liver, white adipose tissue and skeletal and cardiac muscles. Notably, gene and protein expression findings showed that when T1AM is exogenously administered to mice, it modulates the expression of sirtuins involved in energetic metabolic pathways, namely SIRT6, SIRT4 (in liver and adipose tissue) and *Sirt1* gene (in muscle) through modulation of PPARs (*Pparγ* and *Pparβ/δ*). The mammalian sirtuins (SIRT1–7) are a class of NAD$^+$-dependent protein deacetylases and/or ADP-ribosyltransferases that regulate a large variety of cellular, physiological and metabolic processes including cell cycle, apoptosis, energy homeostasis, mitochondrial function and longevity [27,28]. Results from recent studies indicate that reduced sirtuin action, as observed with aging and high-fat feeding, is related with Type 2 diabetes. Consequently, sirtuin activators are rapidly emerging as an effective therapeutic strategy against diabetes [29].

Even though the most well-studied sirtuin protein with effects on metabolism is SIRT1, growing evidence indicates that nuclear SIRT6 plays fundamental roles in the maintenance of glucose and lipid homeostasis [30,31]. It negatively regulates glycolysis, triglyceride synthesis and fat metabolism. Accordingly, SIRT6-overexpressing mice are protected from diet-induced obesity and liver-specific deletion of SIRT6 in mice causes fatty liver formation [18,22,23,31].

SIRT4 is a mitochondrial sirtuin that acts as a negative regulator of mitochondrial metabolism. It has been recently observed that SIRT4 inhibition increases mitochondrial function and fatty acid oxidation in liver and muscle cells, suggesting therapeutic benefits for metabolic diseases such as type 2 diabetes [32].

In our study, transcriptional gene and protein expression analysis revealed a significant increase of GCK and SIRT6 expression associated with reduced SIRT4 expression in liver. Even though the molecular target(s) responsible for T1AM functional effects are still largely unknown, these findings support the hypothesis that SIRT6 and SIRT4 might be the mediators of T1AM's induced shift of metabolism from carbohydrates to lipids through their metabolic intermediates as the principal metabolic activators [15] in this organ. These results suggest that T1AM has great potential to control the balance between glucose and lipid utilization in vivo and open up the way to future pharmacological studies aimed to investigate the hypothesized sirtuin dependency of T1AM metabolic reprogramming in mice. Collectively, our findings are also consistent with the metabolic changes observed by ^{1}H-NMR metabolomics of liver and muscle tissues, two high metabolically active organs. To more exhaustively confirm the potential of T1AM as body-weight loss and anti-obesity drug, future studies will be directed to analyze the effects of chronic T1AM administration to mice in combination with chronic aerobic exercise (CAE) or fasting, which are both known to be strong metabolic triggers [33–35].

In accordance with previous studies [36], the assay of tissue T1AM concentration (Table 2) suggests that liver and adipose tissue might be regarded as T1AM storage sites and support the elevated metabolic and transcriptional activities observed in both tissues especially when T1AM was administered at the highest dose.

4. Materials and Methods

4.1. Chemicals

All reagents, unless otherwise other specified, were from Sigma-Aldrich (St. Louis, MO, USA). 3-Iodothyronamine (T1AM) was kindly provided by Thomas Scanlan, Oregon Health & Science University, Portland, OR, USA.

Purified crystalline T1AM was dissolved in 100 µL dimethyl sulfoxide (DMSO) to increase solubility and then diluted ~1:400 with medical grade 0.9% medical grade saline (Hospira Corp., Lake Forest, IL, USA) to minimize toxicity of DMSO. Two final stock concentration levels were made at 25 and 10 mg/kg in 0.9% saline and each solution type was aliquoted and stored at −80 °C until further use.

A vehicle solution was also made for control animals using 100 µL DMSO and 0.9% saline. This solution was injected in control animals (herein referred as "Saline").

4.2. Animal Study

All animal procedures were approved by the University of Wisconsin, College of Letters and Sciences, Animal Care and Use Committee (Madison, WI protocol # L00408, 12 February 2013), Eighteen out-bred female CD-1 mice obtained from Harlan, (Indianapolis, IN, USA) were used for this study. Female mice are being used in accordance with the previous T_1AM study [15] to increase the relatability of the results between the two experiments.

Mice were fed an AIN-93G diet (17.7% protein, 60.1% carbohydrate and 7.2 % fat) (Harlan, Indianapolis, IN, USA) and water ad libitum during normal housing conditions and until they reached body weight between 40–45 g for ~4 weeks. Mice were age and weight matched initially between control females and treated with T1AM. These animals were divided into three groups ($n = 6$/group), intraperitoneally injected (i.p.) daily from day 1 to day 7 with: (1) saline solution (control), (2) lower dose T1AM (10 mg/kg/day) and (3) higher dose T1AM (25 mg/kg/day). Mice were injected at 12:00, video monitored from 12:00–14:00 and then were weighed at 16:00. On day 7 of the study, at 16:00 animals were anesthetized using 2.5% isoflurane. Blood was drawn using the retro-orbital venus plexus and heparinized capillary tubes. At this point, animals were sacrificed and their organs collected for NMR-metabolomics, real-time qPCR, Western blot and LC/MS-MS analyses. All tissues were flash frozen in liquid nitrogen and kept at −80 °C until further use.

4.3. Plasma NMR Samples Preparation

Blood was drawn on day −3, 4 and 7 of the study through the retro orbital venous plexus using heparinized capillary tubes. Animals were fasted at 08:00 and injected with saline vehicle solution (as described above) on day −3 to collect a baseline plasma sample. On days 4 and 7, food was removed from cages and animals were fasted under the same condition as day −3 at 08:00 and injected with their assigned T1AM treatment at 12:00 and blood was collected at 16:00. Animals had full access to water. Plasma was then separated from blood using centrifugation at 1957× g for 10 min and 4 °C. Plasma samples were stored at −80 °C until sample preparation time for NMR. NMR samples from plasma were prepared as previously described [15].

4.4. Tissue NMR Samples Preparation

Individual 50 mg tissue samples were placed on ice in 10 mM phosphate buffer containing 2 mM sodium orthovanadate, 1 mM NaF, 1 mM phenylmethanesulfonyl fluoride (PMSF) and protease inhibitor cocktail, followed by homogenization by Omni Bead Ruptor Homogenizer (Omni International Inc., Waterbury, CT, USA) for 3 min [37]. Homogenized tissues were transferred to new tubes and were centrifuged for 10 min at $5000\times g$. Supernatant was then transferred to a new tube and ice-cold methanol (2:1, v/v) was quickly added to aliquots of supernatants, vortexed for 30 s to enhance protein precipitation, followed by cooling to $-20\ °C$ for 30 min. After a precipitation period, tubes were vortexed once more for 10 s and centrifuged at $5000\times g$ for 10 min. The supernatant was dried in a speed vacuum overnight. The dried supernatant was then reconstituted in the 10 mM phosphate buffer containing 2 mM PMSF, 2 mM ethylenediaminetetraacetic acid (EDTA) and pH adjusted to 7.4 ± 0.05.

4.5. NMR Data Collection and Analysis

All one-dimensional (1D) ^{1}H-NMR spectra were collected at 25 °C on a 600 MHz Varian VNMRS spectrometer equipped with a cryogenic probe according to our previously published method [38]. Each 1D spectrum was accumulated for 1028 scans, with an acquisition time of ~2.5 s (24,576 complex points) and a 3 s repetition delay for a total collection time of ~2 h [38]. 1D ^{1}H-NMR spectra were referenced to 0.5 mM 4,4-dimethyl-4-silapentane-1-sulfonic acid (DSS). NMR signals arising from small water-soluble metabolites (<1000 Da) were identified and quantified relative to formate (1 mM) as the internal reference by Chenomx software version 6 (http://www.chenomx.com). All metabolite concentrations are reported as values relative to formate.

4.6. Tissue Preparation for Gene and Protein Expression Analyses

Liver, subcutaneous adipose tissue, heart and skeletal muscle from the left leg (50 mg/tissue) were rapidly extracted from 18 mice and kept in RNAlater (Qiagen GmbH, Hilden 40724, Germany) for 24 h and stored at $-80\ °C$ until use. Tissues were homogenized using a Teflon-glass homogenizer in 1 mL of ice-cold buffer (pH = 7.4) containing 50 mM Tris pH = 7.4, 250 mM NaCl, 5 mM EDTA, 20 mM sodium pyrophosphate, 1% Igepal CA-630, 2 mM sodium orthovanadate, 1 mM NaF, 1 mM PMSF and protease inhibitor cocktail. Homogenates were centrifuged at $3914\times g$ for 10 min at 4 °C to pellet cellular debris. The supernatant was collected and frozen to $-80\ °C$. The protein concentration of the supernatant fraction was determined by the Bradford method [39].

4.7. Gene Expression Analysis

Expression of 20 genes (*Acsl5, Apod, Insig2a, Insig2b, Ldlrap1, Me1, Gck, Igfbp2, Cebpb, Abcd2, Abcd3, Abcd4, Pparα, Pparα/δ, Pparγ, Sirt1, Sirt2, Sirt3, Sirt4, Sirt6*) (Table S1) was evaluated in 4 tissues (liver, subcutaneous adipose tissue, skeletal muscle and heart) from the 18 mice by reverse transcription qPCR (RT-PCR). Total RNA was isolated with RNeasy Lipid Tissue Mini kit (Qiagen, GmbH, Hilden, Germany) following the manufacturer's protocol. 50–100mg of tissue was homogenized in 1mL of QiaZol (Qiagen) with TissueRuptor (Qiagen) for 30–40 s. An on-column DNase treatment with RNase-free DNase Set (Qiagen) was included. RNA concentration and purity (260/280 and 260/230 ratios) were analyzed using a Qubit fluorometer (Life Technologies, Carlsbad, CA, USA) and an ND-1000 Spectrophotometer (NanoDrop Technologies, Wilmington, DE, USA). RNA integrity was checked by agarose gel electrophoresis. 1 μg of total RNA was retrotranscribed in 20 μL (5 min at 25 °C, 20 min at 42 °C, 15 min at 46 °C, 15 min at 50 °C and 5 min at 85 °C) using an iScript cDNA Synthesis Kit (Bio-Rad Laboratories, Hercules, CA, USA).

Relative quantity of gene transcripts was measured by real-time PCR on samples' cDNA using a SYBRGreen chemistry and an iQ5 instrument (Bio-Rad). Two μL of 25-fold cDNA dilutions and 8 pmol of each oligonucleotide were added to 10 μL SsoAdvanced SYBRGreen Supermix (Bio-Rad) in a 20 μL total volume reaction. The PCR cycle program consisted of an initial 30 s denaturation at 95 °C followed by 40 cycles of 5 s denaturation at 95 °C and 15 s annealing/extension at 60 °C. A final melting protocol with ramping from 65 °C to 95 °C with 0.5 °C increments of 5 s was performed for verification of amplicon specificity and primer dimer formation.

Primers were designed with Beacon Designer Software v.7.9 (Premier Biosoft International, Palo Alto, CA, USA) with a junction primer strategy. In any case, negative control of retro-transcription was performed to exclude any interference from residual genomic DNA contamination. For quantity data normalization, two to three reference genes were chosen for each tissue type and the values were reported as fold change. Choice was based on testing expression stability of 9 candidate reference genes (*Actb*, *B2m*, *Gusb*, *Hprt*, *Kdm2b*, *Ppia*, *Psmd4*, *Tbp* and *Rpl13*) (Table S2) in tissue specific experiments including all 18 samples. The expression stability of each gene was assessed using geNorm version 3.5 [40]. Efficiency and specificity of primers were tested making standard curves with fivefold serial dilutions of a cDNA obtained from a pool (1 μg) of all mouse liver RNA samples. The first dilution was the two-fold diluted cDNA. All reactions were run in duplicate. Samples were analyzed by the $2^{-\Delta\Delta Ct}$ method as described by Livak and Schmittgen [41].

4.8. Western Blotting Analysis

Western blotting was performed according to manufacturer's instructions (Bio-Rad laboratories, Hercules, CA, USA). In brief, 20–40 μg of proteins was subjected to sodium dodecyl sulfate polyacrylamide gel electrophoresis (SDS-PAGE) (Criterion TGX anykD acrylamide separating gel Bio-Rad). The separated proteins were transferred to a polyvinylidene difluoride (PVDF) membrane (Millipore Corporation, Billerica, MA, USA) according to the manufacturer's instructions. The membranes were dried and then incubated for 1h using primary antibodies for sirtuin proteins (SIRT1, SIRT2, SIRT3, SIRT4, SIRT6), Santa Cruz Biotechnology, Santa Cruz, CA, USA and glucokinase (GCK) Santa Cruz Biotechnology, Santa Cruz, CA, USA) in TBS (20 mM TRIS, pH = 7.6, 137 mM NaCl), 0.04% Tween-20 and 5% low fat milk and then incubated for 30 min with secondary antibodies conjugated with horseradish peroxidase (anti-rabbit, Santa Cruz Biotechnology, Santa Cruz, CA, USA). After washing with TBS, immunoblots were visualized by means of a chemiluminescence reaction (Millipore) by Image Lab™ Software (Bio-Rad) under a luminescent image analyzer (Chemidoc XSR + Bio-Rad, Philadelphia, PA 19103, USA). Only bands below the saturation limit were analyzed and shown. β-actin (Sigma-Aldrich, S.r.l., Milan, Italy) was used as loading control.

4.9. LC/MS-MS Assay

T1AM was assayed in liver, abdominal adipose tissue, skeletal muscle and heart by high performance liquid chromatography (HPLC; LC) coupled with tandem mass spectrometry (MS-MS). Liver, heart and skeletal muscle samples (10–50 mg) were homogenized on ice in 1.5 ml of phosphate buffer (154 mM NaCl, 6.7 mM NaH$_2$PO$_4$, pH 7.4) by 15 + 15 passes in a Potter-Elvejheim homogenizer. The homogenate was centrifuged for 10 min at 18620× *g*, the pellet was discarded and the supernatant was placed in a 15 mL centrifuge tube. After vortexing, 60 mg of NaCl was added; the mixture was equilibrated at room temperature for one hour and then de-proteinized with 2 mL acetone in an ice bath for 30 min. After centrifugation for 15 min the supernatant was evaporated to 1 mL using a Concentrator Plus (Eppendorf, Hamburg, Germany) kept at 30 °C. Subsequent steps included solid phase extraction, HPLC separation and MS-MS assay, which were performed as previously described [4]. Adipose tissue samples (10–50 mg) were extracted for 30 min in 1 mL of acetonitrile and 0.1 M HCl (85:15, *v/v*), in an ultrasound bath (LBS1 3Lt, Falc Instruments, Treviglio, Italy). The material was diluted to 2 mL with acetonitrile and homogenized by 12 + 12 passes in a Potter-Elvejheim homogenizer. The homogenate was further sonicated for 10 min with 1510 Branson and a frequency of 40 kHz, then vortexed for 1 min and centrifuged at 720× *g* for 15 min. The supernatant was subjected for three times to liquid/liquid extraction with 1 mL hexane: the upper phase (hexane) was discarded and the lower phase (acetonitrile) was eventually dried under a gentle stream of nitrogen. The dried samples were reconstituted with 0.1 M HCl/methanol (50:50, *v/v*) and subjected to HPLC separation and MS-MS assay, as previously described [4].

4.10. Biochemical Assays

Plasma samples were prepared at the time of sacrifice as previously described [37]. Plasma total cholesterol (Wako Diagnostics, Richmond, VA, USA) and total triglycerides (Sigma Aldrich, St. Louis, MO, USA) were measured using specific colorimetric kits (Table 1).

4.11. Statistical Analysis

Data analysis was performed by Graph-Pad Prism 6.0 statistical program (GraphPad Software Inc., San Diego, CA, USA). All data were reported as the mean ± SEM (standard error of mean). The threshold of statistical significance was set at $p < 0.05$ unless otherwise specified. For biochemical assays, gene expression, Western blot and LC/MS-MS, statistical significance was assessed by one-way analysis of variance (ANOVA), followed by Dunnett's post hoc test, or Tukey's multiple comparison post hoc test.

Metabolomics

Prior to statistical analyses, all data were examined for assumptions of normality and homogeneity of variance using metaboanalyst procedures (www.metaboanalyst.ca). To evaluate major metabolic changes contributing to separation of metabolome profiles in plasma and tissues under different treatments conditions (T1AM dosages and saline) and their time-dependent changes, we first used a partial least square discriminant analysis (PLSDA) [42–44] (www.metaboanalyst.ca). Heat maps were generated based on the fold changes between metabolome profiles of T1AM-treatment and control groups using R statistical software program (http://www.R-project.org). Fold change is defined by ratio of a given metabolite concentration in the control group, [metabolite]$_{saline}$, and the corresponding metabolite concentration in the T1AM treatment group, [metabolite]$_{T1AM-treatment}$. Color codes were used to indicate increase or decrease of each metabolite concentration. Finally, to determine the statistical significance in plasma metabolite profiles, collected at two time points (i.e., Day4 and Day7), two-way analysis of variance (ANOVA) was used followed by Tukey Honest Significant Differences (HSD) test. For tissue (liver and muscle) metabolomics data set, ANOVA followed by Tukey's multiple comparison post hoc test were performed. In all analyses * indicated $p < 0.05$, control versus treatment; ** indicated $p < 0.01$, control versus treatment; # indicated $p < 0.05$, 10 mg versus 25 mg. Data were presented as standard error measurement (± SEM) for relative concentrations of metabolites to the internal standard (formate).

5. Conclusions

Our multidisciplinary approaches provide consistent multiple perspectives on evidence that treatment with exogenous T1AM affects lipid metabolism in a dose-dependent and tissue-specific manner. Taken together, the identification of SIRT6, as well as SIRT4, as potential targets for T1AM-mediated metabolic regulation, coupled with a better knowledge of T1AM tissue distribution and accumulation, has the potential to open broad new avenues in the treatment of a wide variety of diet- and age-related diseases.

Supplementary Materials: Supplementary materials can be found at http://www.mdpi.com/1422-0067/19/5/1535/s1.

Author Contributions: F.M.A.-P. and G.C. designed the study and wrote the manuscript. M.T. and E.S.S.A. collected NMR metabolomics data. E.S.S.A. and F.M.A.-P. participated in metabolomics data collection and data analysis. H.R. conducted animal studies. M.S. and V.C. conducted gene expression studies. A.S. and M.S. carried out LC/MS-MS data collection. S.G. and L.L. conducted protein expression studies. M.R. collected biochemical data. R.Z. analyzed the data and discussed the results. All authors contributed to the final manuscript.

Acknowledgments: The authors gratefully acknowledge Prof. Thomas S. Scanlan (OHSU, Portland, OR, USA) for providing T1AM as a gift and Prof. Warren Porter for his comments on the manuscript and language editing. This work was supported by UW-Madison Hilldale Undergraduate Scholarship to HR, NIH RC4 EY021357 grant to F.M.A-P and a grant (PRA_2017_55) from the University of Pisa, Italy to G.C. F.M.A-P and E.S.S.A. are cofounders of Metresponse LLC (www.Metresponse.net), F.M.A-P and M.T. are also cofounder of Isomark, LLC (www.isomark.com). This study made use of the National Magnetic Resonance Facility at Madison, which is supported by NIH grant P41GM103399 (NIGMS), old number: P41RR002301. Equipment was purchased with funds from the University of Wisconsin-Madison, the NIH P41GM103399, S10RR02781, S10RR08438, S10RR023438, S10RR025062, S10RR029220) and the NSF (DMB-8415048, OIA-9977486, BIR-9214394).

Conflicts of Interest: The authors declare no conflict of interest. The authors declare no competing financial interests.

References

1. Scanlan, T.S.; Suchland, K.L.; Hart, M.E.; Chiellini, G.; Huang, Y.; Kruzich, P.J.; Frascarelli, S.; Crossley, D.A.; Bunzow, J.R.; Ronca-Testoni, S.; et al. 3-Iodothyronamine is an endogenous and rapid-acting derivative of thyroid hormone. *Nat. Med.* **2004**, *10*, 638–642. [CrossRef] [PubMed]

2. Bunzow, J.R.; Sonders, M.S.; Arttamangkul, S.; Harrison, L.M.; Zhang, G.; Quigley, D.I.; Darland, T.; Suchland, K.L.; Pasumamula, S.; Kennedy, J.L.; et al. Amphetamine, 3,4-methylenedioxymethampheta-mine, lysergic acid diethylamide and metabolites of the catecholamine neurotransmitters are agonists of a rat trace amine receptor. *Mol. Pharmacol.* **2001**, *60*, 1181–1188. [CrossRef] [PubMed]

3. Borowsky, B.; Adham, N.; Jones, K.A.; Raddaz, R.; Artymyshyn, R.; Ogozalek, K.L.; Durkin, M.M.; Lakhlani, P.P.; Bonini, J.A.; Pathirana, S.; et al. Trace amines: Identification of a family of mammalian G protein-coupled receptors. *Proc. Natl. Acad. Sci. USA* **2001**, *98*, 8966–8971. [CrossRef] [PubMed]

4. Saba, A.; Chiellini, G.; Frascarelli, S.; Marchini, M.; Ghelardoni, S.; Raffaelli, A.; Tonacchera, M.; Vitti, P.; Scanlan, T.S.; Zucchi, R. Tissue distribution and cardiac metabolism of 3-iodothyronamine. *Endocrinology* **2010**, *151*, 5063–5073. [CrossRef] [PubMed]

5. Hoefig, C.S.; Köhrle, J.; Brabant, G.; Dixit, K.; Yap, B.; Strasburger, C.J.; Wu, Z. Evidence for extrathyroidal formation of 3-iodothyronamine in humans as provided by a novel monoclonal antibody-based chemiluminescent serum immunoassay. *J. Clin. Endocrinol. Metab.* **2011**, *96*, 1864–1872. [CrossRef] [PubMed]

6. Hackenmueller, S.A.; Marchini, M.; Saba, A.; Zucchi, R.; Scanlan, T.S. Biosynthesis of 3-iodothyronamine (T1AM) is dependent on the sodium-iodide symporter and thyroperoxidase but does not involve extrathyroidal metabolism of T4. *Endocrinology.* **2012**, *153*, 5659–5667. [CrossRef] [PubMed]

7. Dinter, J.; Muhlhaus, J.; Wienchol, C.L.; Cöster, M.; Hermsdorf, T.; Stäubert, C.; Köhrle, J.; Schöneberg, T.; Kleinau, G.; Mergler, S.; et al. The Multitarget Ligand 3-Iodothyronamine Modulates β-Adrenergic Receptor 2 Signaling. *Eur. Thyroid J.* **2015**, *4*, 21–29. [CrossRef] [PubMed]

8. Dinter, J.; Muhlhaus, J.; Jacobi, S.F.; Wienchol, C.L.; Cöster, M.; Meister, J.; Hoefig, C.S.; Müller, A.; Köhrle, J.; Grüters, A.; et al. 3-Iodothyronamine differentially modulates alpha-2A-adrenergic receptor-mediated signaling. *J. Mol. Endocrinol.* **2015**, *54*, 205–216. [CrossRef] [PubMed]

9. Snead, A.N.; Santos, M.S.; Seal, R.P.; Miyakawa, M.; Edwards, R.H.; Scanlan, T.S. Thyronamines inhibit plasma membrane and vesicular monoamine transport. *ACS Chem. Biol.* **2007**, *2*, 390–398. [CrossRef] [PubMed]

10. Chiellini, G.; Frascarelli, S.; Ghelardoni, S.; Carnicelli, V.; Tobias, S.C.; de Barber, A.; Brogioni, S.; Ronca-Testoni, S.; Cerbai, E.; Grandy, D.K.; et al. Cardiac effects of 3-iodothyronamine: A new aminergic system modulating cardiac function. *FASEB J.* **2007**, *21*, 1597–1608. [CrossRef] [PubMed]

11. Ghelardoni, S.; Suffredini, S.; Frascarelli, S.; Brogioni, S.; Chiellini, G.; Ronca-Testoni, S.; Grandy, D.K.; Scanlan, T.S.; Cerbai, E.; Zucchi, R. Modulation of cardiac ionic homeostasis by 3-iodothyronamine. *J. Cell. Mol. Med.* **2009**, *13*, 3082–3090. [CrossRef] [PubMed]

12. Zucchi, R.; Accorroni, A.; Chiellini, G. Update on 3-iodothyronamine and its neurological and metabolic actions. *Front. Physiol.* **2014**, *5*, 402. [CrossRef] [PubMed]

13. Accorroni, A.; Chiellini, G.; Origlia, N. Effects of Thyroid Hormones and their Metabolites on Learning and Memory in Normal and Pathological Conditions. *Curr. Drug Metab.* **2017**, *18*, 225–236. [CrossRef] [PubMed]

14. Braulke, L.J.; Klingenspor, M.; de Barber, A.; Tobias, S.C.; Grandy, D.K.; Scanlan, T.S.; Heldmaier, G. 3-Iodothyronamine: A novel hormone controlling the balance between glucose and lipid utilisation. *J. Comp. Physiol. B.* **2008**, *178*, 167–177. [CrossRef] [PubMed]

15. Haviland, J.A.; Reiland, H.; Butz, D.E.; Tonelli, M.; Porter, W.P.; Zucchi, R.; Scanlan, T.S.; Chiellini, G.; Assadi-Porter, F.M. NMR-based metabolomics and breath studies show lipid and protein catabolism during low dose chronic T1AM treatment. *Obesity* **2013**, *21*, 2538–2544. [CrossRef] [PubMed]

16. Jensen, T.L.; Kiersgaard, M.K.; Sørensen, D.B.; Mikkelsen, L.F. Fasting of mice: A review. *Lab. Anim.* **2013**, *47*, 225–240. [CrossRef] [PubMed]

17. Michishita, E.; McCord, R.A.; Berber, E.; Kioi, M.; Padilla-Nash, H.; Damian, M.; Cheung, P.; Kusumoto, R.; Kawahara, T.L.; Barrett, J.C.; et al. SIRT6 is a histone H3 lysine 9 deacetylase that modulates telomeric chromatin. *Nature* **2008**, *452*, 492–496. [CrossRef] [PubMed]

18. Kim, H.S.; Xiao, C.; Wang, R.H.; Lahusen, T.; Xu, X.; Vassilopoulos, A.; Vazquez-Ortiz, G.; Jeong, W.I.; Park, O.; Ki, S.H.; et al. Hepatic-specific disruption of SIRT6 in mice results in fatty liver formation due to enhanced glycolysis and triglyceride synthesis. *Cell. Metab.* **2010**, *12*, 224–236. [CrossRef] [PubMed]

19. Laurent, G.; German, N.J.; Saha, A.K.; de Boer, V.C.; Davies, M.; Koves, T.R.; Dephoure, N.; Fischer, F.; Boanca, G.; Vaitheesvaran, B.; et al. SIRT4 coordinates the balance between lipid synthesis and catabolism by repressing malonyl CoA decarboxylase. *Mol. Cell* **2013**, *50*, 686–698. [CrossRef] [PubMed]

20. Soupene, E.; Kuypers, F.A. Mammalian Long-Chain Acyl-CoA Synthetases. *Exp. Biol. Med.* **2008**, *233*, 507–521. [CrossRef] [PubMed]

21. Perdomo, G.; Henry Dong, H. Apolipoprotein D in lipid metabolism and its functional implication in atherosclerosis and aging. *Aging* **2009**, *1*, 17–27. [CrossRef] [PubMed]

22. Kanfi, Y.; Peshti, V.; Gil, R.; Naiman, S.; Nahum, L.; Levin, E.; Kronfeld-Schor, N.; Cohen, H.Y. SIRT6 protects against pathological damage caused by diet-induced obesity. *Aging Cell.* **2010**, *9*, 162–173. [CrossRef] [PubMed]

23. Yao, L.; Cui, X.; Chen, Q.; Yang, X.; Fang, F.; Zhang, J.; Liu, G.; Jin, W.; Chang, Y. Cold-Inducible SIRT6 Regulates Thermogenesis of Brown and Beige Fat. *Cell Rep.* **2017**, *20*, 641–654. [CrossRef] [PubMed]

24. Fuentes, E.; Guzmán-Jofre, L.; Moore-Carrasco, R.; Palomo, I. Role of PPARs in inflammatory processes associated with metabolic syndrome. *Mol. Med. Rep.* **2013**, *8*, 1611–1616. [CrossRef] [PubMed]

25. Ghanian, Z.; Maleki, S.; Reiland, H.; Bütz, D.E.; Chiellini, G.; Assadi-Porter, F.M.; Ranji, M. Optical imaging of mitochondrial redox state in rodent models with 3-iodothyronamine. *Exp. Biol. Med.* **2014**, *239*, 151–158. [CrossRef] [PubMed]

26. Mariotti, V.; Melissari, E.; Iofrida, C.; Righi, M.; di Russo, M.; Donzelli, R.; Saba, A.; Frascarelli, S.; Chiellini, G.; Zucchi, R.; et al. Modulation of gene expression by 3-iodothyronamine: Genetic evidence for a lipolytic pattern. *PLoS ONE* **2014**, *9*, e106923. [CrossRef] [PubMed]

27. Haigis, M.C.; Guarente, L.P. Mammalian sirtuins–emerging roles in physiology, aging and calorie restriction. *Genes Dev.* **2006**, *20*, 2913–2922. [CrossRef] [PubMed]

28. Guarente, L. Sirtuins as potential targets for metabolic syndrome. *Nature* **2006**, *444*, 868–874. [CrossRef] [PubMed]

29. Huynh, F.K.; Hershberger, K.A.; Hirschey, M.D. Targeting sirtuins for the treatment of diabetes. *Diabetes Manag.* **2013**, *3*, 245–257. [CrossRef] [PubMed]

30. Zhong, L.; D'Urso, A.; Toiber, D.; Sebastian, C.; Henry, R.E.; Vadysirisack, D.D.; Guimaraes, A.; Marinelli, B.; Wikstrom, J.D.; Nir, T.; et al. The histone deacetylase Sirt6 regulates glucose homeostasis via hif1α. *Cell* **2010**, *140*, 280–293. [CrossRef] [PubMed]

31. Tasselli, L.; Zheng, W.; Chua, K.F. SIRT6: Novel mechanisms and links to aging and disease. *Trends Endocrinol. Metab.* **2017**, *28*, 168–185. [CrossRef] [PubMed]

32. Nasrin, N.; Wu, X.; Fortier, E.; Feng, Y.; Barè, O.C.; Chen, S.; Ren, X.; Wu, Z.; Streeper, R.S.; Bordone, L. SIRT4 Regulates Fatty Acid Oxidation and Mitochondrial Gene Expression in Liver and Muscle Cells. *J. Biol. Chem.* **2010**, *285*, 31995–32002. [CrossRef] [PubMed]

33. Batatinha, H.A.; Lima, E.A.; Teixeira, A.A.; Souza, C.O.; Biondo, L.A.; Silveira, L.S.; Lira, F.S.; Rosa Neto, J.C. Association Between Aerobic Exercise and Rosiglitazone Avoided the NAFLD and Liver Inflammation Exacerbated in PPAR-α Knockout Mice. *J. Cell. Physiol.* **2017**, *232*, 1008–1019. [CrossRef] [PubMed]

34. Geisler, C.E.; Hepler, C.; Higgins, M.R.; Renquist, B.J. Hepatic adaptations to maintain metabolic homeostasis in response to fasting and refeeding in mice. *Nutr. Metab.* **2016**, *13*, 62. [CrossRef] [PubMed]

35. De Lange, P.; Farina, P.; Moreno, M.; Ragni, M.; Lombardi, A.; Silvestri, E.; Burrone, L.; Lanni, A.; Goglia, F. Sequential changes in the signal transduction responses of skeletal muscle following food deprivation. *FASEB J.* **2006**, *20*, 2579–2581. [CrossRef] [PubMed]

36. Chiellini, G.; Erba, P.; Carnicelli, V.; Manfredi, C.; Frascarelli, S.; Ghelardoni, S.; Mariani, G.; Zucchi, R. Distribution of exogenous [125I]-3-iodothyronamine in mouse in vivo: Relationship with trace amine-associated receptors. *J. Endocrinol.* **2012**, *213*, 223–230. [CrossRef] [PubMed]

37. Selen, E.S.; Bolandnazar, Z.; Tonelli, M.; Bütz, D.E.; Haviland, J.A.; Porter, W.P.; Assadi-Porter, F.M. NMR Metabolomics Show Evidence for Mitochondrial Oxidative Stress in a Mouse Model of Polycystic Ovary Syndrome. *J. Proteome Res.* **2015**, *14*, 3284–3291. [CrossRef] [PubMed]

38. Haviland, J.A.; Tonelli, M.; Haughey, D.T.; Porter, W.P.; Assadi-Porter, F.M. Novel diagnostics of metabolic dysfunction detected in breath and plasma by selective isotope-assisted labeling. *Metabolism* **2012**, *61*, 1162–1170. [CrossRef] [PubMed]

39. Bradford, M.M. A Rapid and Sensitive Method for the Quantification of Microgram Quantities of Protein Utilizing the Principle of Protein-Dye Binding. *Anal. Biochem.* **1976**, *72*, 248–254. [CrossRef]

40. Vandesompele, J.; de Preter, K.; Pattyn, F.; Poppe, B.; van Roy, N.; de Paepe, A.; Speleman, F. Accurate normalization of real-time quantitative RT-PCR data by geometric averaging of multiple internal control genes. *Genome Biol.* **2002**, *3*, research0034.1. [CrossRef] [PubMed]

41. Livak, K.J.; Schmittgen, T.D. Analysis of relative gene expression data using real-time quantitative PCR and the $2^{-\Delta\Delta Ct}$ Method. *Methods* **2001**, *25*, 402–408. [CrossRef] [PubMed]

42. Lindgren, F.; Hansen, B.; Karcher, W.; Sjostrom, M.; Eriksson, L. Model validation by permutation tests: Applications to variable selection. *J. Chemometr.* **1996**, *10*, 521–532. [CrossRef]

43. Teng, Q. NMR-Based Metabolomics. In *Structural Biology-Practical NMR Applications*, 2nd ed.; Springer Science + Business Media Inc.: New York, NY, USA, 2013; pp. 311–392. ISBN 978-1-4614-3964-6.

44. Pattini, L.; Mazzara, S.; Conti, A.; Iannaccone, S.; Cerutti, S.; Alessio, M. An integrated strategy in two-dimensional electrophoresis analysis able to identify discriminants between different clinical conditions. *Exp. Biol. Med.* **2008**, *233*, 483–491. [CrossRef] [PubMed]

International Journal of
Molecular Sciences

MDPI

Review

Use of Curcumin, a Natural Polyphenol for Targeting Molecular Pathways in Treating Age-Related Neurodegenerative Diseases

Panchanan Maiti [1,2,3,4,5,6,*] and Gary L. Dunbar [1,2,3,4,*]

1   Field Neurosciences Institute Laboratory for Restorative Neurology, Central Michigan University, Mt.
    Pleasant, MI 48859, USA
2   Program in Neuroscience, Central Michigan University, Mt. Pleasant, MI 48859, USA
3   Department of Psychology, Central Michigan University, Mt. Pleasant, MI 48859, USA
4   Field Neurosciences Institute, St. Mary's of Michigan, Saginaw, MI 48604, USA
5   Department of Biology, Saginaw Valley State University, Saginaw, MI 48610, USA
6   Brain Research Laboratory, Saginaw Valley State University, Saginaw, MI 48610, USA
*   Correspondence: maiti1p@cmich.edu (P.M.); dunba1g@cmich.edu (G.L.D.);
    Tel.: +1-901-246-2649 (P.M.); +1-989-497-3105 (G.L.D.)

Received: 31 March 2018; Accepted: 25 May 2018; Published: 31 May 2018

Abstract: Progressive accumulation of misfolded amyloid proteins in intracellular and extracellular spaces is one of the principal reasons for synaptic damage and impairment of neuronal communication in several neurodegenerative diseases. Effective treatments for these diseases are still lacking but remain the focus of much active investigation. Despite testing several synthesized compounds, small molecules, and drugs over the past few decades, very few of them can inhibit aggregation of amyloid proteins and lessen their neurotoxic effects. Recently, the natural polyphenol curcumin (Cur) has been shown to be a promising anti-amyloid, anti-inflammatory and neuroprotective agent for several neurodegenerative diseases. Because of its pleotropic actions on the central nervous system, including preferential binding to amyloid proteins, Cur is being touted as a promising treatment for age-related brain diseases. Here, we focus on molecular targeting of Cur to reduce amyloid burden, rescue neuronal damage, and restore normal cognitive and sensory motor functions in different animal models of neurodegenerative diseases. We specifically highlight Cur as a potential treatment for Alzheimer's, Parkinson's, Huntington's, and prion diseases. In addition, we discuss the major issues and limitations of using Cur for treating these diseases, along with ways of circumventing those shortcomings. Finally, we provide specific recommendations for optimal dosing with Cur for treating neurological diseases.

Keywords: neurodegenerative diseases; amyloidosis; curcumin; neuroinflammation; anti-amyloid; molecular chaperones; natural polyphenol

1. Introduction

Aggregation of misfolded amyloid proteins and their deposition in intracellular and extracellular spaces of the central nervous system (CNS) are associated with several neurological diseases, including Alzheimer's (AD), Parkinson's (PD), Huntington's (HD) and prion diseases [1,2]. Most of these diseases are age-related, complicated disorders which involve a multitude of causative factors, including neuroinflammation [3], oxidative damage and deposition of misfolded protein aggregates [4]. These events can occur separately or together or in causing neuronal degeneration, which leads to perturbation of neuronal communications, resulting in long-term cognitive and motor dysfunction. The neuropathological onset of these diseases may have occurred long before the manifestation of overt symptoms, which underscores the need for early diagnosis and therapy. Although there have been several studies using therapeutic strategies involving anti-amyloids, anti-inflammatory agents, and small molecule drugs [5], to slow or halt the progression of neurological diseases, none of them have proven effective without serious effects or abbreviated half-lives. As a potent anti-amyloid natural polyphenol, curcumin (Cur) has gained considerable attention as a promising therapeutic agent for AD and other complicated neurological diseases [6,7]. Although Cur has been considered a wonder molecule for use in Indian and Southeast Asian traditional Ayurvedic medicine for a very long time, primarily due to its anti-inflammatory or wound-healing properties [8], its anti-amyloidogenic properties have only been discovered recently [9]. Cur binds to, and inhibits, amyloid-beta protein (Aβ) aggregation, and improves motor coordination and cognition in animal models of AD and other neurodegenerative diseases [9,10]. However, the poor water solubility, instability in body fluids, rapid degradation, and limited bioavailability has curtailed the use of Cur as a therapeutic for neurological diseases [11].

The outlook for using Cur as a therapeutic agent has changed dramatically with the discovery of new formulations for Cur, including liposome-Cur, Cur-conjugated with nanogel, dendrimer-Cur, Cur with silver, or gold nanoparticles and Cur in solid lipid nanoparticles (SLN) [11]. The SLN formula of Cur (nanoCur) has been shown to increase its bioavailability and therapeutic value for neurological diseases [12–14]. This review article addresses the basic understanding of the molecular signaling mechanisms of Cur therapy and its potential impact on major neurological diseases, such as AD, PD, HD, and prion diseases, along with recent findings from our laboratory.

2. Curcumin: The Major Active Polyphenol of Turmeric

2.1. Source

The rhizomes of the *Curcuma longa* (family: *Zingiberaceae*) herb is the source of turmeric (Figure 1A–C). The principal yellow pigment present in the turmeric root is Cur (Figure 1E), which was identified in early 1900 by Lampe and Milobedzka. Its structure and biochemical analyses revealed that about 2.5–6% of turmeric contains pure Cur [15] (Table 1). The commercial turmeric extract contains many other components, including three main types of curcuminoids, such as (a) Cur-I (diferuloylmethane, ~77%); (b) Cur-II (demethoxyCur, DMC, ~17%); and (c) Cur-III (bisdemethoxyCur, BDMC, ~3%) (Figure 1). In addition, four identified turmerones (α-turmerone, β-turmerone, ar-turmerone, and aromatic-turmerone), as well as α-santalene, aromatic-curcumene, curlone, and other compounds were also found in turmeric extract (Figure 1D).

2.2. Chemistry of Cur

Cur is a natural polyphenol, chemically known as diferuloylmethane ($C_{21}H_{20}O_6$), with molecular mass of 368.37 g/mol. Its International Union of Pure and Applied Chemistry (IUPAC) name is 1,7-bis (4-hydroxy-3-methoxy phenyl)-1,6-heptadiene-3,5-dione. There are two aryl rings containing orthomethoxy phenolic OH-groups, which are symmetrically linked to a β-diketone moiety (Figure 1E). The melting point of Cur is ~183 °C. Cur can co-exist with several tautomeric forms, of which two predominant forms are 1,3-diketo form and 1,3-dienol form (Figure 1E). Although in the solid phase or solution the enol form is more stable, their relative concentrations may vary with temperature, polarity of solvent, pH, and substitution of the aromatic rings [16,17].

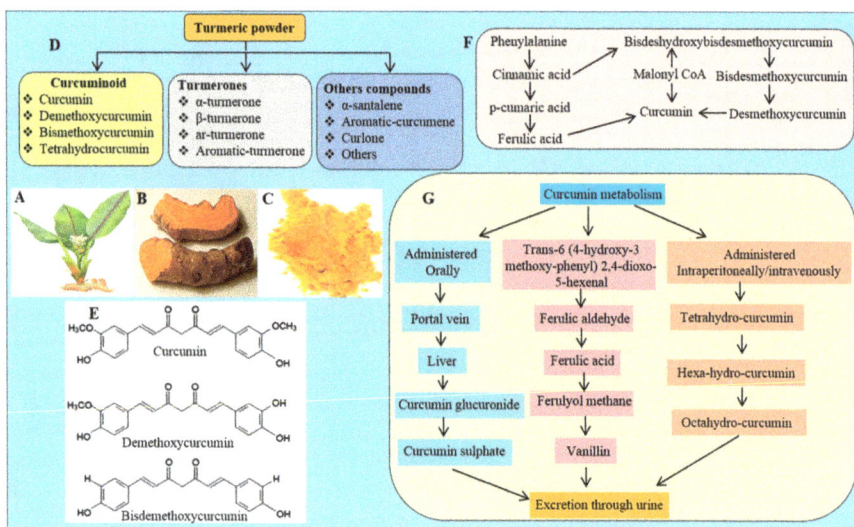

Figure 1. Chemical structure of Cur and its derivatives. (**A–C**) *Curcuma longa*, its rhizomes and turmeric extract; (**D**) Different chemical components of turmeric extract; (**E**) Chemical structure of principal ingredients of curcuminoid; (**F**) pathway of Cur-biosynthesis; and (**G**) Cur metabolism in our body.

Table 1. The chemical and biophysical properties of curcuminoid [15].

Characteristics	Cur-I	Cur-II	Cur-III
Common Name	Cur	DemethoxyCur	BisdemethoxyCur
Chemical Name	Dicinnamoyl methane	4-OH cinnamoyl methane	Bis-4-OH cinnamoyl methane
Color	Bright orange-yellow	Bright orange-yellow	Bright orange-yellow
Amount Present (%)	77	17	3
Molecular Mass (g/mol)	368.4	338.0	308.1
Melting Point (°C)	183.0–186.0	172.5–174.5	224.0
Neutral Solvent (water)	Poorly soluble	Poorly soluble	Poorly soluble
Solubility in Organic Solvents	Soluble	Soluble	Soluble
Solubility in Hexane or Ether	Insoluble	Insoluble	Insoluble
Excitation/Emission in	420/530 nm	420/530 nm	420/530 nm
Excitation/Emission in Alcohol	536–560 nm	Unknown	Unknown

The amounts of keto-enol forms in Cur also play vital roles in the physicochemical properties, biological functions, and anti-oxidant activities [18]. The keto form is predominant in acidic (pH 3) to neutral conditions, while, the enol form is predominant in alkaline solutions (pH > 8), and is a potent free radical-scavenger [15]. Cur is hydrophobic in nature, so has poor solubility in water or hydrophilic solutions, although the solubility can be improved in basic conditions. Cur shows greater solubility in organic solvents (Figure 2), such as ethanol, methanol, isopropanol, acetone and dimethyl-sulfoxide (DMSO), whereas it is moderately soluble in hexane, cyclohexane, tetrahydrofuran and dioxane [15].

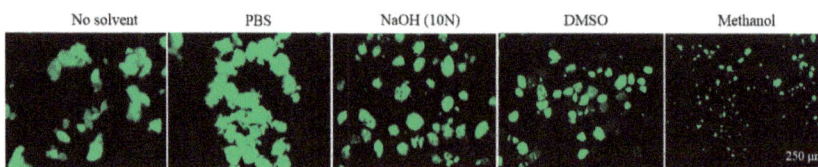

Figure 2. Curcumin solubility in different solvents. Please note that Cur is more soluble in methanol than in phosphate buffer saline (PBS), NaOH or dimethyl-sulfoxide (DMSO).

Interestingly, Cur is a natural fluorophore, with its absorption noted in polar solvents ranging from 408 to 540 nm [16,19]. The maximal fluorescence intensities of Cur is noted in chloroform, acetonitrile, and in acetone to be in the range of 494 to 538 nm, whereas in alcohols and dimethyl formamide (DMF), the fluorescence spectra may shift from 536 to 560 nm [15]. In contrast, in non-polar solvents (e.g., benzene, hexane and cyclohexane), sharper peaks (~460 nm) are observed, because of the blue-shifting of its absorption spectra [15] (Table 1).

2.3. Cur Biosynthesis

Cur may be biosynthesized via two ways (Figure 1F). Phenylalanine is the precursor molecule, and the cinnamic acid is the first byproduct of the Cur biosynthetic pathway (Figure 1F). When cinnamic acid reacts with 5-malonyl CoA, it forms bis-dehydroxybisdesmethoxy-Cur. This compound can then be converted to bisdemethoxy-Cur (BDMC) and demethoxy-Cur (DMC), which can be transformed into Cur (Figure 1F). The second pathway involved in Cur synthesis is the production of cinnamic acid, which is then converted to p-coumaric acid, and ferulic acid. The ferulic acid reacts with 5-malonyl CoA to form Cur [20,21] (Figure 1F).

2.4. Cur Metabolism

The metabolism of Cur, including its pharmacokinetics (PK) and pharmacodynamics (PD) has been studied by several investigators in rodents and in human [22–25]. The profile of Cur metabolites depends on the route of administration. For example, oral administration of Cur immediately reaches the liver, following intestinal absorption, and become sulfated or glucuronidated by liver-specific enzymes, such as sulfatase and glucuronidase, respectively (Figure 1). The major Cur metabolites in animal liver are the glucuronides of tetrahydrocurcumin (THC) and hexahydrocurcumin (HHC), whereas the traces amounts of dihydroferulic acid and ferulic acid are also found as the minor metabolites [26]. However, both these glucuronides, and sulfate conjugates are water soluble, and found in the urine of rats. According to Pan and colleagues, 99% of Cur in plasma was present as glucuronide-conjugates, which suggests that Cur first undergoes extensive reduction by alcohol dehydrogenase, followed by conjugation [27]. In contrast, when Cur is administered intravenously (i.v.), or intraperitonally (i.p.), it can form more stable, and water soluble Cur-derivatives, such as THC, HHC and octahydrocurcumin (OHC), which are easily eliminated from body through urine [27]. In addition, after absorption, Cur is readily catabolized to several degradation products, such as ferulic aldehyde, ferulic acid, ferulyol methane and vanillin (Figure 1G). A pharmacokinetic (PK) study revealed that the maximum concentration of curcuminoid conjugates in plasma was found within 1 h after its oral administration [28], but whether these Cur-metabolites are active, in a manner similar to free Cur, is not yet clear. However, some experimental data demonstrated that Cur-glucuronides and THC are less active than Cur itself [26], but other studies reported that they may be more active than Cur, because of their greater stability in body fluids, [29]. For example, THC shows better anti-diabetic and anti-oxidative effects than Cur in a rat model of type-2 diabetic [30], whereas Sandur and colleagues reported that THC has much lower anti-inflammatory and anti-proliferative activities than Cur [31].

3. Pleotropic Actions of Cur on Nervous System

3.1. Anti-Amyloid Properties

The most promising application of Cur in neurodegenerative diseases therapy is its anti-amyloid property [9,32]. Its preferential binding and potent inhibitory effects on amyloid aggregation has attracted researchers to investigate its beneficial roles for treating neurological diseases [13,33]. It not only binds with Aβ-oligomers and fibrils in AD [9,34], but also binds readily with other amyloid proteins, such as α-synuclein (α-syn) in PD [35], huntingtin (HTT) in HD [36], phosphorylated tau (p-tau) in tauopathies and AD [37], as well as with prion proteins in prion diseases [38] (Table 2).

Most interestingly, the high lipid content of brain tissue allows lipophilic Cur molecules to cross the blood brain barrier (BBB) and inhibit the aggregation of amyloid proteins.

Table 2. Anti-amyloid activities of Cur in major neurodegenerative diseases.

Proteins	Diseases	Nature of Binding of Cur	Outcomes	Ref.
Aβ	AD	With amino acid 16–21 of Aβ	Inhibits oligomer and fibril formation, thus decrease Aβ induced neurotoxicity	[9,13,34]
Tau	Tauopathies, AD	In the microtubule-binding region of tau	Inhibits phosphorylated tau, thus decrease neurofibrillary tangle	[12,37]
α-Syn	PD	In the hydrophobic no Aβ component region	Inhibits α-syn oligomers and fibril formation, thus decrease α-Syn induced oxidative damage	[35,39]
HTT	HD	Unknown	Lower doses (nM) decrease HTT aggregates	[36,40]
Prion	Prion	α-Helical intermediate and to the amyloid form of prion protein	Inhibits PrP^{sc} accumulation	[38,41]

3.2. Potent Antioxidant

Due to the high metabolic rate, increased demand of O_2, large quantities of membrane phospholipids and polyunsaturated fatty acids (PUFA), and lower levels of anti-oxidants relative to other organs, the CNS is particularly vulnerable to oxidative damage (Figure 3). All these factors significantly contribute to increase reactive oxygen species (ROS) and peroxynitrite (ONOO-) levels, which lead to inflammation, mitochondria dysfunction and, ultimately, induce neuronal death. Chronic progressive neurological diseases induce inflammation, oxidative stress, lipid peroxidation, DNA damage, oxidized protein products [42].

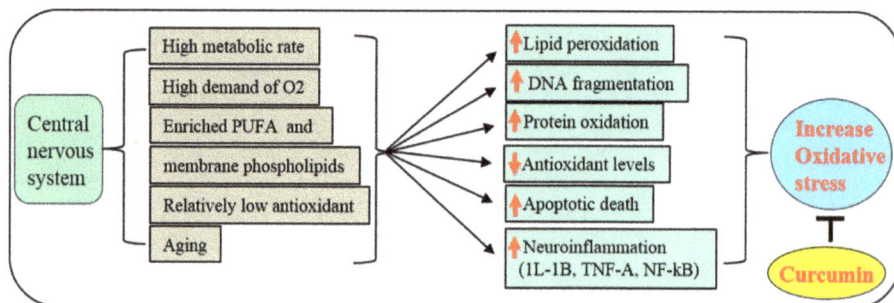

Figure 3. Amelioration of oxidative stress by Cur in brain. The CNS is vulnerable to oxidative stress due to high metabolic rate, which causes higher O_2 demand. This leads to an increase in oxidative stress in the brain tissue. Whereas Cur, as a potent free radical scavenger, can ameliorate these effects.

Chronic oxidative stress has also been associated with induction of misfolded protein aggregates in brain tissues [43]. To counter this, Cur, as a potent anti-oxidant, can scavenge superoxide anions ($O_2{}^-$) and hydroxyl radicals (OH^-), and increase anti-oxidant levels, such as glutathione (GSH) [44]. Cur also can stabilize the brain anti-oxidant enzyme systems, including activation of superoxide dismutase (SOD), glutathione peroxidase (GPx), glutathione S-transferase (GST) [45]. It also protects cells from lipid peroxidation, DNA damage, and protein oxidation or protein carbonylation [46] (Figure 3). Although most researchers describe oxidative stress as an imbalance of pro-oxidants and antioxidants, the actual mechanism involves a disruption of redox signaling and control [47,48]. Therefore, measurement of the signaling proteins associated with oxidative stress, as well as the ROS and anti-oxidant system, are required to investigate the effectiveness of Cur treatments.

3.3. Anti-Inflammatory Agent

Next to its anti-oxidant properties, the second most important reason for interest in Cur as a therapy for neurological diseases is its ability to reduce neuroinflammation [33,49]. Several reports suggest that Cur is a potent anti-inflammatory agent, which can downregulate many neuroinflammatory marker proteins, such as nuclear factor kappa beta (NF-κB) [45]. Cur also inhibits phospholipases and arachidonic acid metabolic enzymes, such as cyclooxygenase-2 (COX-2), 5-lipoxygenase (5-LOX) [50]. In addition, it reduces the levels of several cytokines, such as tumor necrosis factor (TNF), interleukin-1 (IL-1) and interleukin-6 (IL-6) [51,52] (Figure 4). Similarly, Cur is also an agonist for peroxisome proliferator-activated receptor gamma (PPARγ) which can inhibit the pro-inflammatory pathways [53].

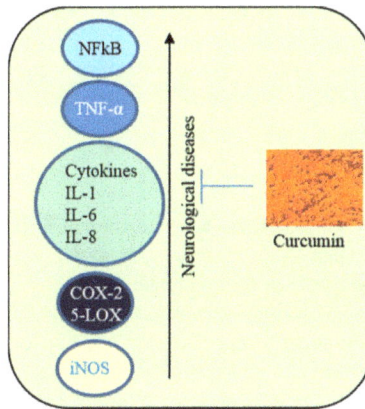

Figure 4. Anti-inflammatory properties of Cur. Curcumin increase levels of anti-inflammatory cytokines, inhibits inflammatory chemokines, iNOS levels and inhibits transcription factors, such as NF-κB.

3.4. Modulate Activity of Molecular Chaperones

Cellular protein quality is control by a set of proteins, called molecular chaperones. They are very important to reverse the unfolded or misfolded proteins to their native form. One of the important molecular chaperones are the heat shock proteins (HSPs). These proteins are downregulated in different neurological diseases [54]. Recently, we, and others, have shown that Cur is neuroprotective through the activation of molecular chaperones, such as HSP70, HSP90, HSP60 and HSP40 and heat shock cognate 70 (HSC70) [55].

3.5. Increase Neurotrophins, Neurogenesis and Synaptogenesis

Mounting evidence indicates that significant declines in neuronal growth factors, such as NGF, BDNF, GDNF, as well as other supporting factors, such as PDGF can lead to synaptic damage and neuronal death. Diets containing Cur have shown to stimulate NGF, BDNF, GDNF, PDGF levels in vivo [56]. Cur also enhances neurogenesis, synaptogenesis and improves cognition in rats [57], which may be due to promoting these neurotrophic factors. Furthermore, improved memory functions in animal models of neurological diseases have been observed after Cur therapy, which may be due to increase levels of these neuronal growth factors. The pre-synaptic and post-synaptic markers, such as synaptophysin and PSD95, are also restored in different animal models of neurodegenerative diseases after Cur treatment [58] (Figure 5).

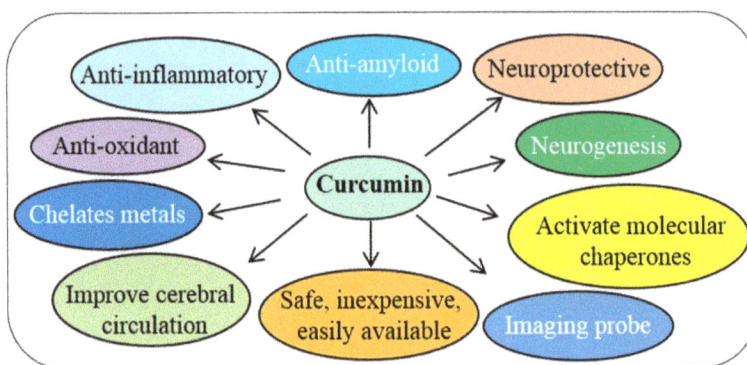

Figure 5. Uses and advantages of Cur for diagnosing and treating neurological diseases. In addition to its pleotropic therapeutic effects Cur is safe, inexpensive, and readily available and can be used to label Aβ deposits in the brain.

3.6. Metal Chelator

Heavy metals, such as aluminum (Al), copper (Cu), cadmium (Cd), iron (Fe), lead (Pb), manganese (Mn), and zinc (Zn) can induce misfolded protein aggregation and production of ROS in different neurological diseases [59,60]. The presence of two phenolic (OH) groups (see Figure 1) and one active methylene (CH2) group in a Cur makes it an excellent ligand for any metal chelation [61]. For example, Cur can interact with Cd and Pb and prevent neurotoxicity caused by these metals [62]. In fact, Cur effectively binds to Cu, Fe, and Zn, and makes them unavailable to induce amyloid protein aggregation. In addition, most of these heavy metals can induce neuroinflammation by increasing the expression of NF-κB levels, whereas it is speculated that Cur suppresses inflammation by inhibiting the NF-κB levels, perhaps via metal chelation [7,63]. The Cur-metal complexes also show greater anti-oxidant properties. For example, Cu-Cur complexes scavenge ROS more efficiently than Cur alone [61]. Similarly, Mn-Cur complex exhibits a more potent neuroprotection than Cur, as shown in both *in vitro* and in vivo experiments [18,61] (Figure 5).

3.7. Cur Regulates Epigenetics

Epigenetics play vital roles in gene expression in different disease conditions. Cur plays significant regulatory roles in modulating the methylation, acetylation, ubiquitination, and phosphorylation status of histone and other DNA-binding proteins [64]. For example, when the lysine at position 4 in histone-3 becomes methylated (H3K4me3), it activates the gene, whereas the lysine methylation at position 27 in the same protein (H3K27me3) silences the gene [65]. Similarly, histone acetylation produces increase in gene expression, whereas deacetylation has opposite effect. The histone acetylation is governed by histone acetyltransferase (HAT) and the enzyme involved deacetylation is the histone deacetylase (HDAC).

However, the epigenetic role of gene expression of Cur has been shown by inhibiting HAT activity and activating HDAC in AD (Figure 6). Cur can directly bind to HAT at a nM levels and can inhibit the catalytic activity of HATs [64], thus inhibiting nuclear histone acetylation. Decreases in histone acetylation reduce the inflammation via NF-κB pathway in some brain diseases [45].

AD	AD + Cur
Decrease HDAC activity	Increase HDAC activity
Increase HAT activity	Decrease HAT activity
Increase acetylated p65	Decrease acetylated p65
Decreased DNA methylation	Increased DNA methylation

Figure 6. Effects of Cur on epigenetics in AD. Cur restores the activity of HDAC and inhibits HAT activity, along with increase DNA methylation in animal models of AD.

3.8. Improving Cerebral Circulation

Decreased cerebral circulation in aging brain causes an increase in risk of cerebral hemorrhage and stroke, whereas Cur has an influential role on cerebral circulation [66]. It can reduce the adhesion of platelets in brain microvascular endothelial cells (BMECs) [67,68], and also can inhibit the inflammation of blood vessels, which may improve overall cerebral circulation [69].

4. Limitations of Cur Delivery

Cur has been delivered in animals and humans by several means, including oral, subcutaneous, intraperitoneal, intravenous, nasal, and topical deliveries to achieve its beneficial effects. The major concerns about Cur delivery involve its instability and poor solubility in most body fluids, which reduces its absorption through the gastrointestinal (GI) tract, and facilitates its metabolism and degradation, as well as rapid elimination from the body, mitigating its bioavailability [70]. For example, researchers were unable to detect free Cur from the plasma of AD patient in a clinical trial in which 2–4 g Cur were delivered daily [71]. It is hypothesized that after absorption, Cur becomes rapidly glucuronidated in the liver by glucuronidase, which makes it water soluble, and, thus, promotes its rapid excretion through the urine [71,72]. Similarly, oral administration of 1 g/kg dose of Cur causes excretion of ~75% of Cur through feces, with negligible amounts in the urine [73]. Similarly, approximately 40% of the Cur was found in the feces, along with Cur-glucuronide and sulfates in the urine when 400 mg per day Cur is administered orally [26]. Most of the Cur is excreted within 72 h when lower doses (10 or 80 mg) are administered, whereas Cur remains in tissues for 12 days after oral administration of higher (400 mg) doses [24]. In contrast, no Cur or its metabolites were found in urine in a clinical trial when 36 and 180 mg was given daily for 4 months by oral administration, but some of these metabolites was excreted in the feces [74]. Clearly, major challenges for successful Cur delivery and its clinical applications for neurological diseases will require a special formula, which can optimize its solubility, stability, and bioavailability. In addition, it is critical to determine the amount of Cur required to prevent further neurodegeneration or to rescue degenerating neurons in neurological diseases.

5. Nano-Technological Approaches for Cur Delivery

To improve the bioavailability of Cur, numerous approaches and many promising novel formulations have been undertaken by several investigators, which included the use of nanoparticles, liposomes, micelles, and phospholipid complexes, nanogels, noisomes, cyclodextrins, dendrimers, silver, gold, and structural analogues of Cur [11] (Table 3). Most of these novel delivery mechanisms increase Cur bioavailability by providing longer circulation, better permeability, and/or resistance to metabolic processes.

Table 3. Different components used to increase Cur solubility and bioavailability. Scientists used different materials, including adjuvants, proteins, lipid nanoparticles, and synthetic materials to increase Cur solubility [75].

Materials	Compounds Used with Cur
Adjuvant	Piperine
Bio-conjugates	Turmeric oil, glycine, alanine, EGCG
Lipids	Phospholipid, liposome, oil body emulsion
Nanoparticles	GMO, Chitosan, cyclodextrin, PLGA, silica, PHEMA, gold, silver, casein, orange gel-based nano emulsion, dendrimer, solid lipid particles
Protein	BSA, soy protein isolated
Others	Hyaluronic acid, hydrogel, polymer, PEG-PEI emulsion, polymer encapsulated, beta-lactoglobulin

5.1. Adjuvants

Conjugation of piperine (extracted from black pepper, a well-known inhibitor of hepatic and intestinal glucuronidation of Cur) with Cur increase free Cur levels in animals and human plasma [76]. For example, co-supplementation with 20 mg of piperine with 2 g of Cur significantly increased the bio-availability of Cur by 2000 folds in a clinical trial [75] (Figure 7).

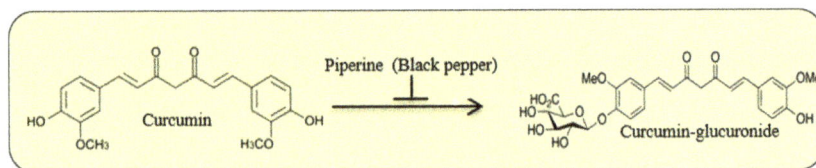

Figure 7. Use of adjuvant piperine to increase bioavailability of free Cur levels. Rapid glucuronidation process reduces bioavailability of Cur, whereas piperine present in black pepper can inhibit this glucuronidation process, thus increasing the amount of free Cur in different tissues.

5.2. Bio-Conjugates

Several bio-conjugates (e.g., turmeric oil, glycine, alanine, and/or piperic acid) can be used to increase the cellular uptake and greater bioavailability of Cur [26] (Table 3). For example, when curcuminoid is combined with turmeric oil (turmerons) in a specific proportion to make Biocurcumax-95 (BCM-95), the Cur-bioavailability was 7-8 times more than that of natural Cur [77] (Table 4).

Table 4. Average values of pharmacokinetic parameters of Cur-lecithin-piperine and BCM-95® (Biocurcumax™).

Parameters	Tmax	Cmax	Ke	t1/2	AUC (o.inf)	Cl (Observed)/F	Vz (observed)/F
Cur	2	149.8	0.296	2.63	461.86	0.006735	0.026362
BCM—95 RCG	3.44	456.88	0.26	4.96	3201.28	0.001682	0.006784

Tmax: Time of peak plasma concentration, Cmax: Peak plasma concentration, Ke: Elimination rate constant, t1/2: Half-life, AUC (0-infinity): Area under curve from '0' h to infinity, Cl/F: Clearance/Bioavailability, Vd/F: Volume of distribution/bioavailability [77].

This formula has better absorption into blood and longer retention time compared to natural Cur. It also increases the activity of Cur up to a 700%, as confirmed by clinical trials [77]. Similarly, when Cur was bio-conjugated with glycine, alanine, and/or piperic acid, these formulae improved the anti-microbial properties over natural Cur, which suggests that these formulae promote cellular uptake, reduced their degradation, and increased Cur activity inside the cells [78]. In another study, when epigallocatechin-3-gallate (EGCG, a polyphenol of green tea) was conjugated with Cur, this formula also increased Cur uptake, as well as its beneficial effects [79].

5.3. Cur-Phospholipid Complex

Cur-phospholipid complex increased aqueous solubility of Cur up to 3-fold and showed greater protection against tissue damage compared to unformulated Cur [80]. Oral administration of Cur–phospholipid complex (100 mg/kg) in rats showed a maximum plasma Cur levels of 600 ng/mL after 2.33 h, whereas administration of the same amount (100 mg/kg) of unformulated Cur, (i.e., the free Cur) was ~267 ng/mL after 1.62 h, indicating that the Cur-phospholipid complex has greater bioavailability than unformulated Cur. Importantly, this formula provides more tissue protection by increasing antioxidant enzyme systems. Similarly, a formula of phosphatidylcholine with Cur, when administered orally (340 mg/kg) in rat, showed an increase in Cur-bioavailability by 5-folds in comparison to unformulated Cur [80].

5.4. Liposomes

Liposomes are spherical, self-assembling, closed colloidal structures, composed of lipid bilayers, with both hydrophilic and hydrophobic characteristics (Table 5). Therefore, they provide an excellent system for delivering hydrophobic compounds, such as Cur. Liposome vesicles are typically 25 nm to 2.5 mm in diameter and can be extracted from natural phospholipids or can be artificially synthesized. Cur can be encapsulated with liposomes, which can be delivered into the cell by membrane fusion or endocytosis, a formulation which has proved to be safe and to enhance Cur solubility and cellular activities. Moreover, liposome-encapsulated Cur is transported without rapid degradation, along with minimum side-effects, and with greater stability. In addition, liposomal-Cur is more stable than free Cur in PBS, but both are equally stable in human plasma and in culture media. Furthermore, the liposome-Cur complex increases the bioavailability and efficacy of Cur after intravenous administration in animals. This formula also possesses anti-cancer effects, both in vitro and in vivo, against osteosarcoma and breast cancer, and inhibits the growth of melanoma cells. It also increases serum creatinine, while decreasing tissue damage, cell death, and inflammation [11].

Table 5. Nanoparticles-conjugated Cur formulae, characteristics and their biological effects [11].

Nanoparticles	Schematic Diagram	Shape	Size (nm)	Methods	Outcome
Liposome		Globular	25–205	In vitro, in vivo (dogs & mice)	Increased solubility, tissue distribution, and stability
Micelles		Spherical	10–100	In vitro, In vivo (mice)	Increased solubility and bioavailability; Improved anti-oxidative properties
Noisome		Lamellar	190–1140	In vitro, In vivo (snake and mice)	Increased skin penetration; Prolonged delivery system

Table 5. *Cont.*

Nanoparticles	Schematic Diagram	Shape	Size (nm)	Methods	Outcome
Nanogel network		Cross-linked polymer	10–200	In vitro	Increased stability, fluorescence effects, developed bioavailability, get better control release; Prolonged half-life
Chitosan		Linear polysaccharide composed	100–250	In vitro, In vivo (rats and mice)	Improved chemical stability, improved antioxidant effects; Prolonged blood circulation
Gold		Globular	200–250	In vitro	Improved solubility; Enhanced antioxidant
Silver		Film layer	~15	In vitro	Improved wound healing; Increased antiviral and anticancer effects
Cyclodextrin		cyclic oligomers of glucose oligosaccharide		In vitro	Improve stability and bioavailability of Cur
Dendrimer		Globular polymer	15–150	In vitro, In vivo (mice)	Improved stability; Increased antitumor and anti-proliferative effects
Solid lipid		Spherical	50–1000	In vitro, In vivo (rat and mice)	Prolonged circulation of blood; Increased anti-inflammatory effects; Improved brain delivery

5.5. Micelles

Unlike liposomes, micelles are monolayered phospholipid complexes (in solution) with various shapes (spherical, vesicles, rod-like, or star-shaped) and sizes. They are amphipathic molecules, form emulsions when and solubilized in water, and act as excellent surfactants. A major advantage of using micelles for Cur delivery is that they are of smaller sizes (~10–100 nm in diameter), making them more stable in biological fluids [81]. Micelles can form a nano-sized core/shell structures in aqueous media, facilitate the permeability of hydrophobic drugs, such as Cur, by burring themselves inside the hydrophobic core and making the polymer more water soluble. Thus, micelles can act as transporters of Cur, and increase their efficiency by targeting specific organs, such as the brain. Using a solid dispersion method, Liu and colleagues [72] prepared bio-degradable, self-assembled polymeric micelles, loaded with Cur, which significantly increased the release of free Cur. In vitro results of studies using spherical Cur-loaded mixed micelles revealed an enhanced solubility and biological activity of Cur [82]. Similarly, ε-poly-lysine micelles coated with curcuminoid also improved their solubility and cellular anti-oxidative activities, in comparison to free curcuminoid [83]. Furthermore, novel biodegradable micelles that were synthesized by conjugating methoxy-polyethylene glycol provided sustained Cur release for 24 h in vitro and enhanced its aqueous solubility and stability with a 3-fold reduction in IC_{50} value of Cur [84]. To increase prolongation of its half-life, higher bio-distribution, and bioavailability, while decreasing total clearance of Cur, Song and colleagues [85] synthesized a poly (D, L-lactide-co-glycolide)-b-poly(ethyleneglycol)-b-poly(D, L-lactide-coglycolide) (PLGA-PEG-PLGA) polymeric micelles which coated Cur. This prolonged its circulation time because of its smaller size and hydrophilic shell that reduced the drug uptake by the mononuclear phagocytic systems [85]. In an attempt to increase the aqueous solubility of hydrophobic drugs, a polymeric micellar formulation containing methoxy poly (ethylene glycol)-block-polycaprolactone diblock copolymers (MePEG-b-PCL) provided a 13 to 105 fold increase in solubility [11]. However, Cur-loaded micelles can boost the efficiency of the drugs by targeting specific cells, resulting in less drug accumulation in healthy tissues and a reduction in toxicity. Therefore, micelle encapsulation of Cur provides an enormous increase in solubility and bioavailability of Cur, making this formulation a very promising avenue for developing clinically effective therapeutic tools.

5.6. Noisome

Chemically, noisomes are alkyl or dialkyl polyglycerols that contain cholesterol, which acts as nonionic surfactant [86]. It is an excellent carrier for all kinds of drug molecules, including those which are hydrophilic, amphiphilic, or lipophilic in nature. Noisomes behave similar to liposomes in vivo, therefore, providing an alternative for liposome-based drug delivery. In addition, they have significant potential for anti-cancer and anti-inflammatory activity [81]. Because, they are very stable, which prolongs their delivery and suppresses the level of degradation, their use can improve oral bioavailability of Cur and can increase its skin penetration [87]. Therefore, noisomes are a potential delivery system for Cur that would increase its stability and bioavailability.

5.7. Nanogels

These are covalently cross-linked polymers with 3-D chain networks making them suitable for delivering bio-compatible drugs to different tissues (Table 5). They provide the perfect reservoir for loading and delivering different amphipathic drugs and prevent them from environmental degradation [88]. The size of nanogels can be customized by manipulating the functional groups used, the density or degree of cross-linking, ionic strength, and pH of the solution [88]. Several chemical interactions, such as salt bonds, hydrogen bonds, or hydrophobic interactions can be used to react with Cur to enhance its stability in biological fluids, as well as its oral and brain bioavailability [89]. For example, a self-assembled dextrin nanogel was used for Cur delivery, which proved to be a suitable carrier for controlling the release of Cur. By using dynamic light scattering (DLS), scanning electron microscope (SEM), and Fourier transform infrared spectroscopy (FTIR), it was shown that Cur-chitin nanogels had higher levels of Cur release at an acidic pH compared to neutral pH, and proved to be a potent toxic agent to cancer cells (0.1–1.0 mg/mL), without harming normal cells [90]. Furthermore, water-dispersible hybrid nanogels have been made by coating silver or gold bimetallic nanoparticles for intracellular delivery of Cur. This formula increases the Cur loading yields and its sustained release, along with its bioavailability, and also prevents Cur from surrounding temperature or exogenous irradiation with near-infrared light [91]. Therefore, nanogels might be an excellent carrier for Cur, due to their smaller particle size (10–200 nm), which significantly enhances their biodegradability, stability, loading efficiency, and/or biocompatibility, while prolonging half-life, increasing transdermal penetration and providing protection against degradation by the immune system [88]. Overall, the use of specifically designed, multifunctional hybrid nanogels appears to be safe and appropriate for Cur delivery in clinical trials aimed at the prevention of neurodegeneration or cancers.

5.8. Chitosan

This is a linear polysaccharide of deacetylated and acetylated units of chitin, present in exoskeleton of crustaceans and the walls of the fungi cells. They contain primary amine groups, which make it cationic in nature and increase its solubility in various media, while possessing its polyelectrolyte behavior, metal chelation, and structural uniqueness [92]. Chitosan-coated nanocarriers, contain Cur particles, and are positive in nature, are 114–125 nm in diameter, and can increase the fluorescence intensity of Cur. Cur-phytosome-loaded chitosan microspheres also improve Cur absorption, prolong the retention of Cur and increase its bioavailability due to the accumulation of nanoparticles in the ER [93]. Furthermore, binding of Cur to chitosan nanoparticles improves its chemical stability and prevents its degradation. Overall, Cur-coated chitosan derivatives can easily enter the cell membrane and release Cur in a controlled manner, and are nontoxic to normal cells, while toxic to tumor cells, at the same time maintaining stronger antioxidant and chelating effects than free Cur [94].

5.9. Gold Particles

Due to their optical and electrochemical uniqueness, stability in biological systems, capability for combining with biomolecules, and their lower cytotoxicity, gold nanoparticles (AuNPs) can be a potent carrier for Cur [95]. These particles can also be easily synthesized and functionalized and have improved longevity in the circulatory system. A formulation of chitosan-Cur nano-capsules with AuNPs formulated by a solvent evaporation method produced 18–20 nm diameter of AuNPs-Cur, which provides a more controlled and steady release of Cur, compared with Cur-encapsulated chitosan nanoparticles. The effect of Cur-conjugated-AuNPs on peripheral blood lymphocytes are those typical characteristics of apoptosis, including chromatin condensation, membrane blebbing, and the occurrence of apoptotic bodies [96]. Therefore, Cur conjugated AuNPs could provide better targeting of cells, sustained release of Cur, and more powerful antioxidant effects than free Cur.

5.10. Silver Particles

As a safe and potent anti-microbial agent, silver (Ag) can be used to improve Cur delivery [97]. Using a diffusion mechanism, researchers loaded Cur into a 15-nm diameter sodium carboxyl methyl cellulose silver nanocomposite (AgNPs) film. This film improved Cur encapsulation and increased its anti-microbial activity. Similarly, a novel hydrogel-AgNPs-Cur composite has been developed which produces greater anti-microbial activity than AgNPs-Cur films that lack the hydrogel. Moreover, there is a sustained release of Cur from the Ag-encapsulated composite, which can increase its bioavailability, as well as therapeutic values [11]. In addition, the use of AgNPs could protect the cells from anti-microbial attack and also act as an anti-inflammatory, anti-viral, and anti-cancer agent, along with its wound-healing properties [11] (Table 5).

5.11. Cyclodextrin

Cyclodextrins (Cds) are the cyclic oligomers of glucose or oligosaccharide residues synthesized from starch molecules, which have wide applications, including use in pharmaceutical, drug-delivery, and food processing industries [81]. Chemically, they are pseudo-amphiphilic molecules, which help their solubility and stability in aqueous solution and can act as vehicles for oral or intravenous delivery of hydrophobic molecules (e.g., Cur) to improve their bioavailability and prevent their degradation without alteration of their pharmacokinetics [81]. A preparation with Cds-Cur complexes (Figure 8) improved the hydrolytic stability of Cur with enhancement of photodecomposition efficiency in organic solvents, thus increasing their stability and reducing their degradation rate compared to the free Cur [98].

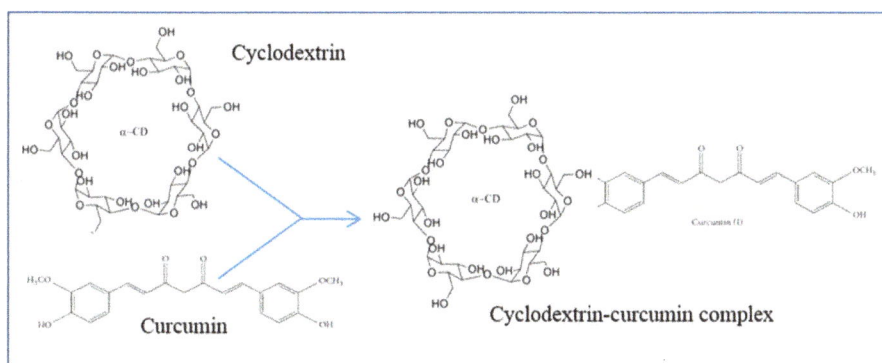

Figure 8. Cyclodextrin-Cur complex. Cyclodextrin is a water-soluble pseudo-amphiphilic starch molecule, which is a potent carrier for Cur and can increase its solubility and bioavailability.

Similarly, Cur tagged with β-Cds nanosponges has more solubilization efficiency and can release Cur more readily, given that the complex is nonhemolytic in comparison to free Cur [99]. Another study showed that the Cds-Cur complex becomes more potent as an anti-inflammatory agent than free Cur by inhibiting the nuclear factor kappa beta (NF-κB), inducing death of cancer cells [100], such as prostate cancer cells, while acting as a telomerase inhibitor. Therefore, Cds could encapsulate Cur and increase its stability and bioavailability, compared to the free Cur, without altering their pharmacokinetics.

5.12. Dendrimer

Dendrimers are a group of small (nM), dense spherical, branched series of polymeric globular polymers (Figure 9), which are considered as synthetic proteins. They consist of a core, branched interiors and numerous surface functional groups (i.e., OH or NH_2), which serve as a platform for carrying and delivering many drugs, (like Cur), small molecules, or DNA [101]. As a safe molecule, dendrimers can be used as a probe for molecular imaging. Dendrimer-Cur formulations are readily dissolved in aqueous solution, can increase cellular uptake, and show more cytotoxic effects in human breast cancer cells than free Cur. The dendrimer surface containing poly-amidoamine group can carry hydrophobic drugs (such as Cur) for their successful delivery. For example, a poly-amidoamine (PAMAM) encapsulated Cur conjugates show significant inhibition of telomerase activity and induce apoptosis by enhancing Cur uptake in human cancer cell lines [102].

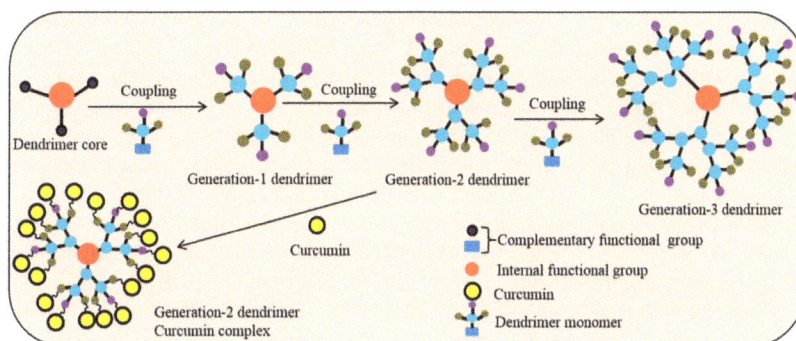

Figure 9. Schematic diagram showing dendrimer structure and construction of dendrimer-curcumin complexes. The Cur can be coated with the outer surface of the branched dendrimer, thus it can carry numerous Cur molecules, depending on the surface groups and charge.

Furthermore, when Cur is conjugated with dendrosomal nanoparticles, which are neutral, amphipathic, and biodegradable nanomaterials, this compound increases the stability and, uptake of Cur, while increasing its antitumor activity through its induction of apoptosis, as demonstrated in both in vitro and in vivo experiments [103]. In addition, the dendrosomal nanoparticle-Cur has chemo-protective and chemotherapeutic effects on colon cancer by inhibiting the cell proliferation and induction of apoptosis [103]. These properties make dendrimers especially attractive as a carrier for Cur, relative to other nanoparticles.

5.13. Solid Lipid Nanoparticles (SLNP)

SLNPs are the spherical and submicron colloidal lipid carriers (50 to 1000 nm) which maintain a solid shape at room temperature (Figure 10). There are several advantages for using of SLNPs for Cur delivery, including improvement of release kinetics, enhancement of bioavailability, increased protection via encapsulation, ease of manufacturing, increased stability, along with versatile applications [104]. Moreover, the size of the Cur-SLNPs is much smaller, ranging from 100 to 300 nm, with a very favorable total drug content of <92% when manufactured by micro-emulsification technique. One of our recent experiments with SLNPs-Cur, using a dose of 555 ppm, showed that Cur level was 250–300 nM in mouse brain tissue, along with improved neurobehavioral outcomes [12,40].

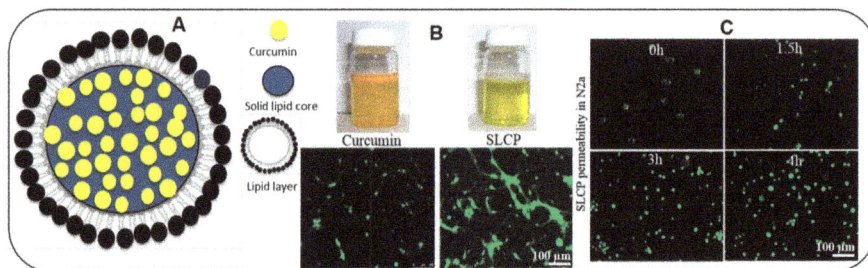

Figure 10. Schematic diagram showing formulation of solid lipid Cur particle (SLCP). (**A**) In this formula, outer layer is composed of long chain fatty acid bilayers, with the inner layer being composed of a solid fatty acid core and on that core that is coated with Cur molecules. (**B**) Comparative solubility (**upper**) and cellular permeability in primary hippocampal neurons (**lower**) and (**C**) permeability of Cur and SLCP in N2a cells. Scale bar = 100 μm

In an experimental set up using an animal model of cerebral ischemia, rats fed with SLNPs-Cur had a 90% improvement in cognitive function, along with a 52% inhibition of acetyl cholinesterase activity [105]. Furthermore, this formula has been shown to increase the levels of superoxide dismutase (SOD), catalase, glutathione (GSH), and the activities of mitochondrial enzymes, while decreasing lipid peroxidation and peroxynitrite levels. This formula also improved the bioavailability of Cur in the brain by 16.4- and 30-fold with oral and intravenous administration, respectively. Similarly, solid lipid microparticles of Cur that were prepared with palmitic acid, stearic acid, and soya lecithin, had more powerful anti-angiogenic and anti-inflammatory activities. As such, the SLNP-Cur formula has several advantages over other nanoparticles, such as: (a) larger carrying capacity of Cur; (b) ease of scaling and sterilization; (c) protection via encapsulation, (d) more favorable kinetics, (e) increased bioavailability, (f) ease of manufacturing, and (g) superior stability with application versatility [11]. For example, Verdure Sciences has developed a SLNP-formulation of Cur, called "Longvida", which achieves a 0.1 to 0.2 μM plasma level with an associated 1–2 μM brain level of free Cur in animals [6,10,13,55]. Later they optimized this formula as "lipidated Cur" which can achieve more than 5 μM in the brains of mice [106,107]. We have been working with this formula and found significant beneficial effects both in vivo and in vitro models of AD [108], and in an in vitro model of glioblastoma [14].

5.14. Derivatives and Analogues of Cur

The biological properties of Cur and its derivatives depend on the chemical structure of Cur. For example, isomeric forms of Cur have better antioxidant properties. Therefore, structural modifications of Cur might be a good strategy to improve its biological activities. Several Cur derivatives and/or analogues have been synthesized and tested by many researchers. Among them, EF-24, a Cur analogue (Figure 11) has shown to possess promising anti-tumor activity in vitro and in vivo, in comparison to natural Cur [109]. Up to 32 mg/kg of this compound was safe in mice after intravenous administration, and the absorption was rapid after both oral and i.p. administration. At this dose, when mice were injected with EF-24 i.p., within 3 min, the peak plasma concentration of Cur reached 1000 nM and the absorption and elimination half-life values were 177 and 219 min, respectively. The bioavailability of oral and i.p. EF-24 was 60% and 35%, respectively [109].

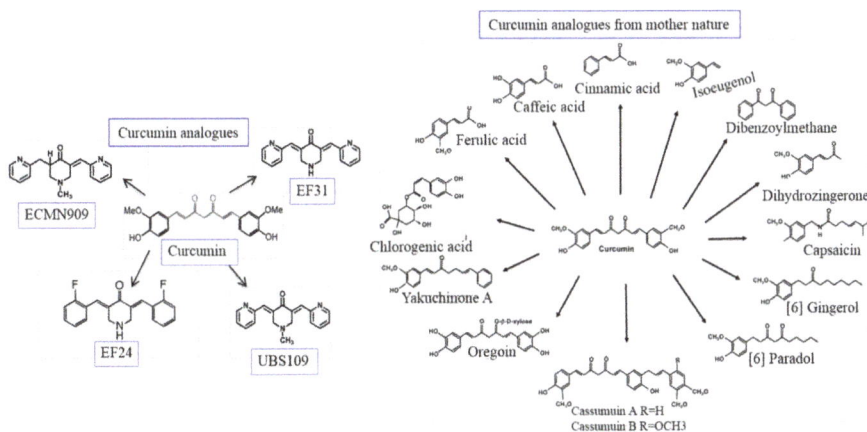

Figure 11. Different Cur analogues and derivatives. By modifying the structure, several analogues and derivative of Cur have been developed by many researchers, which improved its solubility, stability, bio-availability, and biological activities.

These new analogues exhibit no in vivo toxicity and have shown growth suppressive activity that is ~30 times greater than that of natural Cur [110]. Furthermore, synthesized Cur analogues, when complexed with other chemicals, such as sodium dodecyl sulfate and cetyl-trimethyl-ammonium bromide micelles, show anti-oxidative effects against free-radical-induced lipid peroxidation [111], suggesting that synthesized Cur can be used as an antioxidant, as is the case with natural Cur.

6. Rationale for Cur Therapy in Neurodegenerative Diseases

Several experiments have demonstrated that Cur has pleiotropic effects on the nervous system. It is a neuroprotective agent, with potent antioxidant properties, along with the significant anti-inflammatory activity [26,33]. Therefore, its anti-amyloid properties make it a most promising compound for treating different brain diseases caused by amyloid accumulation. In addition, Cur is hydrophobic, as well as lipophilic in nature, and because the brain contains huge amounts of lipids, the absorption, bioavailability, and half-life profiles of Cur are very favorable in the CNS. Several experiments have shown that neuroinflammation, oxidative damage, and deposition of misfolded amyloid proteins synergistically contribute to the pathogenesis of many neurological diseases. Therefore, targeting these processes is a prime strategy for developing therapies for different neurodegenerative diseases. In this context, use of Cur as a treatment for neurodegenerative diseases, has several advantages (Figure 12), including it can: (i) readily cross the blood brain barrier [13,32]; (ii) bind and dis-aggregate amyloid oligomers and fibrils (anti-amyloid) [9,112]; (iii) enhance amyloid clearance similar to vaccine [113]; (iv) reduce chronic inflammation in neurodegenerative diseases; (v) act as a potent antioxidant; (vi) stimulate neurogenesis, as shown in animal models; (vii) chelate metals, including removal of the metals from Aβ; (viii) be taken at relatively high doses (12 g/day) with no negative effects; (ix) be obtained readily and inexpensively; (x) be absorbed into hydrophobic and lipophilic nature, and (xi) produce high fluorescent intensity when it binds to amyloid-plaques, for use in labeling and imaging of amyloid plaques ex vivo and in vivo, or as an imaging probe for non-invasive techniques [13,114] (Figure 12).

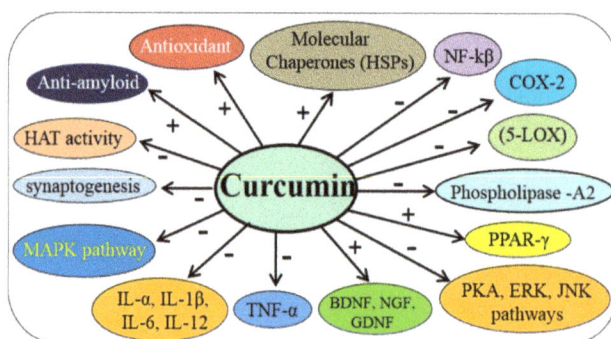

Figure 12. Multiple-reasons for use Cur for treating neurodegenerative diseases. Among these, its anti-amyloid property is particularly attractive as a therapeutic tool.

6.1. Curcumin Therapy in Alzheimer's Disease

Alzheimer's disease is the major age-related neurodegenerative disease, characterized by various neurobehavioral abnormalities, but most prominently by early memory deficits, with a gradual decline of cognitive and intellectual functions that culminate in dementia. It is the leading cause of death in the elderly [115]. The hallmark pathologies of AD are the deposition of Aβ protein as senile plaques in extracellular spaces [116–120] and the phosphorylated tau as neurofibrillary tangles (NFT), intracellularly [121–125]. The accumulation of these abnormal or misfolded proteins are thought to be the principal reasons for the synaptic deficits, neuronal loss, oxidative damage and increase in neuroinflammation in numerous brain regions. Therefore, drugs with pleotropic actions, especially those with anti-amyloid properties and those that reduce oxidative damage, and neuroinflammation should provide the greatest potential for preventing the neuronal loss observed in AD. Unfortunately, numerous drugs or small molecules that have been developed and tested to halt neurodegeneration have not prevented or reduced the symptoms of this disease. Recently, Cur is being considered one of the most potent and promising natural polyphenols for use in AD therapy, due to its pleotropic actions (See Table 6).

Table 6. Pleotropic actions of Cur for AD therapy. Among them, anti-amyloid, anti-oxidant and anti-inflammatory activities are considered the most promising for treating AD [33,126].

Actions	Mechanisms	References
Anti-amyloid properties	Binds with Aβ and prevent its oligomerization and fibril formation	[9,112,127]
Inhibition of Aβ production	Inhibits activities of β-secretase (BACE), inhibiting amyloid precursor protein (APP) processing pathway	[33,128]
Aβ clearance	Stimulates phagocytosis, thus decrease Aβ-plaques	[9,10,51]
Inhibition of NFTs	Binds with NFTs and inhibits tau phosphorylation (pTau)	[129]
Inhibition of other amyloids	Binds with α-synuclein in PD, huntingtin in HD, and prion aggregates in prion diseases	[35,130]
Potent antioxidant	Scavenges ROS/RONS, increase antioxidant levels, decreases lipid peroxidation, chelates toxic metals	[10,51,131]
Anti-inflammatory activities	Downregulates NF-κB, COX-2, 5-LOX, TNFα, IL-1, IL-6.	[10,51]
Regulates activity of molecular chaperones	Restores levels of heat shock proteins (HSP90, 70, 60, 40, HSC70), proteasome system	[132]
Enhance NGF, BDNF, GDNF, neurogenesis and synaptogenesis	Increase expression of BDNF, NGF, GDNF and can promote neurogenesis, and synaptogenesis	[10,133]
Improving cerebral circulation	Inhibits inflammation of brain vasculature leading to improvement of overall blood supply, reduces platelet adhesion in the brain microvascular endothelial cells	[69,134]

Table 7. Curcumin therapy in different animal models and their outcomes [135].

Animal Models	Dose and Duration of Treatment	Disease	Outcomes	Ref.
Sprague-Dawley rat	Diet, 500 and 2000 ppm, 2 months	AD (Aβ ICV infusion)	Decrease spatial memory deficit, oxidative damage, microgliosis	[135]
3XTg-AD mice	Diet, 555 ppm, 2 months	AD (Aβ overexpression)	Decreased Aβ plaque deposition	[12]
APPswe/PS1dE9 mice	Diet, 160 and 5000 ppm, 6 months	AD (Aβ overexpression)	Reduced hippocampal Aβ40/Aβ42 levels	[136]
APPswe/PS1dE9 mice		AD (Aβ overexpression)	Improved spatial memory and decreased Aβ40/Aβ42 levels	[114]
Tg2576 mice	Diet, 500 ppm, 4 m	AD (Aβ overexpression)	Decrease cell death, Aβ-plaques, prevent fibril formation	[9]
PS-1dE9 mice	IV, 7.5 mg/kg/day, 7 days	AD (Aβ overexpression)	Increased restoration of distorted neuritis, plaque disruption	[135]
Kunming mice	PO, 200 mg/kg, 45 days	AD (AlCl3, D-galactose)	Decrease spatial memory deficit	[135]
Sprague-Dawley rat	PO, 50 mg/kg, 4 days	PD (6-OHDA)	Improve TH+ cell numbers	[135]
ICR mice	IP, 50 mg/kg, 3 times	PD (MPTP)	Decreased oxidative damage, increase dopaminergic neurons	[137]
Swiss albino mice	IP, 80 mg/kg, 7 days	PD (MPTP)	Decreased MAO-B	[135,138]
CAG140 mice	Diet, 555 ppm, 2m	HD (knock in)	Decreased huntingtin aggregation, increase rearing, decrease climbing	[40]
5XFAD	IP, 100 mg/kg, 2–5 days	AD (transgenic)	Decreased Aβ plaque, prevent cell death	[108]

6.1.1. Inhibition of Aβ Aggregation

Numerous experiments have demonstrated that Cur can directly bind to the β-pleated sheet structures of Aβ. Interestingly, Cur shows the strongest inhibitory effects on Aβ aggregation among 214 antioxidant compounds tested in vitro [125,135,139,140], indicating it is one of the most potent anti-amyloid compounds investigated so-far. An in vitro study conducted by Ono and colleagues has demonstrated that Cur has a dose-dependent effects on the inhibition of Aβ1−40/1−42 fibrils, with an EC_{50} of 0.09–0.63 µM [9,112]. Several in vitro studies have demonstrated that Cur can attenuate the assembly of both Aβ40 and Aβ42 oligomers and fibril formation [7] (Figure 13).

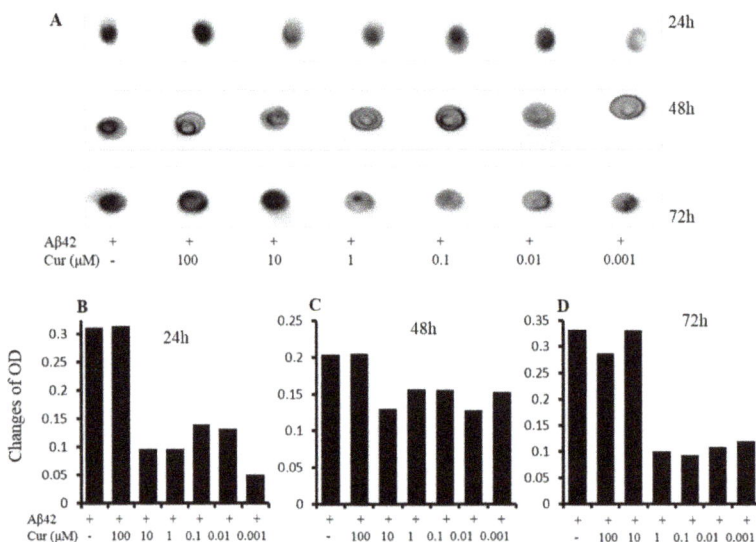

Figure 13. Nanomolar (nM) concentrations of Cur inhibit Aβ42 aggregation in vitro. HFIP-treated Aβ42 was incubated with and without different concentrations of Cur for 24–72 h and a dot blot was performed by probing with Aβ42-fibril specific antibody (OC) and the color was developed with chemiluminescent reagents and the optical density of each dot was measured using Image-J software. Lower concentrations (1–0.001 µM) of Cur inhibited Aβ42 aggregation, whereas higher concentrations had no effect on aggregation.

Following oral or intraperitoneal injections of Cur for 3–7 days in mice, Cur crossed the blood-brain barrier (BBB) and was found in brain tissue, decreasing neuropathology in an animal model of AD (Table 7), as shown by two-photon microscopic imaging [141]. Similarly, significant inhibition of Aβ oligomerization, its plaque formation, and tau phosphorylation, along with behavioral improvements, were observed in a mouse model of AD after oral administration of Cur [9,33,51]. Furthermore, in vivo imaging, using multiphoton microscope, showed a decrease of 30% Aβ plaque size and prevented dystrophic neurites when the animals were injected the Cur via tail vein for one week [141]. In another study, Cur was shown to bind with Aβ-plaques in retina [6,25,139]. In a clinical study, Cur engulfed Aβ effectively and decreased plaque load in AD brain [9]. Though there are no true epidemiological studies that relate Cur intake to the incidence of AD, a trend for reduced incidences of AD is observed among Indian and South Asian countries, in which Cur is consumed everyday as a spice, when compared to the United States and other Western countries in which the intake of Cur is much less [33].

6.1.2. Inhibition of Aβ Production

Aβ is a by-product of a transmembrane protein, called amyloid precursor protein (APP). The production of Aβ is catalyzed by the two successive enzymes, first by β-secretase (BACE), followed by γ-secretase, which contains presenilin-1 (PS-1). It is speculated that during disease progression, induction of inflammatory signals aggravate the expression of Aβ production by increasing the activity of BACE [142], whereas Cur inhibits the activity of BACE, thus reducing the levels of Aβ [9,33]. In addition, Cur is a potent inhibitor of the APP metabolic pathway, thus lowering

Aβ levels [7,143]. Furthermore, it can regulate Aβ production by inhibiting GSK-3β-mediated PS-1 activation [144] (Figure 14).

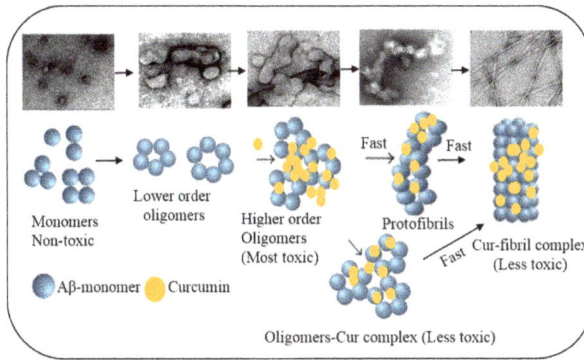

Figure 14. Schematic diagram showing the formation of different Aβ-species during its aggregation process and the inhibitory role of Cur in its assembly process. Cur has been shown to bind with Aβ and attenuate the oligomer formation or slowdown the process. Additionally, it can fasten the transformation of more toxic oligomers into less fibril forms.

6.1.3. Aβ Clearance

The levels of Aβ in the brains of AD patients depend on a balance between production, clearance, and influx of Aβ. When clearance pathways are impaired the levels of Aβ are increased. However, there are several ways in which Aβ is disposed from the cell, including receptor-mediated Aβ transport across the BBB and enzyme-mediated Aβ degradation, as well as the involvement of immune system [145]. Cur can act in a manner that is similar to an amyloid vaccine [33] and can bind with Aβ to enable its removal from the brain by promoting receptor-mediated Aβ efflux [9]. In contrast, Cur could decrease Aβ load by suppressing the Aβ influx across the BBB and by upregulating the enzyme-mediated degradation of Aβ. Furthermore, Cur can stimulate phagocytosis and increase the association of phagocytic cells around Aβ-plaques as observed in a rat AD model [146] and the Tg2576 mouse model of AD, as well as with plaques in post-mortem human brain sections exposed to primary rodent microglia [147] (Figure 15).

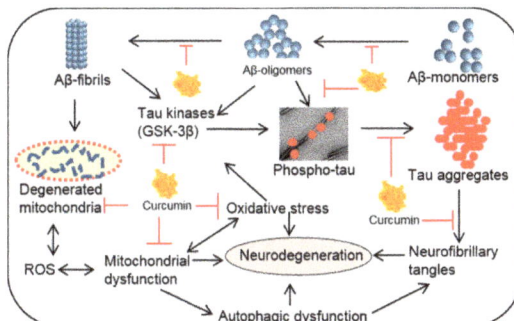

Figure 15. Schematic diagram showing the role of Cur in Aβ clearance. Cur stimulates phagocytosis of Aβ by activating microglia and enzyme-mediated degradation of Aβ. Cur also stimulates B-lymphocytes to activate Aβ-specific antibody, which neutralizes Aβ. In addition, it also inhibits Aβ-influx from the blood stream to the brain and increase Aβ-efflux from brain to the general circulation. "↔"bidirectional; "→" increased; and "⊤" inhibition/decrease.

6.1.4. Inhibition of Tau Phosphorylation

The second most common pathology observed in AD is the tau tangle, which is basically the deposition of phosphorylated tau (pTau) as paired helical filaments (PHF). Tau is a microtubule stabilizing protein, which is abundant in neurons of whole CNS. Hyperphosphorylation of tau causes cytoarchitectural changes, which create oxidative stress, mitochondrial dysfunction, and neurodegeneration [148]. Tau phosphorylation and deposition of NFTs are regulated by several tau kinases, with glycogen synthase kinase-3β (GSK-3β) and mitogen-activated protein kinase (MAPK) being the most common among them [149]. The common tau-kinases, which can phosphorylate the tau protein are cyclin-dependent kinase 5 (Cdk5)/p25, extracellular signal-regulated kinase 2 (ERK2), S6 kinase (S6K), microtubule affinity-regulating kinase (MARK), SAD kinase (SADK), protein kinase A (PKA), calcium/calmodulin-dependent protein kinase II (CaMKII), or Src family kinases, such as Fyn and c-Abl.

Therefore, inhibition of tau kinases could be a viable strategy to prevent NFT-induced neurodegeneration. Cur has been shown to bind to NFTs in human AD brain and mouse models of AD [129] (Figure 16). An in vitro experiment showed that Cur inhibits pTau aggregation by reducing oxidative stress [150]. We have shown that Cur inhibits GSK-3β activity and reduces tau dimer and pTau oligomerization in a human tau transgenic mouse model [12]. In addition, oral administration of Cur (555 ppm) together with DHA, reduced pTau by inhibiting IRS-1 and JNK activities in vivo [12] (Figure 16).

6.1.5. Inhibition of Oxidation and Inflammation

Whether or not Aβ can induce oxidative stress and neuroinflammation is not yet clear, but it is considered one of the primary events involved in neuronal death in AD [151]. However, as a strong antioxidant, Cur can limit the pro-oxidant, pro-inflammatory, and other toxic effects in AD brains [7,33]. Cur can inhibit the inflammatory cytokines, including IL1, IL6, TNF-α, IFN-γ, and COX-2 activity [152]. Several studies have demonstrated that Cur can inhibit activated astrocytes and microglia, as shown by reducing GFAP and Iba-1 levels [108,153] (Figure 17). Therefore, as an anti-inflammatory natural polyphenol, Cur is a promising compound for tackling oxidative stress and inflammation in AD.

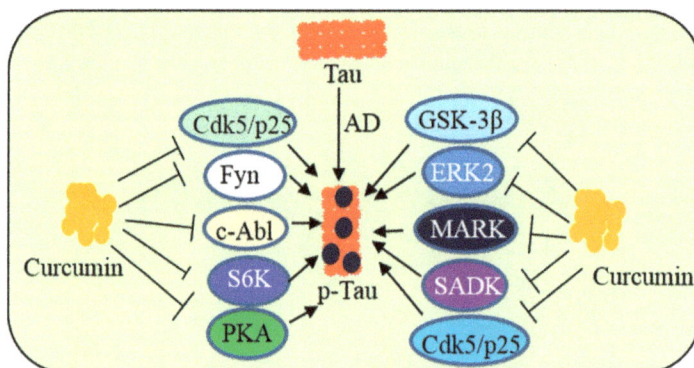

Figure 16. Schematic diagram showing the role of Cur in inhibition of tau phosphorylation in AD. Cur has been shown to inhibit tau-kinases, thus inhibiting phospho-tau formation. It also binds with tau directly to inhibit their aggregation.

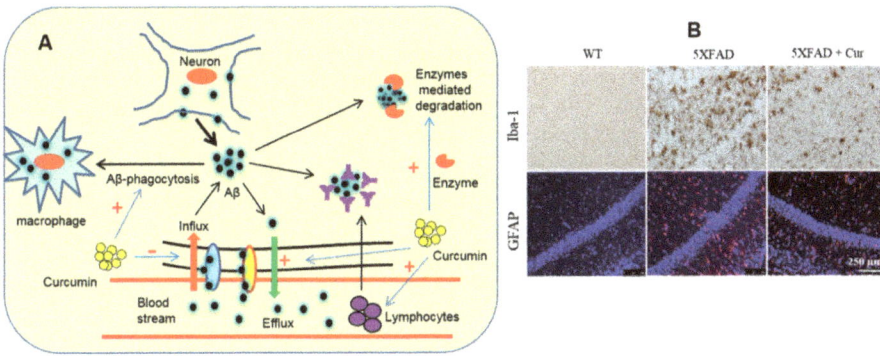

Figure 17. Schematic diagram showing the anti-inflammatory properties of Cur in Alzheimer's disease. (**A**) Cur stimulates B-lymphocytes to produce antibodies, increase anti-inflammatory cytokines and decrease proinflammatory chemokines, increase phagocytosis of Aβ, and increase proteolytic enzymes to degrade Aβ. (**B**) Cur inhibits microglia (Iba-1; **upper**) and astrocyte (GFAP; **lower**) activation in 5xFAD mice brain tissue.

6.1.6. As an Imaging Probe for Aβ-Plaque Detection Ex Vivo and In Vivo

Cur is an ideal fluorophore for Aβ plaque imaging and detection because it is a natural fluorescent molecule and preferentially binds to Aβ plaques [6,9,13,25,141]. Therefore, researchers use Cur for labeling and imaging Aβ-plaques ex vivo and in vivo [13]. Interestingly, it has structural similarities with classical amyloid binding dyes, (such as Thioflavin-S, Congo red, and crysamine-G), which makes it a promising candidate for labeling and imaging of Aβ plaques ex vitro and in vivo [13] (Figure 18). For example, Garcia-Alloza and colleagues have demonstrated that Cur can be used to visualize Aβ-plaques in vivo, as shown in APP-tau transgenic mouse model [141]. Similarly, a strong fluorescent signal was observed when the brain sections both from animal models of AD and from AD patients were incubated with Cur [6,154]. To confirm whether Cur binds to Aβ-plaques, we performed immunohistochemistry in the 5XFAD brain tissue sections with Aβ-specific antibodies (6E10) and then the same sections were stained with Cur, and we observed that Cur was completely co-localized with Aβ-specific antibody (Figure 18), which indicates that Cur has specificity to Aβ similar to Aβ-specific antibody [13].

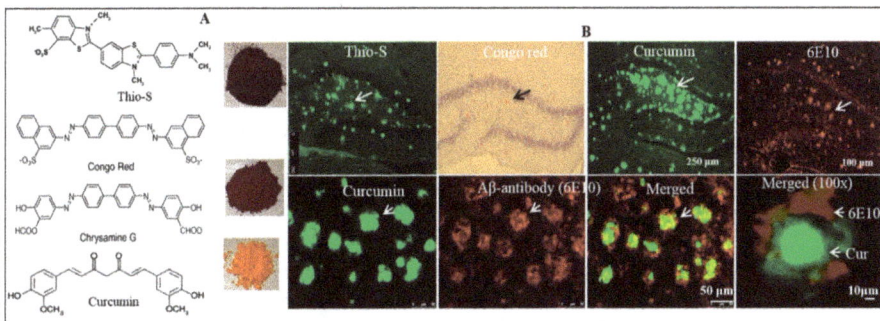

Figure 18. Cur binds to Aβ-plaques greater than classical amyloid binding dyes. (**A**) Curcumin has structural similarities with classical amyloid binding dyes. (**B**) **Upper panel**: Cryostat sections from 5XFAD mouse hippocampus were stained with Thio-S, Congo-red, Cur and Aβ-specific antibody (6E10). Please note that Cur stained Aβ plaques are more visible than other dyes. **Lower panel**: Cur is co-localized with Aβ-specific antibody (6E10) in Aβ-plaques in mouse cortical tissue from 5xFAD.

More recently, researchers have tried to use the fluorescent properties of Cur derivatives for in vivo imaging, such as positron emission tomographic (PET) probes for amyloid imaging or retinal scan for detection of AD in experimental animals and humans [127]. However, it is not a practical probe for in vivo near infrared (NIR) imaging, due to its short emission wavelength (~550 nm), limited bioavailability, and rapid degradation. To overcome these issues, scientists modified the Cur structure to form boro-fluoro-Cur derivatives, which shift the emission wavelength to the NIR range (Figure 19). These derivatives are called CRANAD derivative (e.g., CRANAD-2, CRANAD-44 and CRANAD-28) [155] (Figure 19). These derivatives of the Cur probe significantly increase fluorescence properties upon binding to Aβ-plaques [155,156]. Surprisingly, the binding affinity of Cur for Aβ aggregates is higher (with a Ki of 0.07 nM) for F18-labeled Cur binding of fibrillar Aβ than for other the molecular imaging probes, such as PIB in FDG-PET [127]. Beyond labeling Aβ-plaques, Cur can also help to visualize the distinct morphology of different Aβ-plaques, such as core, neuritic, diffuse, and burned out plaques [13], indicating that it can be used to investigate the overall amyloid plaque loads, as well as an aid to characterize the morphology of Aβ-plaques after anti-amyloid therapy. Therefore, as a potent anti-amyloid polyphenol, Cur has a complete requisite profile for labeling and imaging the amyloid plaques.

6.2. Curcumin Therapy in Parkinson's Disease

Parkinson's disease (PD) is the second most common age-related neurodegenerative disease, and is characterized by bradykinesia, tremor, rigidity, and abnormalities in gait and posture. Gradual and selective degeneration of dopaminergic neurons in the substantia nigra pars compacta (SNpc), with a subsequent decline in dopamine (DA) levels in the nigro-striatal pathway are associated with PD [157,158]. Although most PD cases are sporadic, about 5% of the cases can be inherited. The major pathological hallmarks of PD are the presence of insoluble, fibrous aggregates, composed of α-syn in intraneuronal inclusions of Lewy bodies (LBs). In humans, PD is associated with the α-syn aggregation.

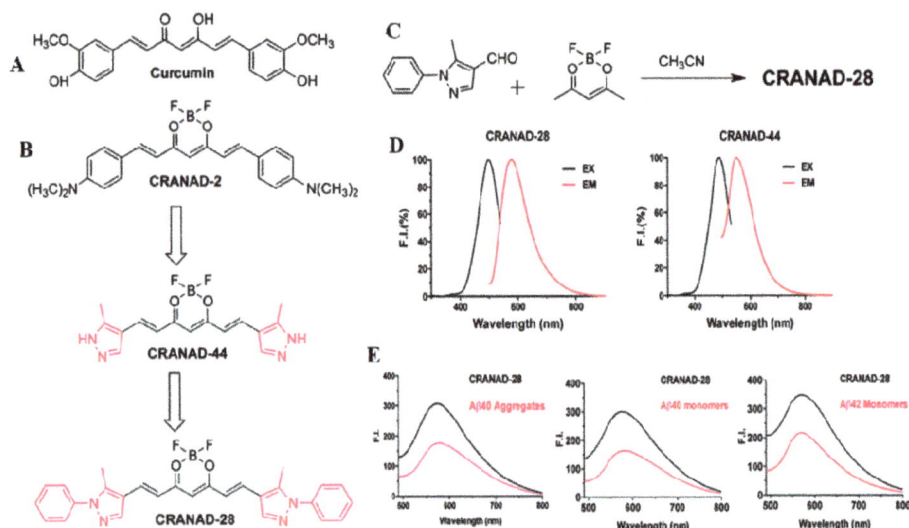

Figure 19. Structural modifications of Cur. (**A**) Structure of natural Cur; (**B**) Design of CRANAD-28 through pyrazole replacement; (**C**) The synthetic route for CRANAD-28 synthesis; (**D**) The excitation/emission spectra of CRANAD-28 and -44; (**E**) fluorescence responses of CRANAD-28 with Aβ40 aggregates, Aβ40 monomers and Aβ42 monomers.

Effects of Cur on PD

Cur has several beneficial effects on PD (Figure 20). First, Cur can inhibit α-syn aggregation and, prevent LB accumulation in vitro, attenuating α-syn oligomer toxicity in cells [35]. First, Cur reduces toxicity by binding to α-syn oligomers and fibrils (but not monomers), modulating the morphology, and reducing their aggregation, as shown by fluorescence and two-dimensional nuclear magnetic resonance (2D-NMR) studies [35]. Secondly, Cur attenuates reduction of DA levels, and degeneration of DA neurons [39]. Thirdly, Cur can reduce oxidative stress, memory deficits, and motor impairments [33]. Fourthly, Cur chelates iron, copper, and other metals, thus preventing α-syn or LB aggregation [159]. Fifthly, Cur promotes the recovery of macroautophagy by activating transcription factor EB, thus reducing cell death and neurotoxicity [160]. Sixthly, Cur can inhibit activity of monoamine oxidase (similar to MAO-inhibitor), thus restoring DA levels [161] and reducing depression [162]. Seventhly, Cur can protect DA neurons in brain by reducing ROS levels, maintaining mitochondrial functions, and attenuating neuroinflammation via CNB001, a Cur-derived compound [163] Finally, Cur can inhibit the JNK pathway and prevent dopaminergic neuronal loss via apoptosis [163] (Figure 20).

- Inhibits α-syn and Lewy Bodies aggregation
- Acts as monoamine oxidase inhibitor
- Prevents reduction in dopamine levels.
- Anti-inflammatory, anti-apoptotic, antioxidant and chelating properties prevent destruction of SNpc DA neurons
- Prevented reduction of antioxidant enzymes
- Prevent mitochondrial dysfunction
- Recovering the macroautophagy process.
- Improve cognition, reduce memory deficits and motor impairment and depression

Figure 20. Pleotropic actions of Cur on Parkinson's disease.

6.3. Curcumin in Huntington's Disease

Huntington's disease (HD), is a poly-glutamine (PolyQ) autosomal dominant genetic disorder characterize by progressive neurodegeneration and impairments of motor, psychiatric and cognitive functions [164]. Degeneration of medium spiny neurons of striatum, and cells in layers V and VI of cortex, SNpc, hippocampus, cerebellum, hypothalamus, and parts of the thalamus are the most commonly observed in HD [165,166]. The hallmark pathology in HD is the abnormal accumulation of misfolded mutated huntingtin protein (mHTT) as intracellular aggregates, which causes selective neuronal loss, primarily in the cortex and medium spiny neurons of striatum [165]. The Huntingtin gene (*HTT*) is present on the short arm of chromosome 4p16.3. HTT is ubiquitously expressed in neurons and is found in many subcellular compartments. Although its exact function is not fully understood, experimental animal studies have revealed that HTT is essential for fetal development and its absence is lethal [167]. HTT is involved in cytoskeleton anchoring, transport, cell signaling, and vesicle trafficking [168]. In addition, HTT upregulates the expression of brain derived neurotrophic factor (BDNF), inhibits caspase-3 and caspase-9, and protects against apoptosis [169]. A mutation in the HTT gene leads to expansion of the CAG (Cystocine-Adenine-Guanine) repeats which leads to elongation of polyQ in HTT protein and results in the accumulation of mHTT [170]. There is a correlation between the number of CAG repeats and the age of onset of the disease [165]. In general, more than 36 CAG repeats could trigger disease symptoms, and it is critical for its toxicity. Increase in the number of CAG repeats will cause greater HTT deposition, which increases neurotoxicity [170]. For example, animal models of HD, such as R6/1 and R6/2 mice or YAC128 mice, which express a larger CAG expansion, progressively develop cognitive, psychiatric and motor symptoms, which are analogous to those observed in HD patients [171–174]. Therefore, both loss of function in the wild type HTT and gain of function in the mutated form of HTT have been proposed to play a role in the development of HD pathology [169].

To date, the exact pathogenesis for the neuronal death in HD is not fully understood. However, glutamate excitotoxicity [175], mitochondrial dysfunction [176], impaired protein degradation, protein misfolding, caspase activation, transcriptional pathways dysregulation, decrease of proteasome function and abnormality in proteolysis are among the major causes for this disease [177].

Beneficial effects of Cur in Huntington Disease

Cur has been shown to neuroprotective effects in animal models of HD. For example, when we fed chow containing 555 ppm Cur to CAG 140 knock-in HD mice, a significant decrease in mHTT aggregates and increased striatal DARPP-32 and D1 receptor mRNAs, as well as an amelioration of rearing deficits was observed [40]. Solid lipid nanoparticles (C-SLNs) given to treated 3-nitropropionic-acid (3-NP) treated rats (a toxin which causes HD-like neuropathology in rodents) resulted in a decrease of HD-like neurodegeneration, as well as significant increase in the activity of mitochondrial complexes and cytochrome levels [178]. Furthermore, C-SLNs also restored glutathione levels and superoxide dismutase (SOD) activity, decreasing lipid peroxidation, protein carbonyl formation, ROS levels, and mitochondrial swelling. In addition, C-SLN-treated rats showed significant improvements in neuromotor coordination, when compared with 3-NP-treated rats [178]. Furthermore, chronic treatment with Cur (10-, 20- and 50-mg/kg, once daily for 8 days, peritoneally) improved the motor and cognitive performances, reduced oxidative stress, and restored succinate dehydrogenase activity, which is inhibited in the in 3-nitropropionic acid (3-NP) HD rat model. An increase in Nrf2 (a key regulator for antioxidants enzyme expression) expression in C-SLN-treated mice, compared to controls, has also been observed [179]. Cur treatment also restored the down-regulated molecular chaperones (e.g. HSP40, HSP70) in HD. In addition, downregulated brain-derived neurotrophic factor (BDNF) in HD was also restored by Cur-treatments [180] (Figure 21).

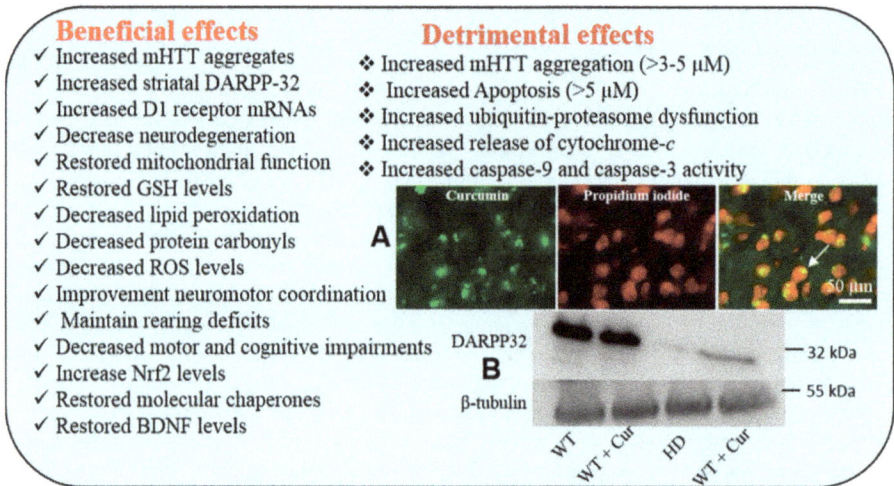

Beneficial effects
✓ Increased mHTT aggregates
✓ Increased striatal DARPP-32
✓ Increased D1 receptor mRNAs
✓ Decrease neurodegeneration
✓ Restored mitochondrial function
✓ Restored GSH levels
✓ Decreased lipid peroxidation
✓ Decreased protein carbonyls
✓ Decreased ROS levels
✓ Improvement neuromotor coordination
✓ Maintain rearing deficits
✓ Decreased motor and cognitive impairments
✓ Increase Nrf2 levels
✓ Restored molecular chaperones
✓ Restored BDNF levels

Detrimental effects
❖ Increased mHTT aggregation (>3-5 µM)
❖ Increased Apoptosis (>5 µM)
❖ Increased ubiquitin-proteasome dysfunction
❖ Increased release of cytochrome-*c*
❖ Increased caspase-9 and caspase-3 activity

Figure 21. Effects of Cur on in vitro and in vivo models of HD. Several experiments have shown beneficial effects of Cur, but higher concentrations (>3–5 µM) of Cur may exacerbate HD symptoms. (**A**) The cortex of YAC128 mice was stained with Cur and images showed that Cur binds with HTT. (**B**) Western blot of DARP32, the marker for medium spiny neuron were partially restored by Cur treatment. Scale bar indicates 50 µm and is applicable to other images.

In contrast, 3–5 μM of Cur can increase mHTT aggregation and mHTT-dependent cell death and promote proteasomal dysfunction in the mHTT expressing cells, in comparison to control cells [40,181,182]. Similarly, Jana and colleagues reported that Cur induced apoptosis through the impairment of the ubiquitin-proteasome system [183]. They have demonstrated that the exposure of Cur to the N2a cells causes a dose-dependent decrease in proteasome activity and an increase in ubiquitinated proteins. They concluded that the Cur can induce more apoptosis in proliferative cells than in differentiated cells, by decreasing the mitochondrial membrane potential, releasing cytochrome-*c* into cytosol, and activating caspase-9 and caspase-3 [183]. Although systemic availability of Cur is very low (~25 nM) after oral administration, it is assumed that low doses of Cur may not increase mHTT aggregation, but rather provides neuroprotective effects, but further study is needed to confirm this hypothesis.

6.4. Curcumin Therapy in Prion Diseases

Transmissible spongiform encephalopathies (TSE) or "prion" diseases are a group of rare, fatal, progressive neurodegenerative disorder which affect mammals [184]. It is a spectrum of diseases, which includes Creutzfeldt-Jakob disease, bovine spongiform encephalopathy, Scrapie, Kuru, Gersmann-Sträussler-Scheinker syndromes and fatal familial insomnia. In all these diseases the "prion" proteins are transmitted from cell to cell and induce accumulation of misfolded "prion" proteins (PrP) [184]. These accumulate as prion plaques, which can be observed throughout the CNS and cause neurodegeneration. Basically, prion aggregates generate several "holes" or "vacuoles" within the CNS, which make for a spongy neuronal architecture [185]. Prion disease can be genetic, sporadic, or infectious [186]. The PrP can take on an α-helix-rich, pathogenic, and cellular forms (PrP^C), which can convert to a β-structure-rich, stable, insoluble, infectious, proteinase K-resistant fibril conformation of prion protein (PrP^{Sc}) [184,185]. Although the mechanism of neuronal cell death in prion disease is unclear, the accumulation of this insoluble PrP^{Sc} is thought to be the principal reasons for dendritic or synaptic loss, as well as neurodegeneration, along with neuroinflammation in the CNS [184].

6.5. Effects of Cur on Prion Disease

Several drugs, small molecules or compounds have been tested for inhibiting PrP^{Sc} aggregation, but none of them have proven to be satisfactory for clinical use. Some recent research suggests that Cur can inhibit PrP^{Sc} fibril formation in scrapie-infected neuroblastoma (scNB) cells [38,41]. It can bind to non-native forms of PrP, thereby inhibiting prion fibril formation, without affecting native PrP [130]. Low doses of Cur (10 nM) can decrease ROS levels and effectively prevent PrP^{Sc}-induced apoptosis [147].

7. Biphasic or Dose-Dependent Effects of Curcumin

Cur shows its biphasic or dose-dependent effects in our body, which means lower doses, show neuroprotection, whereas higher doses may be toxic to cells [147]. Several in vitro and in vivo studies suggest that Cur has beneficial effects, including potent antioxidant, anti-inflammatory, anti-amyloid properties at relatively lower doses (0.1–1 μM). Perhaps due to its limited bioavailability, Cur does not reach high concentrations in all tissues and organs. Therefore, it is assumed that the beneficial effects are due to low levels [147]. In contrast, higher doses (>5 μM) may induce colon or lung cancer or other side effects, such as inhibition of proteasomal function, which can occur at levels above a 3 μM concentration [183]. For example, ≥3 μM of Cur can increase mHTT-induced neurotoxicity in an in vitro model of HD [183]. It has been demonstrated that about 10 μg/mL of turmeric extract caused a dose- and time-dependent induction of chromosome aberrations, as well as DNA damage in several mammalian cell lines [187]. Even doses as small as 2.5–5 μg/mL of Cur have been shown to induce both mitochondrial- and the nuclear-DNA damage [188]. These reports raise a major concern about the safety of Cur therapy in different diseases. A recent report has also shown that Cur can promote lung cancer in mice [189].

Interestingly, several studies also demonstrated that the higher concentration of Cur (5–50 μM) can be used to kill cancer cells, although it requires the cells to be treated with Cur for several hours or days [190]. For example, in one of our recent in vitro experiments, we treated glioblastoma cells (U-87 Mg, derived from human cells) with 25 μM of Cur for 24–72 h, and found a significant increase in DNA fragmentation and apoptotic death [14] (Figure 22). These data support many previous observations and suggest that higher concentrations of Cur can be used to kill cancer cells. Conversely, most therapeutic applications for Cur require low doses. However, when administered orally, achieving therapeutically significant concentrations of Cur outside the GI-tract is practically impossible because of its poor bioavailability. However, the Cur bioavailability in animal models of AD [12,13,108] and HD [40] following i.p. or oral administration was about 250–350 nM in the brain tissue. Also, when we imaged Cur-treated AD mouse brain tissue we observed that Cur was colocalized with Aβ profusely [13] and decreased Aβ plaques. This indicates that at least a few hundred-nM concentrations of unmodified Cur were available following either oral or i.p. administration of lipidated formula of Cur interact with Aβ, p-tau and prevent AD pathologies, such as neuroinflammation, thus reducing cognitive deficits [10–13]. Therefore, achieving optimal concentrations of Cur to the brain by oral administration while reducing the possibility of toxic effects to the periphery, needs to be carefully studied before widespread use of orally administered Cur can be advocated.

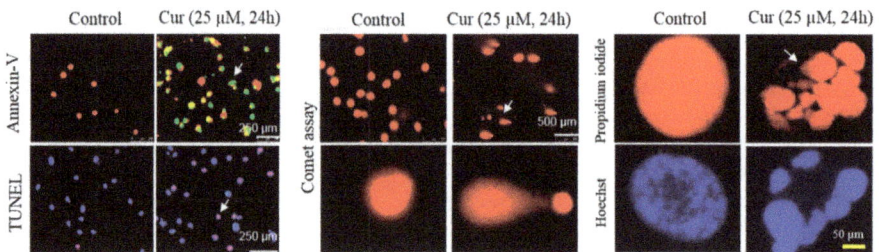

Figure 22. High concentrations of Cur can be used to kill glioblastoma cells. Cur treatment (25 μM) induces apoptosis and DNA fragmentation in human derived glioblastoma cell line (U-87Mg).

8. Recommended Doses and Limitations of Cur Therapy

To understand the beneficial roles of Cur, it is essential to evaluate its pharmacokinetics or pharmacodynamics after its administration in humans. Based on previous clinical studies, it has been demonstrated that 8 g per day of short-term Cur therapy has no significant detrimental effects [70]. Toxicological evaluations revealed that Cur is found to be pharmacologically safe, even up to 12 g per day, as reported by several animal studies and in phase-I clinical trials [32,70]. Similarly, another phase-1 human trial, with 8 g of Cur per day for three months, revealed no toxic effects [7]. However, a few studies have indicated that high doses of Cur can cause highly variable adverse side effects, including gastrointestinal discomfort, chest tightness, skin rashes, and swollen skin, as well as some allergic reactions, such as dermatitis [191] (Table 8).

Table 8. Side effects of Cur therapy when intake orally.

Parameter	Side Effects	Ref.
General effects	Gastrointestinal discomfort, chest tightness, skin rashes, and swollen skin, allergic reactions or dermatitis, nausea, and diarrhea	[192]
Blood clotting	Slow down blood clotting process	[193]
Gall bladder	Increase gallstones contraction and increase bile duct obstruction	[192]
Pregnancy and postnatal complications	Stimulate the uterus or promote a menstrual period. Breast feeding women not recommended	[8]
Stomach problems	Increased stomach acid secretion if taken with antacid drugs	[8]

In human studies with Cur, doses of 0.9–3.6 g per day for 1–4 months caused some adverse effects, including nausea and diarrhea, with an increase in serum alkaline phosphatase and lactate dehydrogenase [194]. Despite reports of possible adverse effects, most studies have revealed significant beneficial effects with Cur [12,30,39,45,46,57,58,69,135,144,150,160,182] (Table 9). For example, one of our previous studies with the 3XTg mouse model of AD showed that 600 nM is sufficient to reduce AD-like pathological symptoms [10,12]. Extrapolation of animal studies to clinical trials also revealed that an oral supplementation of Cur in the range of 80–500 mg per day was recommended to obtain these beneficial effects in humans, which means a daily intake of raw turmeric would be 2–4 g [10,126]. Furthermore, chronic intake of Cur sometimes may be hepatotoxic. Therefore, a person with liver diseases, such as cirrhosis, biliary tract obstruction, gallstones, obstructive jaundice and acute biliary colic, or those are under prescribed medication for hepatic problems are counter-indicated for Cur therapy, because Cur can stimulate bile secretion [74]. In fact, supplementation of even 20–40 mg of Cur per day can increase the gallbladder contractions in healthy people [195,196]. Similarly, alcoholics or heavy drinkers may not receive the benefits of this therapy. Furthermore, the individual taking any blood thinning agents, non-steroidal anti-inflammatory (NSAIDs) drugs, or reserpine are recommended not to take Cur, because, it can interact with these drugs [10,33,126]. However, for therapeutic purposes, dietary Cur is very unstable in most of the body fluids, and because of its poor water solubility and limited tissue bioavailability, it is highly recommended to mix Cur with oil or milk to enhance its absorption and metabolism [8]. Although the vast majority of animal studies and clinical trials using Cur has resulted in more beneficial than adverse effects [12,30,39,45,46,57,58,69,135,144,150,160,182], its use must be tempered by possible toxic effects at high doses.

Table 9. List of a few clinical trials with Cur in Alzheimer's disease.

Study ID	Curcumin Molecule	Cohort	Dose	Duration	Outcomes	Ref.
Baum et al. NCT00164749	Cur + gingko	AD:50+ year, $n= 30$	1, 4 g/day	6 months	No differences in Aβ levels between treatments or MMSE scores	[71]
Ringman et al. (ACT00099710)	Cur C3 complex	Mid/Moderate AD, 49 y+, $n = 30$	2, 4 g/day	24 weeks	No differences detected between treatment groups in biomarkers measured, low bioavailability	[197]
Hishikawa et al.	Tumeric capsule	Severe AD, $n = 3$	100 mg/day	12 months tested after 12 weeks	MMSE and NPIQ: score on NPIQ decreased significantly, MMSE increased in 1/3	[198]
Poncha (NCT01001637)	Longvida	Moderate-severe AD, 50-80 y, $n = 160$	2, 3 g/twice daily	2 months	Efficacy and safety: blood and cognition	[199]
Martin and Goozee (ACTRN12613000681752)	Biocurucmax (BCM-95)	Retirement, healthy, 65-95, $n = 100$	500 mg/thrice/day	12 months	Cognition, blood biomarkers, brain imaging, retinal imaging	[199]
Martin (ACTRN12611000437965)	Biocurucmax (BCM-95)	Community living, healthy, 55-75, $n = 100$	500 mg/thrice/day	12 months	Cognition, blood biomarkers, life style, brain imaging	[199]
Small et al. (NCT01383161)	Tetracurcumin CR-031P™	MCI, normal aging, $n = 132$	90 mg/twice/day	18 months	Cognition, blood genetic profile	[200]
Frautschy (NCT018811381)	Longvida and Yoga	Subjective cognitive complainers, 55-90, $n = 80$	400 mg/twice/daily	6 months	Biochemistry, cognition, brain imaging	[199]
Cox et al. (ACTRN12612001027808)	Longvida™	Healthy and cognitive decline, 65-80, $n = 60$	400, 800 mg/daily	4weeks 8 weeks	Cognition, mood and anxiety, blood biomarkers, MRI	[201]
NCT00595582	Curcumin bioperine	MCI, 55-85, $n = 10$	900 mg/twice/daily	24 months	Cognition and size of metabolic lesion by PET	[199]
ACTRN12614001024639	BCM-95	Healthy and MCI, 65-90 years, $n = 48$	500 mg/twice/daily	3 months	Gene regulation and expression, and cognition	[199]
ACTRN12613000367741	Longvida™	Healthy, MCI, mild/moderate AD, 50 years, $n = 200$	20 g/daily	7 days	Diagnostics, Curcumin fluorescent retinal imaging of Aβ plaques	[199]

9. Future Perspective of Curcumin Research

Numerous studies have been conducted to test the potential of Cur to prevent or treat different neurological diseases. However, several reports have raised questions about its safety and efficacy, especially at high doses, which may be harmful. Some researchers recommend limiting the daily intake of Cur to 1 g per day, whereas other studies have shown no side effects, even up to 12 g /day. However, our review of the studies using Cur for treating neurological disorders underscore a few critical observations. First, the formula of Cur can make a difference. Many researchers used whole turmeric extract, while others used different lipidated formulae of Cur, along with nano-Cur, and their data needs to be directly compared with the effects of natural Cur. In addition, the purity of Cur, along with a detailed description of the extraction method and sources are critical variables that need to be considered when making determination about the efficacy of Cur therapies. Second, the difference between Cur and curcuminoids needs to be considered. Many researchers use the term "curcuminoid" with Cur, but curcuminoid has two additional compounds (bis-methoxy and de-bismethoxy Cur). Therefore, this clarification needs to be considered when interpreting the result of such studies. Third, the mode of administration is important: It is critical to know how Cur is administered, because the PK and PD of Cur can vary with the route of administration, which can significantly affect the efficacy of the treatment. Fourth, the duration of Cur therapy can affect its efficacy: Both short- and long-term Cur therapy have shown mixed effects. Many scientists have reported short-term beneficial effects, whereas the potential of long-term studies to reveal toxic effects may be underreported, which underscores the need for more work using chronic administration of Cur. Overall, most scientists agree that Cur has enormous potential as an effective nutraceutical with advantages of having relatively low toxicity, being quite inexpensive, and easily obtained [12,30,39,45,46,57,58,69,135,144,150,160,182]. However, due to its poor bioavailability, different lipidated forms continue to be developed, which is providing increasingly greater bioavailability and efficacy.

10. Conclusions

Neurodegenerative diseases are age-related complicated disorders with complex neuropathological characteristics. They develop progressively, and oftentimes significant neuropathology precedes any overt clinical symptoms. Neuronal damage and cognitive deficits or impairment of motor coordination are the major problems in these diseases. Because of its pleotropic actions on the nervous system, including anti-amyloid, anti-inflammatory, and anti-oxidant properties, Cur is a promising candidate for targeting protein misfolding neurological diseases. Furthermore, it is safe and inexpensive, readily available and can effectively penetrate the blood-brain barrier and neuronal membranes. We have provided detailed information on the anti-amyloid properties of Cur in major neurodegenerative disorders, such as AD, PD, HD, and prion diseases. Collectively, the information available from reviewing the literature on the therapeutic potential of Cur can provide helpful insights into the potential clinical utility of Cur for treating neurological diseases.

Author Contributions: P.M. designed the study, collected, analyzed, and interpreted data, and wrote the manuscript. G.L.D. directed the study, contributed to the manuscript writing, edited, and approved the manuscript. Both authors read and approved the final manuscript.

Acknowledgments: This work was supported by the Field Neurosciences Institute, St. Mary's of Michigan, and the John G. Kulhavi Professorship in Neurosciences and the Neuroscience Program at Central Michigan University. We thank Verdure Sciences (Noblesville, IN) for donating the solid-lipid curcumin particles to generate some of the data. We also thank Abeer Gharaibeh and Zachary Bowers for their critical comments on portions of this manuscript

Conflicts of Interest: The authors declare no conflict of interest.

Int. J. Mol. Sci. **2018**, *19*, 1637

Abbreviations

Cur	Curcumin
CNS	Central nervous system
AD	Alzheimer disease
SLN	Solid lipid nanoparticles
DMC	Demethoxycurucmin
BDMC	Bisdemethoxycurucmin
IUPAC	International Union of Pure and Applied Chemistry
DMSO	Dimethyl-sulfoxide
DMF	Dimethyl formamide
PK	Pharmacokinetics
PD	Pharmacodynamics
THC	Tetrahydrocurcumin
HHC	Hexahydrocurcumin
i.v.	Intravenously
i.p.	Intraperitonally
OHC	Octahydrocurcumin
Aβ	Amyloid β-protein
α-syn	Alfa-synuclein
HTT	Huntingtin
PD	Parkinson's disease
HD	Huntington's disease
p-tau	Phosphorylated tau
BBB	Blood brain barrier
PUFA	Polyunsaturated fatty acids
ROS	Reactive oxygen species
GSH	Glutathione (reduced)
SOD	Superoxide dismutase
GPx	Glutathione peroxidase
GST	Glutathione S-transferase
NF-κB	Nuclear factor κ-light-chain-enhancer of activated B cells transcription factor
COX	Cyclooxygenase
LOX	Lipoxygenase
TNF	Tumor necrosis factor
IL	Interleukin
PPARγ	Peroxisome proliferator-activated receptor gamma
iNOS	Induced nitric oxide synthase
HSP	Heat shock protein
NGF	Nerve growth factor
BDNF	Brain derived neurotropic factor
GDNF	Glial derived neurotropic factor
PDGF	Platelet derived neurotropic factor
PSD	Post-synaptic density protein
HAT	Histone acetyltransferase
BMECs	Brain microvascular endothelial cells
GI	Gastrointestinal
EGCG	Epigallocatechin gallate
GMO	Genetically modified organism
PLGA	Poly lactic-*co*-glycolic acid
BCM-95	Biocurcumax-95
IC	Inhibitory concentration
DLS	Dynamic light scattering

SEM	Scanning electron microscope
FTIR	Fourier transforms infrared spectroscopy
ER	Endoplasmic reticulum
AuNPs	Gold nanoparticles
AgNPs	Silver nanocomposite
Cds	Cyclodextrins
SLNP	Solid lipid nanoparticles
nm	Nanometer
nM	Nanomolar
ppm	Parts-per-million
SLCP	Solid lipid Cur particle
NFT	Neurofibrillary tangle
BACE	β-secretase
APP	Amyloid precursor protein
PHF	Pair helical filaments
GSK-3β	Glycogen synthase kinase-3β
MAPK	Mitogen-activated protein kinase
Cdk5	Cyclin-dependent kinase 5
ERK2	Extracellular signal-regulated kinase 2
MARK	Microtubule affinity-regulating kinase
SADK	SAD-kinase
PKA	Protein kinase A
CaMKII	Calcium/calmodulin-dependent protein kinase II
JNKc	Jun N-terminal kinase
IRS	Insulin receptor substrate
GFAP	Glial fibrillary acidic protein
Iba-1	Ionized calcium-binding adapter molecule 1
PET	Positron emission tomography
NIR	Near infrared
PIB	Pittsburgh compound B
FDG	2-Deoxy-2-[18F] fluoroglucose
SNpc	Substantia nigra pars compacta
DA	Dopamine
LBs	Lewy bodies
2D-NMR	Two-dimensional nuclear magnetic resonance
MAO	Monoamine oxidase
PolyQ	Poly-glutamine
mHTT	Mutated huntingtin protein
CAG	Cytosine-adenine-guanine
YAC	Yeast artificial chromosome
DARPP	Dopamine-and cAMP-regulated neuronal phosphoprotein
TSE	Transmissible spongiform encephalopathies
PrP	Prion protein
PrP^C	Cellular form of prion protein
NSAIDs	Non-steroidal anti-inflammatory drugs
NPI-Q	Neuropsychiatric Inventory–Questionnaire
PBS	Phosphate Buffer Saline

References

1. Kaytor, M.D.; Warren, S.T. Aberrant protein deposition and neurological disease. *J. Biol. Chem.* **1999**, *274*, 37507–37510. [CrossRef] [PubMed]
2. Selkoe, D.J. Cell biology of protein misfolding: The examples of Alzheimer's and Parkinson's diseases. *Nat. Cell Biol.* **2004**, *6*, 1054–1061. [CrossRef] [PubMed]

3. Amor, S.; Puentes, F.; Baker, D.; van der Valk, P. Inflammation in neurodegenerative diseases. *Immunology* **2010**, *129*, 154–169. [CrossRef] [PubMed]

4. Chen, X.; Guo, C.; Kong, J. Oxidative stress in neurodegenerative diseases. *Neural Regen. Res.* **2012**, *7*, 376–385. [PubMed]

5. Cummings, J.; Lee, G.; Mortsdorf, T.; Ritter, A.; Zhong, K. Alzheimer's disease drug development pipeline: 2017. *Alzheimer's Dement.* **2017**, *3*, 367–384. [CrossRef] [PubMed]

6. Koronyo, Y.; Salumbides, B.C.; Black, K.L.; Koronyo-Hamaoui, M. Alzheimer's disease in the retina: Imaging retinal abeta plaques for early diagnosis and therapy assessment. *Neurodegener. Dis.* **2012**, *10*, 285–293. [CrossRef] [PubMed]

7. Mishra, S.; Palanivelu, K. The effect of curcumin (turmeric) on Alzheimer's disease: An overview. *Ann. Indian Acad. Neurol.* **2008**, *11*, 13–19. [CrossRef] [PubMed]

8. Prasad, S.; Aggarwal, B.B. Turmeric, the Golden Spice: From Traditional Medicine. In *Herbal Medicine: Biomolecular and Clinical Aspects*, 2nd ed.; Benzie, I.F.F., Wachtel-Galor, S., Eds.; CRC Press: Boca Raton, FL, USA, 2011.

9. Yang, F.; Lim, G.P.; Begum, A.N.; Ubeda, O.J.; Simmons, M.R.; Ambegaokar, S.S.; Chen, P.P.; Kayed, R.; Glabe, C.G.; Frautschy, S.A.; et al. Curcumin inhibits formation of amyloid beta oligomers and fibrils, binds plaques, and reduces amyloid in vivo. *J. Biol. Chem.* **2005**, *280*, 5892–5901. [CrossRef] [PubMed]

10. Begum, A.N.; Jones, M.R.; Lim, G.P.; Morihara, T.; Kim, P.; Heath, D.D.; Rock, C.L.; Pruitt, M.A.; Yang, F.; Hudspeth, B.; et al. Curcumin structure-function, bioavailability, and efficacy in models of neuroinflammation and Alzheimer's disease. *J. Pharmacol. Exp. Ther.* **2008**, *326*, 196–208. [CrossRef] [PubMed]

11. Ghalandarlaki, N.; Alizadeh, A.M.; Ashkani-Esfahani, S. Nanotechnology-applied curcumin for different diseases therapy. *Biomed. Res. Int.* **2014**, *2014*. [CrossRef] [PubMed]

12. Ma, Q.L.; Zuo, X.; Yang, F.; Ubeda, O.J.; Gant, D.J.; Alaverdyan, M.; Teng, E.; Hu, S.; Chen, P.P.; Maiti, P.; et al. Curcumin suppresses soluble tau dimers and corrects molecular chaperone, synaptic, and behavioral deficits in aged human tau transgenic mice. *J. Biol. Chem.* **2013**, *288*, 4056–4065. [CrossRef] [PubMed]

13. Maiti, P.; Hall, T.C.; Paladugu, L.; Kolli, N.; Learman, C.; Rossignol, J.; Dunbar, G.L. A comparative study of dietary curcumin, nanocurcumin, and other classical amyloid-binding dyes for labeling and imaging of amyloid plaques in brain tissue of 5x-familial Alzheimer's disease mice. *Histochem. Cell Biol.* **2016**, *146*, 609–625. [CrossRef] [PubMed]

14. Maiti, P.; Al-Gharaibeh, A.; Kolli, N.; Dunbar, G.L. Solid Lipid Curcumin Particles Induce More DNA Fragmentation and Cell Death in Cultured Human Glioblastoma Cells than Does Natural Curcumin. *Oxid. Med. Cell. Longev.* **2017**, *2017*. [CrossRef] [PubMed]

15. Lee, W.H.; Loo, C.Y.; Bebawy, M.; Luk, F.; Mason, R.S.; Rohanizadeh, R. Curcumin and its derivatives: Their application in neuropharmacology and neuroscience in the 21st century. *Curr. Neuropharmacol.* **2013**, *11*, 338–378. [CrossRef] [PubMed]

16. Priyadarsini, K.I. Chemical and structural features influencing the biological activity of curcumin. *Curr. Pharm. Des.* **2013**, *19*, 2093–2100. [PubMed]

17. Galer, P.; Golobic, A.; Koller, J.; Kosmrlj, B.; Sket, B. Structures in solid state and solution of dimethoxy curcuminoids: Regioselective bromination and chlorination. *Chem. Cent. J.* **2013**, *7*, 107. [CrossRef] [PubMed]

18. Amalraj, A.; Pius, A.; Gopi, S.; Gopi, S. Biological activities of curcuminoids, other biomolecules from turmeric and their derivatives—A review. *J. Tradit. Complement. Med.* **2017**, *7*, 205–233. [CrossRef] [PubMed]

19. Priyadarsini, K.I. The chemistry of curcumin: From extraction to therapeutic agent. *Molecules* **2014**, *19*, 20091–20112. [CrossRef] [PubMed]

20. Katsuyama, Y.; Kita, T.; Funa, N.; Horinouchi, S. Curcuminoid biosynthesis by two type III polyketide synthases in the herb Curcuma longa. *J. Biol. Chem.* **2009**, *284*, 11160–11170. [CrossRef] [PubMed]

21. Katsuyama, Y.; Kita, T.; Horinouchi, S. Identification and characterization of multiple curcumin synthases from the herb Curcuma longa. *FEBS Lett.* **2009**, *583*, 2799–2803. [CrossRef] [PubMed]

22. Sharma, R.A.; McLelland, H.R.; Hill, K.A.; Ireson, C.R.; Euden, S.A.; Manson, M.M.; Pirmohamed, M.; Marnett, L.J.; Gescher, A.J.; Steward, W.P. Pharmacodynamic and pharmacokinetic study of oral Curcuma extract in patients with colorectal cancer. *Clin. Cancer Res.* **2001**, *7*, 1894–1900. [PubMed]

23. Heger, M.; van Golen, R.F.; Broekgaarden, M.; Michel, M.C. The molecular basis for the pharmacokinetics and pharmacodynamics of curcumin and its metabolites in relation to cancer. *Pharmacol. Rev.* **2014**, *66*, 222–307. [CrossRef] [PubMed]

24. DiSilvestro, R.A.; Joseph, E.; Zhao, S.; Bomser, J. Diverse effects of a low dose supplement of lipidated curcumin in healthy middle aged people. *Nutr. J.* **2012**, *11*, 79. [CrossRef] [PubMed]

25. Koronyo-Hamaoui, M.; Koronyo, Y.; Ljubimov, A.V.; Miller, C.A.; Ko, M.K.; Black, K.L.; Schwartz, M.; Farkas, D.L. Identification of amyloid plaques in retinas from Alzheimer's patients and noninvasive in vivo optical imaging of retinal plaques in a mouse model. *Neuroimage* **2011**, *54*, S204–S217. [CrossRef] [PubMed]

26. Anand, P.; Kunnumakkara, A.B.; Newman, R.A.; Aggarwal, B.B. Bioavailability of curcumin: Problems and promises. *Mol. Pharm.* **2007**, *4*, 807–818. [CrossRef] [PubMed]

27. Pan, M.H.; Huang, T.M.; Lin, J.K. Biotransformation of curcumin through reduction and glucuronidation in mice. *Drug Metab. Dispos.* **1999**, *27*, 486–494. [PubMed]

28. Vareed, S.K.; Kakarala, M.; Ruffin, M.T.; Crowell, J.A.; Normolle, D.P.; Djuric, Z.; Brenner, D.E. Pharmacokinetics of curcumin conjugate metabolites in healthy human subjects. *Cancer Epidemiol. Biomarkers Prev.* **2008**, *17*, 1411–1417. [CrossRef] [PubMed]

29. Jager, R.; Lowery, R.P.; Calvanese, A.V.; Joy, J.M.; Purpura, M.; Wilson, J.M. Comparative absorption of curcumin formulations. *Nutr. J.* **2014**, *13*, 11. [CrossRef] [PubMed]

30. Ghosh, S.; Bhattacharyya, S.; Rashid, K.; Sil, P.C. Curcumin protects rat liver from streptozotocin-induced diabetic pathophysiology by counteracting reactive oxygen species and inhibiting the activation of p53 and MAPKs mediated stress response pathways. *Toxicol. Rep.* **2015**, *2*, 365–376. [CrossRef] [PubMed]

31. Sandur, S.K.; Pandey, M.K.; Sung, B.; Ahn, K.S.; Murakami, A.; Sethi, G.; Limtrakul, P.; Badmaev, V.; Aggarwal, B.B. Curcumin, demethoxycurcumin, bisdemethoxycurcumin, tetrahydrocurcumin and turmerones differentially regulate anti-inflammatory and anti-proliferative responses through a ROS-independent mechanism. *Carcinogenesis* **2007**, *28*, 1765–1773. [CrossRef] [PubMed]

32. Aggarwal, B.B.; Sundaram, C.; Malani, N.; Ichikawa, H. Curcumin: The Indian solid gold. *Adv. Exp. Med. Biol.* **2007**, *595*, 1–75. [PubMed]

33. Cole, G.M.; Teter, B.; Frautschy, S.A. Neuroprotective effects of curcumin. *Adv. Exp. Med. Biol.* **2007**, *595*, 197–212. [PubMed]

34. Yanagisawa, D.; Taguchi, H.; Yamamoto, A.; Shirai, N.; Hirao, K.; Tooyama, I. Urcuminoid binds to amyloid-beta1-42 oligomer and fibril. *J. Alzheimer's Dis.* **2011**, *24*, 33–42. [CrossRef] [PubMed]

35. Singh, P.K.; Kotia, V.; Ghosh, D.; Mohite, G.M.; Kumar, A.; Maji, S.K. Curcumin modulates alpha-synuclein aggregation and toxicity. *ACS Chem. Neurosci.* **2013**, *4*, 393–407. [CrossRef] [PubMed]

36. Chongtham, A.; Agrawal, N. Curcumin modulates cell death and is protective in Huntington's disease model. *Sci. Rep.* **2016**, *6*, 18736. [CrossRef] [PubMed]

37. Mohorko, N.; Repovs, G.; Popovic, M.; Kovacs, G.G.; Bresjanac, M. Curcumin labeling of neuronal fibrillar tau inclusions in human brain samples. *J. Neuropathol. Exp. Neurol.* **2010**, *69*, 405–414. [CrossRef] [PubMed]

38. Hafner-Bratkovic, I.; Gaspersic, J.; Smid, L.M.; Bresjanac, M.; Jerala, R. Curcumin binds to the alpha-helical intermediate and to the amyloid form of prion protein—A new mechanism for the inhibition of PrP(Sc) accumulation. *J. Neurochem.* **2008**, *104*, 1553–1564. [CrossRef] [PubMed]

39. Mythri, R.B.; Bharath, M.M. Curcumin: A potential neuroprotective agent in Parkinson's disease. *Curr. Pharm. Des.* **2012**, *18*, 91–99. [CrossRef] [PubMed]

40. Hickey, M.A.; Zhu, C.; Medvedeva, V.; Lerner, R.P.; Patassini, S.; Franich, N.R.; Maiti, P.; Frautschy, S.A.; Zeitlin, S.; Levine, M.S.; et al. Improvement of neuropathology and transcriptional deficits in CAG 140 knock-in mice supports a beneficial effect of dietary curcumin in Huntington's disease. *Mol. Neurodegener.* **2012**, *7*, 12. [CrossRef] [PubMed]

41. Lin, C.F.; Yu, K.H.; Jheng, C.P.; Chung, R.; Lee, C.I. Curcumin reduces amyloid fibrillation of prion protein and decreases reactive oxidative stress. *Pathogens* **2013**, *2*, 506–519. [CrossRef] [PubMed]

42. Mosley, R.L.; Benner, E.J.; Kadiu, I.; Thomas, M.; Boska, M.D.; Hasan, K.; Laurie, C.; Gendelman, H.E. Neuroinflammation, Oxidative Stress and the Pathogenesis of Parkinson's Disease. *Clin. Neurosci. Res.* **2006**, *6*, 261–281. [CrossRef] [PubMed]

43. Gregersen, N.; Bross, P. Protein misfolding and cellular stress: An overview. *Methods Mol. Biol.* **2010**, *648*, 3–23. [PubMed]

44. Menon, V.P.; Sudheer, A.R. Antioxidant and anti-inflammatory properties of curcumin. *Adv. Exp. Med. Biol.* **2007**, *595*, 105–125. [PubMed]

45. Biswas, S.K.; McClure, D.; Jimenez, L.A.; Megson, I.L.; Rahman, I. Curcumin induces glutathione biosynthesis and inhibits NF-kappaB activation and interleukin-8 release in alveolar epithelial cells: Mechanism of free radical scavenging activity. *Antioxid. Redox Signal.* **2005**, *7*, 32–41. [CrossRef] [PubMed]

46. Jat, D.; Parihar, P.; Kothari, S.C.; Parihar, M.S. Curcumin reduces oxidative damage by increasing reduced glutathione and preventing membrane permeability transition in isolated brain mitochondria. *Cell Mol. Biol. (Noisy-le-grand)* **2013**, *59*, 1899–1905.

47. Jones, D.P. Radical-free biology of oxidative stress. *Am. J. Physiol. Cell Physiol.* **2008**, *295*, C849–C868. [CrossRef] [PubMed]

48. Sies, H. Oxidative stress: A concept in redox biology and medicine. *Redox Biol.* **2015**, *4*, 180–183. [CrossRef] [PubMed]

49. He, Y.; Yue, Y.; Zheng, X.; Zhang, K.; Chen, S.; Du, Z. Curcumin, inflammation, and chronic diseases: How are they linked? *Molecules* **2015**, *20*, 9183–9213. [CrossRef] [PubMed]

50. Hong, J.; Bose, M.; Ju, J.; Ryu, J.H.; Chen, X.; Sang, S.; Lee, M.J.; Yang, C.S. Modulation of arachidonic acid metabolism by curcumin and related beta-diketone derivatives: Effects on cytosolic phospholipase A(2), cyclooxygenases and 5-lipoxygenase. *Carcinogenesis* **2004**, *25*, 1671–1679. [CrossRef] [PubMed]

51. Lim, G.P.; Chu, T.; Yang, F.; Beech, W.; Frautschy, S.A.; Cole, G.M. The curry spice curcumin reduces oxidative damage and amyloid pathology in an Alzheimer transgenic mouse. *J. Neurosci.* **2001**, *21*, 8370–8377. [CrossRef] [PubMed]

52. Gulcubuk, A.; Altunatmaz, K.; Sonmez, K.; Haktanir-Yatkin, D.; Uzun, H.; Gurel, A.; Aydin, S. Effects of curcumin on tumour necrosis factor-alpha and interleukin-6 in the late phase of experimental acute pancreatitis. *J. Vet. Med. A Physiol. Pathol. Clin. Med.* **2006**, *53*, 49–54. [CrossRef] [PubMed]

53. Mazidi, M.; Karimi, E.; Meydani, M.; Ghayour-Mobarhan, M.; Ferns, G.A. Potential effects of curcumin on peroxisome proliferator-activated receptor-gamma in vitro and in vivo. *World J. Methodol.* **2016**, *6*, 112–117. [CrossRef] [PubMed]

54. Leak, R.K. Heat shock proteins in neurodegenerative disorders and aging. *J. Cell Commun. Signal.* **2014**, *8*, 293–310. [CrossRef] [PubMed]

55. Maiti, P.; Dunbar, G.L. Comparative Neuroprotective Effects of Dietary Curcumin and Solid Lipid Curcumin Particles in Cultured Mouse Neuroblastoma Cells after Exposure to Abeta42. *Int. J. Alzheimer's Dis.* **2017**, *2017*. [CrossRef]

56. Gupta, S.C.; Prasad, S.; Kim, J.H.; Patchva, S.; Webb, L.J.; Priyadarsini, I.K.; Aggarwal, B.B. Multitargeting by curcumin as revealed by molecular interaction studies. *Nat. Prod. Rep.* **2011**, *28*, 1937–1955. [CrossRef] [PubMed]

57. Dong, S.; Zeng, Q.; Mitchell, E.S.; Xiu, J.; Duan, Y.; Li, C.; Tiwari, J.K.; Hu, Y.; Cao, X.; Zhao, Z. Curcumin enhances neurogenesis and cognition in aged rats: Implications for transcriptional interactions related to growth and synaptic plasticity. *PLoS ONE* **2012**, *7*, e31211. [CrossRef] [PubMed]

58. Ahmed, T.; Enam, S.A.; Gilani, A.H. Curcuminoids enhance memory in an amyloid-infused rat model of Alzheimer's disease. *Neuroscience* **2010**, *169*, 1296–1306. [CrossRef] [PubMed]

59. Perry, G.; Sayre, L.M.; Atwood, C.S.; Castellani, R.; Cash, A.D.; Rottkamp, C.A.; Smith, M.A. The role of iron and copper in the aetiology of neurodegenerative disorders: Therapeutic implications. *CNS Drugs* **2002**, *16*, 339–352. [CrossRef] [PubMed]

60. Liu, G.; Huang, W.; Moir, R.D.; Vanderburg, C.R.; Lai, B.; Peng, Z.; Tanzi, R.E.; Rogers, J.T.; Huang, X. Metal exposure and Alzheimer's pathogenesis. *J. Struct. Biol.* **2006**, *155*, 45–51. [CrossRef] [PubMed]

61. Wanninger, S.; Lorenz, V.; Subhan, A.; Edelmann, F.T. Metal complexes of curcumin—Synthetic strategies, structures and medicinal applications. *Chem. Soc. Rev.* **2015**, *44*, 4986–5002. [CrossRef] [PubMed]

62. Eybl, V.; Kotyzova, D.; Bludovska, M. The effect of curcumin on cadmium-induced oxidative damage and trace elements level in the liver of rats and mice. *Toxicol. Lett.* **2004**, *151*, 79–85. [CrossRef] [PubMed]

63. Shakibaei, M.; John, T.; Schulze-Tanzil, G.; Lehmann, I.; Mobasheri, A. Suppression of NF-kappaB activation by curcumin leads to inhibition of expression of cyclo-oxygenase-2 and matrix metalloproteinase-9 in human articular chondrocytes: Implications for the treatment of osteoarthritis. *Biochem. Pharmacol.* **2007**, *73*, 1434–1445. [CrossRef] [PubMed]

64. Reuter, S.; Gupta, S.C.; Park, B.; Goel, A.; Aggarwal, B.B. Epigenetic changes induced by curcumin and other natural compounds. *Genes Nutr.* **2011**, *6*, 93–108. [CrossRef] [PubMed]
65. Rose, N.R.; Klose, R.J. Understanding the relationship between DNA methylation and histone lysine methylation. *Biochim. Biophys. Acta* **2014**, *1839*, 1362–1372. [CrossRef] [PubMed]
66. Awad, A.S. Effect of combined treatment with curcumin and candesartan on ischemic brain damage in mice. *J. Stroke Cerebrovasc. Dis.* **2011**, *20*, 541–548. [CrossRef] [PubMed]
67. Zhang, L.; Gu, Z.L.; Qin, Z.H.; Liang, Z.Q. Effect of curcumin on the adhesion of platelets to brain microvascular endothelial cells in vitro. *Acta Pharmacol. Sin.* **2008**, *29*, 800–807. [CrossRef] [PubMed]
68. Vachharajani, V.; Wang, S.W.; Mishra, N.; El Gazzar, M.; Yoza, B.; McCall, C. Curcumin modulates leukocyte and platelet adhesion in murine sepsis. *Microcirculation* **2010**, *17*, 407–416. [CrossRef] [PubMed]
69. Avci, G.; Kadioglu, H.; Sehirli, A.O.; Bozkurt, S.; Guclu, O.; Arslan, E.; Muratli, S.K. Curcumin protects against ischemia/reperfusion injury in rat skeletal muscle. *J. Surg. Res.* **2012**, *172*, e39–e46. [CrossRef] [PubMed]
70. Gupta, S.C.; Patchva, S.; Aggarwal, B.B. Therapeutic roles of curcumin: Lessons learned from clinical trials. *AAPS J.* **2013**, *15*, 195–218. [CrossRef] [PubMed]
71. Baum, L.; Lam, C.W.; Cheung, S.K.; Kwok, T.; Lui, V.; Tsoh, J.; Lam, L.; Leung, V.; Hui, E.; Ng, C.; et al. Six-month randomized, placebo-controlled, double-blind, pilot clinical trial of curcumin in patients with Alzheimer disease. *J. Clin. Psychopharmacol.* **2008**, *28*, 110–113. [CrossRef] [PubMed]
72. Liu, W.; Zhai, Y.; Heng, X.; Che, F.Y.; Chen, W.; Sun, D.; Zhai, G. Oral bioavailability of curcumin: Problems and advancements. *J. Drug Target* **2016**, *24*, 694–702. [CrossRef] [PubMed]
73. Wahlstrom, B.; Blennow, G. A study on the fate of curcumin in the rat. *Acta Pharmacol. Toxicol.* **1978**, *43*, 86–92. [CrossRef]
74. Dulbecco, P.; Savarino, V. Therapeutic potential of curcumin in digestive diseases. *World J. Gastroenterol.* **2013**, *19*, 9256–9270. [CrossRef] [PubMed]
75. Shoba, G.; Joy, D.; Joseph, T.; Majeed, M.; Rajendran, R.; Srinivas, P.S. Influence of piperine on the pharmacokinetics of curcumin in animals and human volunteers. *Planta Med.* **1998**, *64*, 353–356. [CrossRef] [PubMed]
76. Moorthi, C.; Krishnan, K.; Manavalan, R.; Kathiresan, K. Preparation and characterization of curcumin-piperine dual drug loaded nanoparticles. *Asian Pac. J. Trop. Biomed.* **2012**, *2*, 841–848. [CrossRef]
77. Antony, B.; Merina, B.; Iyer, V.S.; Judy, N.; Lennertz, K.; Joyal, S. A Pilot Cross-Over Study to Evaluate Human Oral Bioavailability of BCM-95CG (Biocurcumax), A Novel Bioenhanced Preparation of Curcumin. *Indian J. Pharm. Sci.* **2008**, *70*, 445–449. [CrossRef] [PubMed]
78. Mishra, S.; Narain, U.; Mishra, R.; Misra, K. Design, development and synthesis of mixed bioconjugates of piperic acid-glycine, curcumin-glycine/alanine and curcumin-glycine-piperic acid and their antibacterial and antifungal properties. *Bioorg. Med. Chem.* **2005**, *13*, 1477–1486. [CrossRef] [PubMed]
79. Comblain, F.; Sanchez, C.; Lesponne, I.; Balligand, M.; Serisier, S.; Henrotin, Y. Curcuminoids extract, hydrolyzed collagen and green tea extract synergically inhibit inflammatory and catabolic mediator's synthesis by normal bovine and osteoarthritic human chondrocytes in monolayer. *PLoS ONE* **2015**, *10*, e0121654. [CrossRef] [PubMed]
80. Maiti, K.; Mukherjee, K.; Gantait, A.; Saha, B.P.; Mukherjee, P.K. Curcumin-phospholipid complex: Preparation, therapeutic evaluation and pharmacokinetic study in rats. *Int. J. Pharm.* **2007**, *330*, 155–163. [CrossRef] [PubMed]
81. Aqil, F.; Munagala, R.; Jeyabalan, J.; Vadhanam, M.V. Bioavailability of phytochemicals and its enhancement by drug delivery systems. *Cancer Lett.* **2013**, *334*, 133–141. [CrossRef] [PubMed]
82. Ma, Z.; Haddadi, A.; Molavi, O.; Lavasanifar, A.; Lai, R.; Samuel, J. Micelles of poly(ethylene oxide)-b-poly(epsilon-caprolactone) as vehicles for the solubilization, stabilization, and controlled delivery of curcumin. *J. Biomed. Mater. Res. A* **2008**, *86*, 300–310. [CrossRef] [PubMed]
83. Yu, H.; Li, J.; Shi, K.; Huang, Q. Structure of modified epsilon-polylysine micelles and their application in improving cellular antioxidant activity of curcuminoids. *Food Funct.* **2011**, *2*, 373–380. [CrossRef] [PubMed]
84. Podaralla, S.; Averineni, R.; Alqahtani, M.; Perumal, O. Synthesis of novel biodegradable methoxy poly(ethylene glycol)-zein micelles for effective delivery of curcumin. *Mol. Pharm.* **2012**, *9*, 2778–2786. [CrossRef] [PubMed]

85. Song, Z.; Feng, R.; Sun, M.; Guo, C.; Gao, Y.; Li, L.; Zhai, G. Curcumin-loaded PLGA-PEG-PLGA triblock copolymeric micelles: Preparation, pharmacokinetics and distribution in vivo. *J. Colloid Interface Sci.* **2011**, *354*, 116–123. [CrossRef] [PubMed]

86. Kazi, K.M.; Mandal, A.S.; Biswas, N.; Guha, A.; Chatterjee, S.; Behera, M.; Kuotsu, K. Niosome: A future of targeted drug delivery systems. *J. Adv. Pharm. Technol. Res.* **2010**, *1*, 374–380. [PubMed]

87. Jain, S.; Singh, P.; Mishra, V.; Vyas, S.P. Mannosylated niosomes as adjuvant-carrier system for oral genetic immunization against hepatitis B. *Immunol. Lett.* **2005**, *101*, 41–49. [CrossRef] [PubMed]

88. Yallapu, M.M.; Ebeling, M.C.; Chauhan, N.; Jaggi, M.; Chauhan, S.C. Interaction of curcumin nanoformulations with human plasma proteins and erythrocytes. *Int. J. Nanomed.* **2011**, *6*, 2779–2790.

89. Kabanov, A.V.; Vinogradov, S.V. Nanogels as pharmaceutical carriers: Finite networks of infinite capabilities. *Angew. Chem. Int. Ed. Engl.* **2009**, *48*, 5418–5429. [CrossRef] [PubMed]

90. Mangalathillam, S.; Rejinold, N.S.; Nair, A.; Lakshmanan, V.K.; Nair, S.V.; Jayakumar, R. Curcumin loaded chitin nanogels for skin cancer treatment via the transdermal route. *Nanoscale* **2012**, *4*, 239–250. [CrossRef] [PubMed]

91. Wu, W.; Shen, J.; Banerjee, P.; Zhou, S. Water-dispersible multifunctional hybrid nanogels for combined curcumin and photothermal therapy. *Biomaterials* **2011**, *32*, 598–609. [CrossRef] [PubMed]

92. Salomon, C.; Goycoolea, F.M.; Moerschbacher, B. Recent Trends in the Development of Chitosan-Based Drug Delivery Systems. *AAPS PharmSciTech* **2017**, *18*, 933–935. [CrossRef] [PubMed]

93. Zhang, J.; Tang, Q.; Xu, X.; Li, N. Development and evaluation of a novel phytosome-loaded chitosan microsphere system for curcumin delivery. *Int. J. Pharm.* **2013**, *448*, 168–174. [CrossRef] [PubMed]

94. Karewicz, A.; Bielska, D.; Loboda, A.; Gzyl-Malcher, B.; Bednar, J.; Jozkowicz, A.; Dulak, J.; Nowakowska, M. Curcumin-containing liposomes stabilized by thin layers of chitosan derivatives. *Colloids Surf. B Biointerfaces* **2013**, *109*, 307–316. [CrossRef] [PubMed]

95. Omidfar, K.; Khorsand, F.; Darziani Azizi, M. New analytical applications of gold nanoparticles as label in antibody based sensors. *Biosens. Bioelectron.* **2013**, *43*, 336–347. [CrossRef] [PubMed]

96. Sindhu, K.; Indra, R.; Rajaram, A.; Sreeram, K.J.; Rajaram, R. Investigations on the interaction of gold-curcumin nanoparticles with human peripheral blood lymphocytes. *J. Biomed. Nanotechnol.* **2011**, *7*, 56. [CrossRef] [PubMed]

97. Sweet, M.J.; Singleton, I. Silver nanoparticles: A microbial perspective. *Adv. Appl. Microbiol.* **2011**, *77*, 115–133. [PubMed]

98. Tonnesen, H.H.; Masson, M.; Loftsson, T. Studies of curcumin and curcuminoids. XXVII. Cyclodextrin complexation: Solubility, chemical and photochemical stability. *Int. J. Pharm.* **2002**, *244*, 127–135. [CrossRef]

99. Torne, S.; Darandale, S.; Vavia, P.; Trotta, F.; Cavalli, R. Cyclodextrin-based nanosponges: Effective nanocarrier for tamoxifen delivery. *Pharm. Dev. Technol.* **2013**, *18*, 619–625. [CrossRef] [PubMed]

100. Yadav, V.R.; Prasad, S.; Kannappan, R.; Ravindran, J.; Chaturvedi, M.M.; Vaahtera, L.; Parkkinen, J.; Aggarwal, B.B. Cyclodextrin-complexed curcumin exhibits anti-inflammatory and antiproliferative activities superior to those of curcumin through higher cellular uptake. *Biochem. Pharmacol.* **2010**, *80*, 1021–1032. [CrossRef] [PubMed]

101. Longmire, M.; Choyke, P.L.; Kobayashi, H. Dendrimer-based contrast agents for molecular imaging. *Curr. Top. Med. Chem.* **2008**, *8*, 1180–1186. [CrossRef] [PubMed]

102. Ramachandran, C.; Fonseca, H.B.; Jhabvala, P.; Escalon, E.A.; Melnick, S.J. Curcumin inhibits telomerase activity through human telomerase reverse transcritpase in MCF-7 breast cancer cell line. *Cancer Lett.* **2002**, *184*, 1–6. [CrossRef]

103. Babaei, E.; Sadeghizadeh, M.; Hassan, Z.M.; Feizi, M.A.; Najafi, F.; Hashemi, S.M. Dendrosomal curcumin significantly suppresses cancer cell proliferation in vitro and in vivo. *Int. Immunopharmacol.* **2012**, *12*, 226–234. [CrossRef] [PubMed]

104. Ekambaram, P.; Abdul, H.S. Formulation and evaluation of solid lipid nanoparticles of ramipril. *J. Young Pharm.* **2011**, *3*, 216–220. [CrossRef] [PubMed]

105. Kakkar, V.; Muppu, S.K.; Chopra, K.; Kaur, I.P. Curcumin loaded solid lipid nanoparticles: An efficient formulation approach for cerebral ischemic reperfusion injury in rats. *Eur. J. Pharm. Biopharm.* **2013**, *85*, 339–345. [CrossRef] [PubMed]

106. Gota, V.S.; Maru, G.B.; Soni, T.G.; Gandhi, T.R.; Kochar, N.; Agarwal, M.G. Safety and pharmacokinetics of a solid lipid curcumin particle formulation in osteosarcoma patients and healthy volunteers. *J. Agric. Food Chem.* **2010**, *58*, 2095–2099. [CrossRef] [PubMed]

107. Dadhaniya, P.; Patel, C.; Muchhara, J.; Bhadja, N.; Mathuria, N.; Vachhani, K.; Soni, M.G. Safety assessment of a solid lipid curcumin particle preparation: Acute and subchronic toxicity studies. *Food Chem. Toxicol.* **2011**, *49*, 1834–1842. [CrossRef] [PubMed]

108. Xie, X.; Tu, J.; You, H.; Hu, B. Design, synthesis, and biological evaluation of novel EF24 and EF31 analogs as potential IkappaB kinase beta inhibitors for the treatment of pancreatic cancer. *Drug Des. Dev. Ther.* **2017**, *11*, 1439–1451. [CrossRef] [PubMed]

109. Zhou, H.; Wu, X.; Xu, W.; Yang, J.; Yang, Q. Fluorescence enhancement of the silver nanoparticales—Curcumin-cetyltrimethylammonium bromide-nucleic acids system and its analytical application. *J. Fluoresc.* **2010**, *20*, 843–850. [CrossRef] [PubMed]

110. Mondal, S.; Ghosh, S. Role of curcumin on the determination of the critical micellar concentration by absorbance, fluorescence and fluorescence anisotropy techniques. *J. Photochem. Photobiol. B* **2012**, *115*, 9–15. [CrossRef] [PubMed]

111. Ono, K.; Hasegawa, K.; Naiki, H.; Yamada, M. Curcumin has potent anti-amyloidogenic effects for Alzheimer's beta-amyloid fibrils in vitro. *J. Neurosci. Res.* **2004**, *75*, 742–750. [CrossRef] [PubMed]

112. Zhang, L.; Fiala, M.; Cashman, J.; Sayre, J.; Espinosa, A.; Mahanian, M.; Zaghi, J.; Badmaev, V.; Graves, M.C.; Bernard, G.; et al. Curcuminoids enhance amyloid-beta uptake by macrophages of Alzheimer's disease patients. *J. Alzheimer's Dis.* **2006**, *10*, 1–7. [CrossRef]

113. Zhang, X.; Tian, Y.; Zhang, H.; Kavishwar, A.; Lynes, M.; Brownell, A.L.; Sun, H.; Tseng, Y.H.; Moore, A.; Ran, C. Curcumin analogues as selective fluorescence imaging probes for brown adipose tissue and monitoring browning. *Sci. Rep.* **2015**, *5*, 13116. [CrossRef] [PubMed]

114. Jack, C.R., Jr.; Albert, M.S.; Knopman, D.S.; McKhann, G.M.; Sperling, R.A.; Carrillo, M.C.; Thies, B.; Phelps, C.H. Introduction to the recommendations from the National Institute on Aging-Alzheimer's Association workgroups on diagnostic guidelines for Alzheimer's disease. *Alzheimer's Dement. J. Alzheimer's Assoc.* **2011**, *7*, 257–262. [CrossRef] [PubMed]

115. Glenner, G.G.; Wong, C.W. Alzheimer's disease: Initial report of the purification and characterization of a novel cerebrovascular amyloid protein. *Biochem. Biophys. Res. Commun.* **1984**, *120*, 885–890. [CrossRef]

116. Masters, C.L.; Simms, G.; Weinman, N.A.; Multhaup, G.; McDonald, B.L.; Beyreuther, K. Amyloid plaque core protein in Alzheimer disease and Down syndrome. *Proc. Natl. Acad. Sci. USA* **1985**, *82*, 4245–4249. [CrossRef] [PubMed]

117. Haass, C. Take five—BACE and the gamma-secretase quartet conduct Alzheimer's amyloid beta-peptide generation. *EMBO J.* **2004**, *23*, 483–488. [CrossRef] [PubMed]

118. Haass, C.; Selkoe, D.J. Soluble protein oligomers in neurodegeneration: Lessons from the Alzheimer's amyloid beta-peptide. *Nat. Rev. Mol. Cell Biol.* **2007**, *8*, 101–112. [CrossRef] [PubMed]

119. Lichtenthaler, S.F.; Haass, C.; Steiner, H. Regulated intramembrane proteolysis—Lessons from amyloid precursor protein processing. *J. Neurochem.* **2011**, *117*, 779–796. [CrossRef] [PubMed]

120. Kosik, K.S.; Joachim, C.L.; Selkoe, D.J. Microtubule-associated protein tau (tau) is a major antigenic component of paired helical filaments in Alzheimer disease. *Proc. Natl. Acad. Sci. USA* **1986**, *83*, 4044–4048. [CrossRef] [PubMed]

121. Nukina, N.; Ihara, Y. One of the antigenic determinants of paired helical filaments is related to tau protein. *J. Biochem.* **1986**, *99*, 1541–1544. [CrossRef] [PubMed]

122. Dickson, D.W. Apoptotic mechanisms in Alzheimer neurofibrillary degeneration: Cause or effect? *J. Clin. Investig.* **2004**, *114*, 23–27. [CrossRef] [PubMed]

123. Spires-Jones, T.L.; de Calignon, A.; Matsui, T.; Zehr, C.; Pitstick, R.; Wu, H.Y.; Osetek, J.D.; Jones, P.B.; Bacskai, B.J.; Feany, M.B.; et al. In vivo imaging reveals dissociation between caspase activation and acute neuronal death in tangle-bearing neurons. *J. Neurosci. Off. J. Soc. Neurosci.* **2008**, *28*, 862–867. [CrossRef] [PubMed]

124. Wakasaya, Y.; Kawarabayashi, T.; Watanabe, M.; Yamamoto-Watanabe, Y.; Takamura, A.; Kurata, T.; Murakami, T.; Abe, K.; Yamada, K.; Wakabayashi, K.; et al. Factors responsible for neurofibrillary tangles and neuronal cell losses in tauopathy. *J. Neurosci. Res.* **2011**, *89*, 576–584. [CrossRef] [PubMed]

125. Hu, S.; Maiti, P.; Ma, Q.; Zuo, X.; Jones, M.R.; Cole, G.M.; Frautschy, S.A. Clinical development of curcumin in neurodegenerative disease. *Expert Rev. Neurother.* **2015**, *15*, 629–637. [CrossRef] [PubMed]

126. Ryu, E.K.; Choe, Y.S.; Lee, K.H.; Choi, Y.; Kim, B.T. Curcumin and dehydrozingerone derivatives: Synthesis, radiolabeling, and evaluation for beta-amyloid plaque imaging. *J. Med. Chem.* **2006**, *49*, 6111–6119. [CrossRef] [PubMed]

127. Xiong, Z.; Hongmei, Z.; Lu, S.; Yu, L. Curcumin mediates presenilin-1 activity to reduce beta-amyloid production in a model of Alzheimer's Disease. *Pharmacol. Rep.* **2011**, *63*, 1101–1108. [CrossRef]

128. Mutsuga, M.; Chambers, J.K.; Uchida, K.; Tei, M.; Makibuchi, T.; Mizorogi, T.; Takashima, A.; Nakayama, H. Binding of curcumin to senile plaques and cerebral amyloid angiopathy in the aged brain of various animals and to neurofibrillary tangles in Alzheimer's brain. *J. Vet. Med. Sci.* **2012**, *74*, 51–57. [CrossRef] [PubMed]

129. Caughey, B.; Raymond, L.D.; Raymond, G.J.; Maxson, L.; Silveira, J.; Baron, G.S. Inhibition of protease-resistant prion protein accumulation in vitro by curcumin. *J. Virol.* **2003**, *77*, 5499–5502. [CrossRef] [PubMed]

130. Anand, P.; Thomas, S.G.; Kunnumakkara, A.B.; Sundaram, C.; Harikumar, K.B.; Sung, B.; Tharakan, S.T.; Misra, K.; Priyadarsini, I.K.; Rajasekharan, K.N.; et al. Biological activities of curcumin and its analogues (Congeners) made by man and Mother Nature. *Biochem. Pharmacol.* **2008**, *76*, 1590–1611. [CrossRef] [PubMed]

131. Maiti, P.; Manna, J.; Veleri, S.; Frautschy, S. Molecular chaperone dysfunction in neurodegenerative diseases and effects of curcumin. *Biomed. Res. Int.* **2014**, *2014*. [CrossRef] [PubMed]

132. Gibson, S.A.; Gao, G.D.; McDonagh, K.; Shen, S. Progress on stem cell research towards the treatment of Parkinson's disease. *Stem Cell Res. Ther.* **2012**, *3*, 11. [CrossRef] [PubMed]

133. Dos Santos Jaques, J.A.; Ruchel, J.B.; Schlemmer, K.B.; Pimentel, V.C.; Bagatini, M.; Souza Vdo, C.; Moretto, M.B.; Morsch, V.M.; Schetinger, M.R.; Leal, D.B. Effects of curcumin on the activities of the enzymes that hydrolyse adenine nucleotides in platelets from cigarette smoke-exposed rats. *Cell Biochem. Funct.* **2011**, *29*, 630–635. [CrossRef] [PubMed]

134. Monroy, A.; Lithgow, G.J.; Alavez, S. Curcumin and neurodegenerative diseases. *Biofactors* **2013**, *39*, 122–132. [CrossRef] [PubMed]

135. Feng, H.L.; Fan, H.; Dang, H.Z.; Chen, X.P.; Ren, Y.; Yang, J.D.; Wang, P.W. Neuroprotective effect of curcumin to Abeta of double transgenic mice with Alzheimer's disease. *China J. Chin. Mater. Med.* **2014**, *39*, 3846–3849.

136. Wang, X.S.; Zhang, Z.R.; Zhang, M.M.; Sun, M.X.; Wang, W.W.; Xie, C.L. Neuroprotective properties of curcumin in toxin-base animal models of Parkinson's disease: A systematic experiment literatures review. *BMC Complement. Altern. Med.* **2017**, *17*, 412. [CrossRef] [PubMed]

137. Rajeswari, A.; Sabesan, M. Inhibition of monoamine oxidase-B by the polyphenolic compound, curcumin and its metabolite tetrahydrocurcumin, in a model of Parkinson's disease induced by MPTP neurodegeneration in mice. *Inflammopharmacology* **2008**, *16*, 96–99. [CrossRef] [PubMed]

138. Maiti, P.; Paladugu, L.; Dunbar, G.L. Solid lipid curcumin particles provide greater anti-amyloid, anti-inflammatory and neuroprotective effects than curcumin in the 5xFAD mouse model of Alzheimer's disease. *BMC Neurosci.* **2018**, *19*, 7. [CrossRef] [PubMed]

139. Koronyo, Y.; Biggs, D.; Barron, E.; Boyer, D.S.; Pearlman, J.A.; Au, W.J.; Kile, S.J.; Blanco, A.; Fuchs, D.T.; Ashfaq, A.; et al. Retinal amyloid pathology and proof-of-concept imaging trial in Alzheimer's disease. *JCI Insight* **2017**, *2*. [CrossRef] [PubMed]

140. Kim, H.; Park, B.S.; Lee, K.G.; Choi, C.Y.; Jang, S.S.; Kim, Y.H.; Lee, S.E. Effects of naturally occurring compounds on fibril formation and oxidative stress of beta-amyloid. *J. Agric. Food Chem.* **2005**, *53*, 8537–8541. [CrossRef] [PubMed]

141. Garcia-Alloza, M.; Borrelli, L.A.; Rozkalne, A.; Hyman, B.T.; Bacskai, B.J. Curcumin labels amyloid pathology in vivo, disrupts existing plaques, and partially restores distorted neurites in an Alzheimer mouse model. *J. Neurochem.* **2007**, *102*, 1095–1104. [CrossRef] [PubMed]

142. Yan, R.; Vassar, R. Targeting the beta secretase BACE1 for Alzheimer's disease therapy. *Lancet Neurol.* **2014**, *13*, 319–329. [CrossRef]

143. Narlawar, R.; Baumann, K.; Schubenel, R.; Schmidt, B. Curcumin derivatives inhibit or modulate beta-amyloid precursor protein metabolism. *Neurodegener. Dis.* **2007**, *4*, 88–93. [CrossRef] [PubMed]

144. Huang, H.C.; Xu, K.; Jiang, Z.F. Curcumin-mediated neuroprotection against amyloid-beta-induced mitochondrial dysfunction involves the inhibition of GSK-3beta. *J. Alzheimer's Dis.* **2012**, *32*, 981–996. [CrossRef] [PubMed]

145. Thal, D.R. Clearance of amyloid beta-protein and its role in the spreading of Alzheimer's disease pathology. *Front. Aging Neurosci.* **2015**, *7*, 25. [CrossRef] [PubMed]

146. Frautschy, S.A.; Hu, W.; Kim, P.; Miller, S.A.; Chu, T.; Harris-White, M.E.; Cole, G.M. Phenolic anti-inflammatory antioxidant reversal of Abeta-induced cognitive deficits and neuropathology. *Neurobiol. Aging* **2001**, *22*, 993–1005. [CrossRef]

147. Cole, G.M.; Morihara, T.; Lim, G.P.; Yang, F.; Begum, A.; Frautschy, S.A. NSAID and antioxidant prevention of Alzheimer's disease: Lessons from in vitro and animal models. *Ann. N. Y. Acad. Sci.* **2004**, *1035*, 68–84. [CrossRef] [PubMed]

148. Mondragon-Rodriguez, S.; Perry, G.; Zhu, X.; Moreira, P.I.; Acevedo-Aquino, M.C.; Williams, S. Phosphorylation of tau protein as the link between oxidative stress, mitochondrial dysfunction, and connectivity failure: Implications for Alzheimer's disease. *Oxid. Med. Cell. Longev.* **2013**, *2013*. [CrossRef] [PubMed]

149. Stoothoff, W.H.; Johnson, G.V. Tau phosphorylation: Physiological and pathological consequences. *Biochim. Biophys. Acta* **2005**, *1739*, 280–297. [CrossRef] [PubMed]

150. Rane, J.S.; Bhaumik, P.; Panda, D. Curcumin Inhibits Tau Aggregation and Disintegrates Preformed Tau Filaments in vitro. *J. Alzheimer's Dis.* **2017**, *60*, 999–1014. [CrossRef] [PubMed]

151. Gella, A.; Durany, N. Oxidative stress in Alzheimer disease. *Cell Adhes. Migr.* **2009**, *3*, 88–93. [CrossRef]

152. Venigalla, M.; Gyengesi, E.; Munch, G. Curcumin and Apigenin—Novel and promising therapeutics against chronic neuroinflammation in Alzheimer's disease. *Neural Regen. Res.* **2015**, *10*, 1181–1185. [PubMed]

153. Liu, Z.J.; Li, Z.H.; Liu, L.; Tang, W.X.; Wang, Y.; Dong, M.R.; Xiao, C. Curcumin Attenuates Beta-Amyloid-Induced Neuroinflammation via Activation of Peroxisome Proliferator-Activated Receptor-Gamma Function in a Rat Model of Alzheimer's Disease. *Front. Pharmacol.* **2016**, *7*, 261. [CrossRef] [PubMed]

154. Zhang, X.; Tian, Y.; Zhang, C.; Tian, X.; Ross, A.W.; Moir, R.D.; Sun, H.; Tanzi, R.E.; Moore, A.; Ran, C. Near-infrared fluorescence molecular imaging of amyloid beta species and monitoring therapy in animal models of Alzheimer's disease. *Proc. Natl. Acad. Sci. USA* **2015**, *112*, 9734–9739. [CrossRef] [PubMed]

155. Ran, C.; Xu, X.; Raymond, S.B.; Ferrara, B.J.; Neal, K.; Bacskai, B.J.; Medarova, Z.; Moore, A. Design, synthesis, and testing of difluoroboron-derivatized curcumins as near-infrared probes for in vivo detection of amyloid-beta deposits. *J. Am. Chem. Soc.* **2009**, *131*, 15257–15261. [CrossRef] [PubMed]

156. Zhang, X.; Tian, Y.; Yuan, P.; Li, Y.; Yaseen, M.A.; Grutzendler, J.; Moore, A.; Ran, C. A bifunctional curcumin analogue for two-photon imaging and inhibiting crosslinking of amyloid beta in Alzheimer's disease. *Chem. Commun.* **2014**, *50*, 11550–11553. [CrossRef] [PubMed]

157. Alexander, G.E. Biology of Parkinson's disease: Pathogenesis and pathophysiology of a multisystem neurodegenerative disorder. *Dialogues Clin. Neurosci.* **2004**, *6*, 259–280. [PubMed]

158. Dauer, W.; Przedborski, S. Parkinson's disease: Mechanisms and models. *Neuron* **2003**, *39*, 889–909. [CrossRef]

159. Perez, C.A.; Tong, Y.; Guo, M. Iron Chelators as Potential Therapeutic Agents for Parkinson's Disease. *Curr. Bioact. Compd.* **2008**, *4*, 150–158. [CrossRef] [PubMed]

160. Jiang, T.F.; Zhang, Y.J.; Zhou, H.Y.; Wang, H.M.; Tian, L.P.; Liu, J.; Ding, J.Q.; Chen, S. Curcumin ameliorates the neurodegenerative pathology in A53T alpha-synuclein cell model of Parkinson's disease through the downregulation of mTOR/p70S6K signaling and the recovery of macroautophagy. *J. Neuroimmune Pharmacol.* **2013**, *8*, 356–369. [CrossRef] [PubMed]

161. Khatri, D.K.; Juvekar, A.R. Kinetics of Inhibition of Monoamine Oxidase Using Curcumin and Ellagic Acid. *Pharmacogn. Mag.* **2016**, *12*, S116–S120. [PubMed]

162. Nam, S.M.; Choi, J.H.; Yoo, D.Y.; Kim, W.; Jung, H.Y.; Kim, J.W.; Yoo, M.; Lee, S.; Kim, C.J.; Yoon, Y.S.; et al. Effects of curcumin (Curcuma longa) on learning and spatial memory as well as cell proliferation and neuroblast differentiation in adult and aged mice by upregulating brain-derived neurotrophic factor and CREB signaling. *J. Med. Food* **2014**, *17*, 641–649. [CrossRef] [PubMed]

163. Jayaraj, R.L.; Elangovan, N.; Manigandan, K.; Singh, S.; Shukla, S. CNB-001 a novel curcumin derivative, guards dopamine neurons in MPTP model of Parkinson's disease. *Biomed. Res. Int.* **2014**, *2014*. [CrossRef] [PubMed]

164. Landles, C.; Bates, G.P. Huntingtin and the molecular pathogenesis of Huntington's disease. Fourth in molecular medicine review series. *EMBO Rep.* **2004**, *5*, 958–963. [CrossRef] [PubMed]

165. Walker, F.O. Huntington's Disease. *Semin. Neurol.* **2007**, *27*, 143–150. [CrossRef] [PubMed]

166. Han, I.; You, Y.; Kordower, J.H.; Brady, S.T.; Morfini, G.A. Differential vulnerability of neurons in Huntington's disease: The role of cell type-specific features. *J. Neurochem.* **2010**, *113*, 1073–1091. [CrossRef] [PubMed]

167. Metzler, M.; Chen, N.; Helgason, C.D.; Graham, R.K.; Nichol, K.; McCutcheon, K.; Nasir, J.; Humphries, R.K.; Raymond, L.A.; Hayden, M.R. Life without huntingtin: Normal differentiation into functional neurons. *J. Neurochem.* **1999**, *72*, 1009–1018. [CrossRef] [PubMed]

168. Schulte, J.; Littleton, J.T. The biological function of the Huntingtin protein and its relevance to Huntington's Disease pathology. *Curr. Trends Neurol.* **2011**, *5*, 65–78. [PubMed]

169. Zuccato, C.; Marullo, M.; Vitali, B.; Tarditi, A.; Mariotti, C.; Valenza, M.; Lahiri, N.; Wild, E.J.; Sassone, J.; Ciammola, A.; et al. Brain-derived neurotrophic factor in patients with Huntington's disease. *PLoS ONE* **2011**, *6*, e22966. [CrossRef] [PubMed]

170. Aronin, N.; Chase, K.; Young, C.; Sapp, E.; Schwarz, C.; Matta, N.; Kornreich, R.; Landwehrmeyer, B.; Bird, E.; Beal, M.F.; et al. CAG expansion affects the xpression of mutant Huntingtin in the Huntington's disease brain. *Neuron* **1995**, *15*, 1193–1201. [CrossRef]

171. Miller, B.R.; Bezprozvanny, I. Corticostriatal circuit dysfunction in Huntington's disease: Intersection of glutamate, dopamine and calcium. *Future Neurol.* **2010**, *5*, 735–756. [CrossRef] [PubMed]

172. Mangiarini, L.; Sathasivam, K.; Seller, M.; Cozens, B.; Harper, A.; Hetherington, C.; Lawton, M.; Trottier, Y.; Lehrach, H.; Davies, S.W.; et al. Exon 1 of the HD gene with an expanded CAG repeat is sufficient to cause a progressive neurological phenotype in transgenic mice. *Cell* **1996**, *87*, 493–506. [CrossRef]

173. Spires, T.L.; Grote, H.E.; Garry, S.; Cordery, P.M.; Van Dellen, A.; Blakemore, C.; Hannan, A.J. Dendritic spine pathology and deficits in experience-dependent dendritic plasticity in R6/1 Huntington's disease transgenic mice. *Eur. J. Neurosci.* **2004**, *19*, 2799–2807. [CrossRef] [PubMed]

174. Nithianantharajah, J.; Barkus, C.; Murphy, M.; Hannan, A.J. Gene-environment interactions modulating cognitive function and molecular correlates of synaptic plasticity in Huntington's disease transgenic mice. *Neurobiol. Dis.* **2008**, *29*, 490–504. [CrossRef] [PubMed]

175. Taylor-Robinson, S.D.; Weeks, R.A.; Sargentoni, J.; Marcus, C.D.; Bryant, D.J.; Harding, A.E.; Brooks, D.J. Evidence for glutamate excitotoxicity in Huntington's disease with proton magnetic resonance spectroscopy. *Lancet* **1994**, *343*, 1170. [CrossRef]

176. Quintanilla, R.A.; Johnson, G.V. Role of mitochondrial dysfunction in the pathogenesis of Huntington's disease. *Brain Res. Bull.* **2009**, *80*, 242–247. [CrossRef] [PubMed]

177. Rubinsztein, D.C.; Carmichael, J. Huntington's disease: Molecular basis of neurodegeneration. *Expert Rev. Mol. Med.* **2003**, *5*, 1–21. [CrossRef] [PubMed]

178. Sandhir, R.; Yadav, A.; Mehrotra, A.; Sunkaria, A.; Singh, A.; Sharma, S. Curcumin nanoparticles attenuate neurochemical and neurobehavioral deficits in experimental model of Huntington's disease. *Neuromol. Med.* **2014**, *16*, 106–118. [CrossRef] [PubMed]

179. Liu, Z.; Dou, W.; Zheng, Y.; Wen, Q.; Qin, M.; Wang, X.; Tang, H.; Zhang, R.; Lv, D.; Wang, J.; et al. Curcumin upregulates Nrf2 nuclear translocation and protects rat hepatic stellate cells against oxidative stress. *Mol. Med. Rep.* **2016**, *13*, 1717–1724. [CrossRef] [PubMed]

180. Bekinschtein, P.; Cammarota, M.; Katche, C.; Slipczuk, L.; Rossato, J.I.; Goldin, A.; Izquierdo, I.; Medina, J.H. BDNF is essential to promote persistence of long-term memory storage. *Proc. Natl. Acad. Sci. USA* **2008**, *105*, 2711–2716. [CrossRef] [PubMed]

181. Dikshit, P.; Goswami, A.; Mishra, A.; Nukina, N.; Jana, N.R. Curcumin enhances the polyglutamine-expanded truncated N-terminal huntingtin-induced cell death by promoting proteasomal malfunction. *Biochem. Biophys. Res. Commun.* **2006**, *342*, 1323–1328. [CrossRef] [PubMed]

182. Dikshit, P.; Goswami, A.; Mishra, A.; Chatterjee, M.; Jana, N.R. Curcumin induces stress response, neurite outgrowth and prevent NF-kappaB activation by inhibiting the proteasome function. *Neurotox. Res.* **2006**, *9*, 29–37. [CrossRef] [PubMed]

183. Jana, N.R.; Dikshit, P.; Goswami, A.; Nukina, N. Inhibition of proteasomal function by curcumin induces apoptosis through mitochondrial pathway. *J. Biol. Chem.* **2004**, *279*, 11680–11685. [CrossRef] [PubMed]

184. Prusiner, S.B. The prion diseases. *Brain Pathol.* **1998**, *8*, 499–513. [CrossRef] [PubMed]

185. Brundin, P.; Melki, R.; Kopito, R. Prion-like transmission of protein aggregates in neurodegenerative diseases. *Nat. Rev. Mol. Cell Biol.* **2010**, *11*, 301–307. [CrossRef] [PubMed]

186. Gambetti, P.; Kong, Q.; Zou, W.; Parchi, P.; Chen, S.G. Sporadic and familial CJD: Classification and characterisation. *Br. Med. Bull.* **2003**, *66*, 213–239. [CrossRef] [PubMed]

187. Shang, H.S.; Chang, C.H.; Chou, Y.R.; Yeh, M.Y.; Au, M.K.; Lu, H.F.; Chu, Y.L.; Chou, H.M.; Chou, H.C.; Shih, Y.L.; et al. Curcumin causes DNA damage and affects associated protein expression in HeLa human cervical cancer cells. *Oncol. Rep.* **2016**, *36*, 2207–2215. [CrossRef] [PubMed]

188. Cao, J.; Jia, L.; Zhou, H.M.; Liu, Y.; Zhong, L.F. Mitochondrial and nuclear DNA damage induced by curcumin in human hepatoma G2 cells. *Toxicol. Sci.* **2006**, *91*, 476–483. [CrossRef] [PubMed]

189. Dance-Barnes, S.T.; Kock, N.D.; Moore, J.E.; Lin, E.Y.; Mosley, L.J.; D'Agostino, R.B., Jr.; McCoy, T.P.; Townsend, A.J.; Miller, M.S. Lung tumor promotion by curcumin. *Carcinogenesis* **2009**, *30*, 1016–1023. [CrossRef] [PubMed]

190. Ravindran, J.; Prasad, S.; Aggarwal, B.B. Curcumin and cancer cells: How many ways can curry kill tumor cells selectively? *AAPS J.* **2009**, *11*, 495–510. [CrossRef] [PubMed]

191. Liddle, M.; Hull, C.; Liu, C.; Powell, D. Contact urticaria from curcumin. *Dermatitis* **2006**, *17*, 196–197. [CrossRef] [PubMed]

192. Burgos-Moron, E.; Calderon-Montano, J.M.; Salvador, J.; Robles, A.; Lopez-Lazaro, M. The dark side of curcumin. *Int. J. Cancer* **2010**, *126*, 1771–1775. [CrossRef] [PubMed]

193. Kim, D.C.; Ku, S.K.; Bae, J.S. Anticoagulant activities of curcumin and its derivative. *BMB Rep.* **2012**, *45*, 221–226. [CrossRef] [PubMed]

194. Sharma, R.A.; Euden, S.A.; Platton, S.L.; Cooke, D.N.; Shafayat, A.; Hewitt, H.R.; Marczylo, T.H.; Morgan, B.; Hemingway, D.; Plummer, S.M.; et al. Phase I clinical trial of oral curcumin: Biomarkers of systemic activity and compliance. *Clin. Cancer Res.* **2004**, *10*, 6847–6854. [CrossRef] [PubMed]

195. Rasyid, A.; Lelo, A. The effect of curcumin and placebo on human gall-bladder function: An ultrasound study. *Aliment. Pharmacol. Ther.* **1999**, *13*, 245–249. [CrossRef] [PubMed]

196. Rasyid, A.; Rahman, A.R.; Jaalam, K.; Lelo, A. Effect of different curcumin dosages on human gall bladder. *Asia Pac. J. Clin. Nutr.* **2002**, *11*, 314–318. [CrossRef] [PubMed]

197. Ringman, J.M.; Frautschy, S.A.; Teng, E.; Begum, A.N.; Bardens, J.; Beigi, M.; Gylys, K.H.; Badmaev, V.; Heath, D.D.; Apostolova, L.G.; et al. Oral curcumin for Alzheimer's disease: Tolerability and efficacy in a 24-week randomized, double blind, placebo-controlled study. *Alzheimer's Res. Ther.* **2012**, *4*, 43. [CrossRef] [PubMed]

198. Hishikawa, N.; Takahashi, Y.; Amakusa, Y.; Tanno, Y.; Tuji, Y.; Niwa, H.; Murakami, N.; Krishna, U.K. Effects of turmeric on Alzheimer's disease with behavioral and psychological symptoms of dementia. *Ayu* **2012**, *33*, 499–504. [CrossRef] [PubMed]

199. Goozee, K.G.; Shah, T.M.; Sohrabi, H.R.; Rainey-Smith, S.R.; Brown, B.; Verdile, G.; Martins, R.N. Examining the potential clinical value of curcumin in the prevention and diagnosis of Alzheimer's disease. *Br. J. Nutr.* **2016**, *115*, 449–465. [CrossRef] [PubMed]

200. Small, G.W.; Siddarth, P.; Li, Z.; Miller, K.J.; Ercoli, L.; Emerson, N.D.; Martinez, J.; Wong, K.P.; Liu, J.; Merrill, D.A.; et al. Memory and Brain Amyloid and Tau Effects of a Bioavailable Form of Curcumin in Non-Demented Adults: A Double-Blind, Placebo-Controlled 18-Month Trial. *Am. J. Geriatr. Psychiatry* **2018**, *26*, 266–277. [CrossRef] [PubMed]

201. Cox, K.H.; Pipingas, A.; Scholey, A.B. Investigation of the effects of solid lipid curcumin on cognition and mood in a healthy older population. *J. Psychopharmacol.* **2015**, *29*, 642–651. [CrossRef] [PubMed]

International Journal of
Molecular Sciences

MDPI

Article

Comparative Transcriptome Analysis Identifies Genes Putatively Involved in 20-Hydroxyecdysone Biosynthesis in *Cyanotis arachnoidea*

Xiu Yun Lei [1], Jing Xia [1], Jian Wen Wang [1] and Li Ping Zheng [2,*]

[1]   College of Pharmaceutical Sciences, Soochow University, Suzhou 215123, China;
      18862239302@163.com (X.Y.L.); jxia@stu.suda.edu.cn (J.X.); jwwang@suda.edu.cn (J.W.W.)
[2]   Department of Horticultural Sciences, Soochow University, Suzhou 215123, China
*   Correspondence: lpzheng@suda.edu.cn; Tel.: +86-521-6588-0195

Received: 9 May 2018; Accepted: 23 June 2018; Published: 27 June 2018

Abstract: *Cyanotis arachnoidea* contains a rich array of phytoecdysteroids, including 20-hydroxyecdysone (20E), which displays important agrochemical, medicinal, and pharmacological effects. To date, the biosynthetic pathway of 20E, especially the downstream pathway, remains largely unknown. To identify candidate genes involved in 20E biosynthesis, the comparative transcriptome of *C. arachnoidea* leaf and root was constructed. In total, 86.5 million clean reads were obtained and assembled into 79,835 unigenes, of which 39,425 unigenes were successfully annotated. The expression levels of 2427 unigenes were up-regualted in roots with a higher accumulation of 20E. Further assignments with Kyoto Encyclopedia of Genes and Genomes (KEGG) pathways identified 49 unigenes referring to the phytoecdysteroid backbone biosynthesis (including 15 mevalonate pathway genes, 15 non-mevalonate pathway genes, and 19 genes for the biosynthesis from farnesyl pyrophosphate to cholesterol). Moreover, higher expression levels of mevalonate pathway genes in roots of *C. arachniodea* were confirmed by real-time quantitative PCR. Twenty unigenes encoding CYP450s were identified to be new candidate genes for the bioreaction from cholesterol to 20E. In addition, 90 transcription factors highly expressed in the roots and 15,315 unigenes containing 19,158 simple sequence repeats (SSRs) were identified. The transcriptome data of our study provides a valuable resource for the understanding of 20E biosynthesis in *C. arachnoidea*.

Keywords: *Cyanotis arachnoidea*; roots; leaves; transcriptome; 20-hydroxyecdysone; biosynthesis

---

1. Introduction

Cyanotis arachnoidea C. B. Clarke (Commelinaceae), a Chinese traditional medicine herb, has been use for treatment of limb numbness and rheumatoid arthritis, as well as for promoting blood circulation and muscle relaxing [1]. Rich phytoecdysteroids, including 20-hydroxyecdysone (20E), dihydroxyrubrosterone, rubrosterone, poststerone, and cyanosterone B, were found in this plant [1,2]. 20E was reported to have agricultural applications in enhancing the synchronous development of *Bombyx mori*, elevating silk yield [3] and reducing the time of the molting cycle of *Alpheus heterochelis* [4]. 20E and its derivatives exhibited a wide variety of pharmacological effects, including anti-depression, antioxidation, anti-diabetes, and neuron protection [5,6]. 20E is primarily obtained from ecdysteroid-rich plants, such as *C. arachnoidea*, *Ajuga turkestanica*, and *Serratula wolffi* [6]. Dried roots and aerial parts from *C. arachnoidea* have been reported to contain 20E at 5.50% and 0.52%, respectively [7]. However, due to the limitation of the wild resources for *C. arachnoidea*, the supply of 20E is in great shortage and its application is always restricted [6].

Originally, the ecdysteroid biosynthetic pathway has been described as a cytosolic pathway, starting from mevalonate (MVA) as a precursor [8]. Subsequently, it has been proven that isopentenyl diphosphate (IPP) and dimethylallyl diphosphate (DMAPP) required for the triterpenoid synthesis can be produced not only from the MVA pathway in cytosol but also the 2-C-methyl-D-erythritol 4-phosphate (MEP) pathway in plastids [9]. However, the studies to date on phytoecdysteroid biosynthesis are limited to the parts mediated by the MVA pathway [8]. In many plants, ecdysteroids were synthesized from cholesterol [10,11]. However, Adler and Grebenok [12] identified lathosterol as a precursor of 20E in spinach. It is obviously controversial whether cholesterol or lathosterol is the preferred substrate for phytoecdysteriod biosynthesis [8]. Conversion of a sterol to ecdysteroid requires several structural modifications, such as the hydroxylation of the sterol nucleus and side chain [12]. In insects, cytochrome P-450 monooxygenases (CYP450s), which catalyze the bioreaction from cholesterol to 20E, have been cloned and identified. However, only one CYP450 enzyme in plants has been identified to be involved in 20E biosynthesis in the hairy roots of *Ajuga reptans* [13]. To date, the biosynthesis of 20E, especially the origin of the precursor IPP and its downstream pathway, remains largely unknown.

The accumulation of 20E in *C. arachnoidea* was different among various tissues. Mu et al. [7] reported that the 20E content in roots was 4–20 times higher than that in leaves. Biosynthesis and accumulation of secondary metabolites are often tissue-specific [14]. Tomás et al. [6] reported that the ecdysteroid content was related to organized structures of whole plants and in vitro propagated plantlets in *A. reptans*, while calluses from leaves or roots could not produce ecdysteroids [15]. The cultured plantlets of *A. reptans* produced seven ecdysteroids with the tissue differentiation. Tomas et al. has revealed that ecdysteroid production is root-specific in *A. reptans* [16]. In recent years, comparative transcriptome analysis has been applied to investigate the biosynthesis of secondary metabolites. Yang et al. compared the transcriptome of leaves and roots from *Salvia miltiorrhiza* and identified candidate genes involved in the biosynthesis of tanshinones [17]. Biosynthesis of withanolide A, a medicinal component synthesized specifically in roots of *Withania somnifrea*, was also investigated by comparative transcriptome analysis [18]. Despite the important pharmacological value of 20E, there is little genetic information revealing the biosynthesis of 20E in *C. arachnoidea*. Thus, as a follow-up to our previous works on the biotechnological production of 20E [19] and cloning of key enzyme genes in 20E biosynthesis [20], we conducted a comparative transcriptome analysis of the leaves and roots of *C. arachnoidea* to identify genes putatively involved in 20E biosynthesis. As we know, this study reported the de novo sequencing of *C. arachnoidea* for the first time. Meanwhile, transcription factors (TFs), which have been found to regulate the secondary metabolism in plants, were also searched for in the transcriptome database. Furthermore, we provided valuable information for developing the important molecular marker of simple sequence repeats (SSRs). This study was designed to characterize the transcriptome of *C. arachnoidea* and provide a valuable basis for further elucidating the biosynthetic pathway of 20E.

2. Results and Discussion

2.1. Quantification of 20E from C. arachnoidea

As shown in Figure 1, 20E was found in both *C. arachnoidea* leaves and roots at different levels. The 20E contents were much (i.e., 6-fold) higher in roots (18.0 mg/g DW) than in leaves (3.0 mg/g DW). It has been reported that the ecdysteroid content varies in different organs of the plants [6]. 20E content in roots of *C. arachnoidea* was reported to be 4–20 times higher than that in leaves [7]. Zhu et al. [21] demonstrated that the 20E content of *C. arachnoidea* roots was the highest in air-dried whole plants, amounting to 2.9% (w/w). Although 20E production is not root-specific in *C. arachnoidea*, such spatial distribution of 20E suggests a root preferential expression pattern of biosynthetic pathway genes.

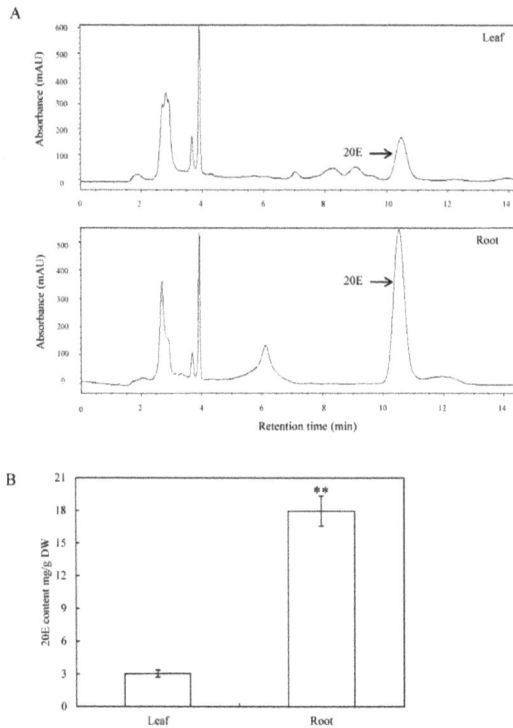

Figure 1. (**A**) Typical high-performance liquid chromatography (HPLC) chromatograms and (**B**) 20E contents in leaves and roots of *C. arachnoidea*. 20E was isolated from four-month old *C. arachnoidea*. Data are expressed as means ± standard deviation of three replicates. Statistical significance in comparison with the corresponding control values is indicated by ** $p < 0.01$.

2.2. Library Sequencing and De Novo Assembly and Annotation

To determine the transcriptome of *C. arachnoidea* leaves and roots, two cDNA libraries were established and sequenced using a HiSeq™2500 platform (Illumina, San Diego, CA, USA). Finally, 45,368,044 and 41,095,316 clean reads were obtained from leaves and roots, respectively (Table 1). We generated a unique transcript library for *C. arachnoidea* using both sets of reads and obtained a total of 79,835 unigenes with an average length of 894 bp and a N_{50} length of 1268 bp. The length distribution of unigenes was presented in Figure S1. Clean reads from leaves and roots were separately mapped to the *C. arachnoidea* transcript library. Finally, 69,782 and 70,556 unigenes were obtained for the leaves and roots, respectively (Table 1).

Table 1. Summary of the sequencing data of the leaf and root transcriptome of *C. arachnoidea*.

The Sequencing Data	Leaf	Root
Clean bases	5,671,005,500	5,136,914,500
Clean reads	45,368,044	41,095,316
Q30 (%)	90.63	89.90
GC content (%)	43.50	44.50
Unigenes ($\geq$ 300 bp)	69,782	70,556
All unigenes ($\geq$ 300 bp)	79,835	
N_{50} (bp)	1268	
Average length (bp)	894	

For sequence annotation, BLASTX alignment was performed against non-redundant (NR), Swiss-Prot, Gene Ontology (GO) and Kyoto Encyclopedia of Genes and Genomes (KEGG) databases. As a result (Figure S2), 39,057 (48.92%) unigenes received gene descriptions through comparisons with known proteins in the NR database, and 49.55% of annotated unigenes exhibited strong homology (E value < $1.0 \times e^{-60}$). Most of these annotated unigenes (30.23%) were best matched to sequences from *Musa acuminata* subsp. Malaccensis, followed by *Elaeis guineensis* (19.74%), and *Phoenix doctylifera* (16.04%). A sum of 28,791 unigenes (36.06%) was annotated using the Swiss-Prot database. Additionally, 27,111 (33.96%) and 9922 (12.43%) unigenes were annotated using the GO and KEGG databases, respectively. Figure S3a showed the number of unigenes annotated by all the databases.

2.3. Different Expression Genes (DEGs) Identified Between Leaves and Roots

The transcript abundance of each unigene was calculated through the fragments per kilobase per million reads (FPKM) method and those with $|\log_2 \text{FoldChange}| \geq 1$, with a p value ≤ 0.05, were considered as DEGs. Finally, a sum of 4811 DEGs was identified in which 2427 unigenes were up-regulated while 2384 unigenes were down-regulated in *C. arachnoidea* roots (Table S2). In addition, we identified 1180 DEGs which were only expressed in leaves and 1250 DEGs expressed specifically in roots (Figure S3b).

2.4. Identification of Unigenes Involved in Secondary Metabolism

The biological function and pathway assignment of unigenes were conducted through KEGG pathway annotation. In the end, 9922 (12.43%) unigenes were mapped to five categories (i.e., organismal systems, metabolism, genetic information processing, environmental information processing, and cellular processes) and 32 sub-categories (Figure S4). The higher proportion of unigenes belonged to the pathway of signal transduction, carbohydrate metabolism, and translation. The majority of unigenes were classified into the metabolism category, and the number of unigenes related to different secondary metabolisms was listed in Table 2. Out of all unigenes involved in secondary metabolite pathways, unigenes in "Phenylpropanoid biosynthesis (ko00940)" had the highest proportion, followed by "Stilbenoid, diarylheptanoid, and gingerol biosynthesis (ko00945)", "Terpenoid backbone biosynthesis (ko00900)", and "Flavonoid biosynthesis (ko00941)". Additionally, 59 unigenes were mapped to the "Steroid biosynthesis (ko00100)" pathway (Figure S5). Among 1070 unigenes related to secondary metabolism, 192 unigenes were differentially expressed in *C. arachnoidea* leaves and roots.

Table 2. Numbers of unigene related to the secondary metabolism biosynthetic pathway in *C. arachnoidea*.

Secondary Metabolites Biosynthesis Pathway	Total	Up-Regulated Unigenes	Down-Regulated Unigenes
Phenylpropanoid biosynthesis (ko00940)	289	62	17
Stilbenoid, diarylheptanoid, and gingerol biosynthesis (ko00945)	85	9	5
Terpenoid backbone biosynthesis (ko00900)	79	1	1
Flavonoid biosynthesis (ko00941)	79	2	12
Limonene and pinene degradation (ko00903)	68	4	8
Ubiquinone and other terpenoid-quinone biosynthesis (ko00130)	67	4	12
Isoquinoline alkaloid biosynthesis (ko00950)	63	8	2
Steroid biosynthesis (ko00100)	59	4	2
Carotenoid biosynthesis (ko00906)	43	1	6
Tropane, piperidine and pyridine alkaloid biosynthesis (ko00960)	39	1	2
Brassinosteroid biosynthesis (ko00905)	37	7	0
Diterpenoid biosynthesis (ko00904)	35	3	2
Zeatin biosynthesis (ko00908)	34	3	1
Flavone and flavonol biosynthesis (ko00944)	24	0	1
Isoflavonoid biosynthesis (ko00943)	16	0	1
Monoterpenoid biosynthesis (ko00902)	14	1	4

Table 2. *Cont.*

Secondary Metabolites Biosynthesis Pathway	Total	Up-Regulated Unigenes	Down-Regulated Unigenes
Sesquiterpenoid and triterpenoid biosynthesis (ko00909)	12	3	1
Anthocyanin biosynthesis (ko00942)	8	0	0
Betalain biosynthesis (ko00965)	6	0	0
Geraniol degradation (ko00281)	5	0	0
Glucosinolate biosynthesis (ko00966)	4	1	0
Caffeine metabolism (ko00232)	2	0	0
Polyketide sugar unit biosynthesis (ko00523)	1	0	0
Indole alkaloid biosynthesis (ko00901)	1	1	0
Total	1070	115	77

The transcript abundance of each unigene was calculated through the FPKM method, and those with $|\log_2$ FoldChange$| \geq 1$, with a p value ≤ 0.05, were considered as DEGs. Positive and negative $\log_2$ FoldChange values indicates up- and down-regulated unigenes, respectively.

2.5. Identification of Unigenes Related to 20E Biosynthesis

20E is an isoprenoid-derived compound using five-carbon building units of IPP or its possible isomer, DMAPP, which can be synthesized by both the MVA and MEP pathway in plants. Previous reports showed that phytoecdysteroid biosynthesis is limited to parts mediated via the MVA pathway [8]. In the *C. arachnoidea* transcriptome, unigenes encoding all the known structural enzymes in the upstream pathways up to IPP were found, including 11 unigenes for six enzymes in the MVA pathway, 15 unigenes for seven enzymes in the MEP pathway, and four unigenes for isopentenyl diphosphate isomerase (IDI) (Step 1 in Table 3 and Figure 2). According to the FPKM value, unigenes involved in the MVA pathway mostly showed higher expression levels in *C. arachnoidea* roots, such as *AACT* (comp70319_c0_seq8), *PMK* (comp65552_c0_seq7), and *PMD* (comp59477_c0_seq1). On the contrary, unigenes involved in the MEP pathway had higher expression level in leaves, such as *DXR* (comp47417_c0_seq1, comp74054_c0_seq1), *MCS* (comp30840_c0_seq1), *HDS* (comp33211_c0_seq1, comp67980_c0_seq1), and *HDR* (comp73401_c0_seq1). Moreover, the expression profile of unigenes involved in the MVA and MEP pathways were validated through real-time quantitative PCR (RT-qPCR) (Figure 3). Unigenes with the |relative expression level| ≥ 2 were presented in Figure 3A. The expression level of *IDI* was up-regulated 3.82-fold in roots. Expression of *ACTT* was up-regulated in roots by 3.80-fold. Additionally, unigenes encoding HMGR, HMGS, PKM, and PMD involved in the MVA pathway all had relative higher expression levels in *C. arachniodea* roots than in leaves. Except for DXS, unigenes encoding DXR and HDS involved in the MEP pathway were down-regulated in *C. arachnoidea* roots (Figure 3A). Because 20E is highly accumulated in roots, the genes encoding enzymes involved in 20E biosynthesis are expected to show a root-preferential expression pattern. These results implied that the backbone of 20E could be synthesized mainly through the MVA pathway rather than the MEP pathway in roots.

In many plants, ecdysteroids are synthesized from the C27 sterol cholesterol [22]. Additionally, lathosterol was also proved as a precursor for ecdysteroid biosynthesis [12]. The result of the pathway assignment mentioned above showed that 59 unigenes were mapped to the "Steroid biosynthesis (ko00100)" pathway. According to the map showed in Figure S5, the downstream steps of 20E biosynthesis were presented in Step 2 of Figure 2. Unigenes encoding most of the enzymes involved in lathosterol and cholesterol biosynthesis were found in our transcriptome database (Step 2 in Table 3). These unigenes showed higher expression levels in roots compared to leaves. For example, *SQLE* (comp35298_c0_seq1), *ERG2* (comp52043_c0_seq5), *DHCR24* (comp29143_c0_seq1, comp48248_c0_seq1), and *EBP* (comp47084_c0_seq3) all had higher FPKM values in roots. To validate the expression level of these unigenes, RT-qPCR analysis was conducted. As shown in Figure 3B, the expression level of *DHCR24* was 673.45-fold higher in roots than in leaves. The expression of *EBP* and *ERG24* was up-regulated in roots by 14.8- and 11.4-fold, respectively. In all, we identified 49 unigenes encoding 24 enzymes involved in 20E biosynthesis in *C. arachniodea*.

Int. J. Mol. Sci. **2018**, 19, 1885

Table 3. The FPKM value of unigenes which encode enzymes involved in 20E biosynthesis.

Enzymes	Enzyme Commission Number	Unigenes ID	FPKM Value	
			Leaf	Root
Step 1: MVA pathway				
acetyl CoA acetyltransferase (AACT)	2.3.1.9	comp60224_c0_seq7	1.35	4.97
		comp70319_c0_seq8	36.31	114.60
3-hydroxy-3-methyl-glutaryl coenzyme A synthase (HMGS)	2.3.3.10	comp69323_c0_seq6	15.40	26.53
3-hydroxy-3-methyl-glutaryl coenzyme A reductase (HMGR)	1.1.1.34	comp68996_c0_seq2	59.92	55.43
		comp52491_c0_seq3	0.29	2.98
		comp31434_c0_seq1	0	1.82
mevalonate kinase (MVK)	2.7.1.36	comp69732_c0_seq13	10.58	5.18
		comp120320_c0_seq1	0	1.10
phosphomevalonate kinase (PMK)	2.7.4.2	comp65552_c0_seq7	35.22	69.05
diphosphomevalonate decarboxylase (PMD)	4.1.1.33	comp33484_c0_seq1	29.43	28.81
		comp59477_c0_seq1	30.56	72.50
MEP pathway				
1-deoxy-D-xylulose-5-phosphate synthase (DXS)	2.2.1.7	comp66526_c0_seq5	17.96	9.56
		comp24131_c0_seq1	2.04	0.25
		comp71989_c1_seq3	6.86	6.63
		comp59095_c0_seq1	3.92	12.75
1-deoxy-D-xylulose-5-phosphate reductoisomerase (DXR)	1.1.1.267	comp47417_c0_seq1	41.16	0
		comp71987_c0_seq5	31.46	22.10
		comp74054_c0_seq1	59.57	35.26
2-C-methyl-D-erythritol4-phosphate cytidylyi transferase (MCT)	2.7.7.60	comp71928_c2_seq77	21.02	15.51
4-diphosphocytidyl-2-C-methyl-D-erythritol kinase (CMK)	2.7.1.148	comp74100_c0_seq1	24.09	10.73
2-C-methyl-D-erythritol-2,4-cyclodiphosphate synthase (MCS)	4.6.1.12	comp30840_c0_seq1	88.79	22.58
hydroxy-2-methyl-2-(E)-butenyl 4-diphosphate synthase (HDS)	1.17.7.1	comp33211_c0_seq1	30.23	4.26
		comp53197_c0_seq1	7.95	2.19
		comp67980_c0_seq1	91.56	39.75
hydroxy-2-methyl-2-(E)-butenyl 4-diphosphate reductase (HDR)	1.17.1.2	comp90900_c0_seq1	0.82	0.69
		comp73401_c0_seq1	331.58	174.81
isopentenyl diphosphate isomerase (IDI)	5.3.3.2	comp64868_c0_seq4	24.51	13.51
		comp61897_c0_seq2	8.51	10.42
		comp29765_c0_seq1	106.24	337.84
		comp74567_c0_seq1	19.86	25.58

Table 3. *Cont.*

Enzymes	Enzyme Commission Number	Unigenes ID	FPKM Value	
			Leaf	Root
Step 2				
farnesyl diphosphate synthase (FDS)	2.5.1.10	comp96291_c0_seq1	0.47	0.57
		comp60714_c0_seq9	6.56	3.03
		comp28732_c0_seq1	71.68	87.98
squalene synthase (SQS)	2.5.1.21	comp68019_c0_seq1	105.92	144.18
squalene monooxygenase (SQLE)	1.14.13.132	comp35298_c0_seq1	201.88	387.62
sterol 14-demethylase (CYP51)	1.14.13.72	comp67692_c0_seq1	33.34	24.69
		comp67692_c1_seq3	335.08	250.89
δ14-sterol reductase (ERG24)	1.3.1.70	comp52043_c0_seq5	21.53	69.73
sterol-4α-carboxylate 3-dehydrogenase (ERG26)	1.1.1.170	comp40656_c0_seq1	1.25	1.91
		comp63129_c0_seq5	8.84	18.50
		comp48599_c0_seq1	22.32	23.31
		comp69306_c0_seq26	7.47	29.19
δ24-sterol reductase (DHCR24)	1.3.1.72	comp29143_c0_seq1	136.31	215.60
		comp48248_c0_seq1	1.45	272.54
cholestenol δ-isomerase (EBP)	5.3.3.5	comp47084_c0_seq3	10.87	59.19
		comp60181_c0_seq1	40.63	54.68
δ7-sterol 5-desaturase (ERG3)	1.14.19.20	comp50441_c0_seq1	5.89	5.25
		comp61289_c0_seq4	28.21	24.24
7-dehydrocholesterol reductase (DHCR7)	1.3.1.21	comp71639_c2_seq1	55.04	56.70

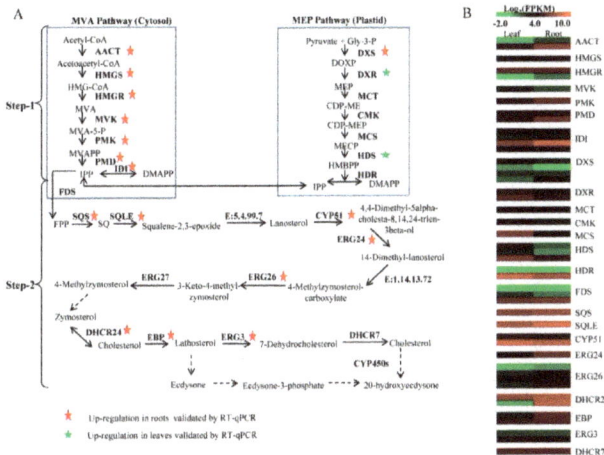

Figure 2. (**A**) The putative pathway of 20E biosynthesis in *C. arachnoidea*. (**B**) The gene expression was validated by RT-qPCR. The solid arrows denote known steps while the dashed arrows denote multi-step and incomplete reactions. Enzymes related to the different steps are shown between the reactions catalyzed. The expression level of unigenes related to these enzymes in leaf and root is shown by a heatmap. Names of the enzymes are provided in Table 3.

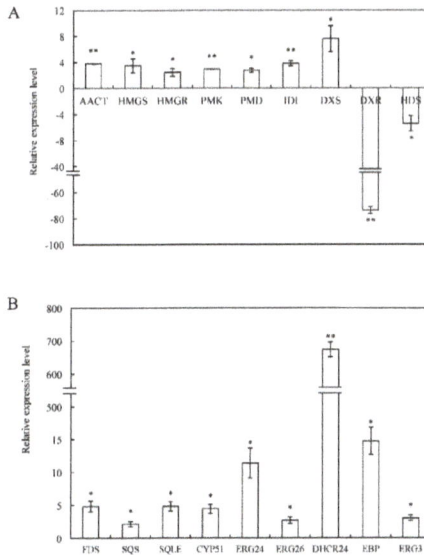

Figure 3. RT-qPCR analysis of selected genes involved in 20E biosynthesis. (**A**) Unigenes involved in terpenoid backbone biosynthesis. (**B**) Unigenes involved in steroid biosynthesis. Leaves and roots of four-month old *C. arachnoidea* were collected and their total RNAs were used for RT-qPCR analysis. Relative expression level of unigenes in roots was calculated by the $2^{-\Delta\Delta Ct}$ method with actin as an internal control and was compared to those in leaves, which were all set to be 1. The negative value represents higher expression levels of unigenes in leaves than in roots. The error bar indicated standard deviations of three biological replicates. Asterisks represented significant differences: $* p < 0.05$ and $** p < 0.01$. The full names of the abbreviations (unigenes) are provided in Table 3.

2.6. Validation of Differentially Expressed CYP450s

Conversion of a sterol to ecdysteroid requires several structural modifications, such as the hydroxylation of the sterol nucleus and side chain [12]. In insects, the biosynthesis of 20E from cholesterol was catalyzed by CYP450 enzymes [23]. The hydroxylation steps in plants are presumably mediated by CYP450s as they are in insects [12]. Tsukagoshi et al. [13] identified a CYP450 enzyme CYP71D443, which catalyzes the C-22 hydroxylation of 20E in *Ajuga* hairy roots. There was only one reported CYP450 enzyme to be involved in the 20E biosynthesis of plants so far [13]. In order to identify the possible involvement of CYP450s in 20E biosynthesis, up-regulated CYP450s unigenes in roots were selected from DEGs. As a result, 20 unigenes encoding CYP450s were identified and their up-regulated expression levels in roots were validated using RT-qPCR (Table 4). For example, the expression level of unigene comp36620_c0_seq1 was 1666.05-fold higher in roots than in leaves. The expression level of comp67837_c0_seq4 in roots was 597.69-fold higher than in leaves. Much work needs be done to narrow down these candidate CYP450 genes to reveal the genes responsible for the bioreaction from cholesterol to 20E.

Table 4. Validation of the expression levels of CYP450 unigenes by using RT-qPCR

Unigene ID	Description	Foldchange	RT-qPCR
comp36620_c0_seq1	cytochrome P450 94C1-like	Inf	$(1.67 \pm 0.13) \times 10^3$ **
comp56632_c0_seq1	cytochrome P450 90B1-like	Inf	38.29 ± 10.74 *
comp60467_c1_seq1	cytochrome P450 710A1-like	Inf	49.69 ± 0.24 **
comp70580_c0_seq1	cytochrome P450 704C1	Inf	32.05 ± 7.01 *
comp28108_c0_seq1	cytochrome P450 CYP736A12	1013.73	$(1.26 \pm 0.27) \times 10^2$ *
comp30002_c0_seq1	cytochrome P450 90A1 isoform X2	388.30	$(1.83 \pm 0.63) \times 10^2$
comp67837_c0_seq4	cytochrome P450 86B1-like isoform X2	147.93	$(5.98 \pm 0.71) \times 10^2$ **
comp57563_c0_seq2	cytochrome P450 71A1-like	124.00	19.39 ± 1.42 **
comp61518_c0_seq1	cytochrome P450 90A1-like	115.05	$(4.67 \pm 0.34) \times 10^2$ **
comp71681_c0_seq1	cytochrome P450 71A1-like	101.70	$(4.15 \pm 0.10) \times 10^2$ **
comp69723_c0_seq2	cytochrome P450 86B1-like	82.30	43.66 ± 8.29 *
comp60312_c0_seq1	cytochrome P450 734A1-like	56.02	$(1.12 \pm 0.07) \times 10^2$ **
comp60108_c0_seq3	cytochrome P450 90B1-like	47.42	73.56 ± 3.60 **
comp60517_c0_seq1	cytochrome P450 734A6-like	39.25	2.10 ± 0.19 **
comp66356_c0_seq1	cytochrome P450 71A1-like	33.79	18.72 ± 2.78 **
comp70580_c1_seq4	cytochrome P450 704C1-like	22.36	24.68 ± 3.93 **
comp60316_c0_seq2	cytochrome P450 714B3-like	21.86	33.65 ± 4.75 **
comp66583_c0_seq3	cytochrome P450 714D1-like	19.16	19.30 ± 7.73
comp67012_c0_seq1	cytochrome P450 734A6-like	16.32	17.48 ± 4.38 **
comp70760_c4_seq11	Cytochrome P450 86A1	12.32	$(2.99 \pm 0.07) \times 10^2$ **

Foldchange equals the ratio of S1/S2. S1 means the FPKM value of a unigene in roots while S2 means the FPKM value of a unigene in leaves. RT-qPCR represents the relative expression level of unigenes in roots measured by RT-qPCR compared to those in leaves. Values are reported as mean $\pm$ SD from three independent experiments. Asterisks represented significant differences: * $p < 0.05$ and ** $p < 0.01$.

2.7. Transcription Factors Predicted and Statistics of Simple Sequence Repeats

In our research, TFs were searched for in the *C. arachnoidea* transcriptome, and a total of 1312 TFs were found (Table 5). Among all the TFs identified, basic helix-loop-helix (bHLH), ethylene response factor (ERF), and C2H2 -type zinc finger family members were of a higher proportion. Furthermore, 90 TFs were up-regulated while 58 TFs were down-regulated in *C. arachnoidea* roots. Among the up-regulated TFs, MYBranked the highest (15), followed by bHLH (13), and WRKY (10). In plants, TFs have been found to regulate secondary metabolism pathways [14]. For example, an ERF transcription factor named JRE4 stimulates the biosynthesis of steroidal glycoalkaloids (SGAs), cholesterol-derived metabolites, through activating the transcription of SGA biosynthetic genes, such as *HMGS, HMGR, IDI,* and *SQS* [24]. Considering the root-preferential accumulation of 20E, these up-regulated TFs were worth further investigating for their function in 20E biosynthesis.

Table 5. Summary of transcription factor (TF) unigenes in *C. arachnoidea*

TF Family	Number of Unigenes Detected	Up-Regulated in Roots	Down-Regulated in Roots
bHLH	146	13	10
ERF	80	3	1
C2H2	79	6	3
NAC	78	4	4
WRKY	75	10	5
MYB-related	74	8	4
bZIP	63	7	0
G2-like	60	4	2
MYB	60	15	1
HD-ZIP	49	2	7
Others	548	18	21
Total	1312	90	58

The transcript abundance of each unigene was calculated through the FPKM method, and those with $|\log_2$ FoldChange$| \geq 1$ with a p value ≤ 0.05 were considered as DEGs. Positive and negative $\log_2$ FoldChange values indicates up- and down-regulated unigenes, respectively.

SSRs are one of the most important molecular markers and have various applications in genetics and plant breeding [25]. Thus, SSRs in *C. arachnoidea* were analyzed in our present study. Out of 79,835 unigenes, 15,315 unigenes containing 19,158 SSRs were identified (Table 6). Within these SSRs, 1194 SSRs presented in a compound formation. Mononucleotide repeats were the most abundant (14,598), followed by tri-nucleotide repeats (2270), and di-nucleotide repeats (2100). Penta-nucleotide repeats were the least (17). We also checked the SSRs in unigenes involved in 20E biosynthesis. Unigenes comp68996_c0_seq2 (*HMGR*) and comp63129_c0_seq5 (*ERG26*) contained compound SSRs. Comp71987_c0_seq5 encoding *DXR* had a di-nucleotide repeat. Additionally, the eight unigenes encoding AACT, HMGS, DXS, DXR, MCT, HDS, DHCR24, and EBP contained mononucleotide repeats. SSRs at different positions in a gene can help determine the regulation of gene expression and the function of the protein produced [25]. Sharopova [26] reported that genes containing five or more SSRs had the highest average level of expression in *Arabidopsis*. In rice, amylase content was correlated with a variation in the number of SSRs in the 5′-untranslated region of the waxy gene [27]. These SSRs identified in the *C. arachnoidea* transcriptome were valuable genetic resources for future studies of this species.

Table 6. Statistics of simple sequence repeats (SSRs) identified from *C. arachnoidea* transcriptome data.

SSR Statistics	Number
Total number of sequences examined	79,835
Total size of examined sequences (bp)	71,342,686
Total number of identified SSRs	19,158
Number of SSR containing sequences	15,313
Sequences containing more than 1 SSR	3067
SSRs present in compound formation	1194
Mononucleotide repeats	14,598
Di-nucleotide repeat	2100
Tri-nucleotide repeat	2270
Tetra-nucleotide repeat	155
Penta-nucleotide repeat	17
Hexa-nucleotide repeat	18

3. Materials and Methods

3.1. Plant Materials

C. arachnoidea seeds were collected in September 2011 from the suburbs of Luquan County, Yunnan Province of China with its voucher specimen (SCU-110923), identified by C.Y. Liu. They were deposited in the herbarium of Soochow University. Germination and transplantation were performed in the horticultural nursery of Soochow University, Suzhou, China. One month after germination, the seedlings were transplanted into a plastic pot (35 × 25 × 7 cm in length, width, and height, respectively) containing sand and vermiculite in a 1:2 (v/v) ratio. Cultures were maintained in a growth chamber at 25 ± 2 °C under a 14/10-h light/dark cycle photoperiod with white fluorescent light at 1500 lux. Leaf and root tissues of four-month old C. arachnildea were collected separately, frozen in liquid nitrogen, and stored at −80 °C until use. At least three biological replicates were used for subsequent studies, and each replicate contained leaf or root tissues from at least 15 seedlings.

3.2. 20E Extraction and Analysis

The extraction and quantification of 20E in C. arachnoidea leaves and roots were conducted as described by our previous report [20] with slight modifications. Dry leaves or roots (0.5 g) were ground into a powder and extracted with 30 mL methanol under sonication for 90 min. The extract was then evaporated to dryness and dissolved in 1 mL methanol. High performance liquid chromatography (HPLC) conditions are as follows: an Aglient 1280 HPLC system equipped with 250 × 4.6 mm Aglient HC-C18 column, samples were eluted with 20:80 (v/v) acetonitrile/water at a flow rate 1 mL/min, and monitored at 242 nm. 20E was quantified with a genuine standard (Sigma, St. Louis, CA, USA). Figure 1A presented a typical chromatogram of 20E in C. arachnoidea leaves and roots under the condition.

3.3. cDNA Library Construction, Sequencing and Quality Control

For cDNA library construction, total RNA was firstly extracted from leaf or root tissues of four-month old C. arachnildea using a mirVana™ RNA isolation kit (Applied Biosystems, Foster City, CA, USA) and then treated with DNase I for 30 min at 37 °C. RNA integrity and purity was confirmed using the Agilent 2100 Bioanalyzer (Agilent Technologies, Palo Alto, CA, USA). For each pool, total RNA (10 μg) was prepared for the cDNA library. The mRNA was purified from total RNA using magnetic beads with Oligo (dT). Subsequently, the mRNA was sheared into small fragments for cDNA synthesis. Double-stranded cDNA was synthesized using random hexamers. These cDNA fragments were subjected to an end repair process, the ligation of adapters, and were enriched by PCR to create the final libraries. Paired-end sequencing at 125 bp was performed using a HiSeq™2500 platform (Illumina, San Diego, CA, USA). High-quality reads were obtained by removing adaptor fragments, reads containing more than 5% ambiguous bases, and low-quality reads containing more than 20% of bases with a Q value ≤ 20.

3.4. De Novo Assembly and Sequence Annotation

De novo assembly of all clean reads was performed using the Trinity program (version: trinityrnseq_r20131110). The raw RNA-seq data were submitted to NCBI's Gene Expression Omnibus (GEO) repository under accession number SRP144398 (http://www.ncbi.nlm.nih.gov). Further processing to form longer sequences was undertaken using the software TGICL (http://compbio.dfci.harvard.edu/tgi/software/) and, finally, the unigenes were acquired. All unigenes were assigned a putative gene description following BLASTX alignment to the NR (ftp://ftp.ncbi.nih.gov/blast/db), Swiss-Prot (http://www.uniprot.org/downloads), GO (http://www.geneontology.org/), and KEGG (http://www.genome.jp/kegg/pathway.html) databases, with a cut off E value of ≤ 1 × e^{-5}. To gain an overview of gene pathway networks, all unigenes were mapped to KEGG pathways using the

KEGG Automatic Annotation Server. The number of unigenes corresponding to different KEGG pathways was calculated.

3.5. Identification of DEGs

Clean reads from *C. arachnoidea* leaves and roots were separately mapped to the assembled transcripts, and the transcript abundance of each unigene was calculated using the FPKM method. Bowtie2 (http://bowtie-bio.sourceforge.net/bowtie2/manual.shtml) and eXpress (http://www.rna-seqblog.com/express-a-tool-for-quantification-of-rna-seq-data/) were used for mapping the sequencing reads to calculate the FPKM values. In this work, the significance of gene expression differences was assessed using the $|\log_2 \text{FoldChange}| \geq 1$ and a p value ≤ 0.05.

3.6. Real-Time Quantitative PCR Analysis

The expression levels of selected unigenes were analyzed through RT-qPCR. Total RNA of *C. arachnoidea* leaves and roots were isolated with the RNAprep Pure Plant Kit (Tiangen, Beijing, China). The first cDNA strand was synthesized using the RevertAid First Strand cDNA Synthesis Kit (Thermo Scientific, San Jose, CA, USA) according to the manufacturer's instructions. RT-qPCR was performed using the CFX96™ Real-Time System (Bio-Red, Hercules, CA, USA). The reaction mixture included 2 μL five-fold diluted cDNA template, 1 μL of 10 mM forward primer, 1 μL of 10 mM reverse primer, 10 μL FS Universal SYBR Green Master (Roche, Indianapolis, IN, USA), and 6 μL ddH$_2$O. Amplification conditions were 94 °C for 4 min, and then 40 cycles of 94 °C for 1 min, 56 °C for 30 s, and 72 °C for 15 s. All the primers used were listed in Table S1.

3.7. Identification of Simple Sequence Repeats

The identification of SSRs in the *C. arachnoidea* transcriptome was conducted using a microsatellite program (MISA) (http://pgrc.ipk-gatersleben.de/misa/). SSRs from the mononucleotides to the hexa-nucleotides were searched for in all unigenes. Both perfect and compound repeats were identified.

3.8. Statistical Analysis

To examine significant differences statistically between the means of two groups, we used Microsoft Excel software to conduct Student's *t*-test. Values are reported as the mean ± SD from three independent experiments. Asterisks represented significant differences: * $p < 0.05$ and ** $p < 0.01$.

4. Conclusions

Our research established the transcriptome database of the important medicinal plant, *C. arachniodea*, for the first time. In total, 79,835 unigenes were assembled and 39,425 unigenes were successfully annotated. The comparative analysis of the *C. arachniodea* leaf and root transcriptome demonstrated that the expression levels of 2427 unigenes were up-regulated in roots with a higher accumulation of 20E. Forty-nine unigenes encoding enzymes in the MVA and MEP pathways and in the downstream of phytoecdysteroid backbone biosynthesis were identified. The higher expression levels of MVA pathway genes in roots of *C. arachniodea* implied that 20E biosynthesis could be mainly mediated by the MVA pathway. Moreover, twenty unigenes encoding CYP450s were identified to be the new candidate genes for the bioreaction from cholesterol to 20E. In addition, 90 TFs up-regulated in roots and molecular marker SSRs were analyzed for further research on gene regulation and plant breeding. Our work will be helpful in understanding phytoecdysteroid biosynthesis and providing important genetic information on *C. arachniodea*.

Supplementary Materials: The following are available online at http://www.mdpi.com/1422-0067/19/7/1885/s1.

Author Contributions: X.Y.L. and J.X. undertook experiments and transcriptomic analysis. J.W.W., X.Y.L. and L.P.Z. prepared the manuscript. L.P.Z. planned and designed the research. All authors discussed the results and commented on the manuscript.

Funding: This work was supported by the National Science Foundation of China (No. 81273487, 81473183). and the Postgraduate Research & Practice Innovation Program of Jiangsu Province (KYCX17_2042) financially supported this work.

Conflicts of Interest: The authors declare no conflict of interest.

References

1. Tan, C.Y.; Wang, J.H.; Xiao, W.; Li, X. A new phytosterone from *Cyanotis arachnoidea*. *J. Asian Nat. Prod. Res.* **2002**, *4*, 7–11. [CrossRef] [PubMed]

2. Tan, C.Y.; Wang, J.H.; Li, X. Phytoecdysteroid constituents from *Cyanotis arachnoidea*. *J. Asian Nat. Prod. Res.* **2003**, *5*, 237–240. [CrossRef] [PubMed]

3. Trivedy, K.; Nair, K.S.; Ramesh, M.; Gopal, N.; Kumar, S.N. Effect of phytoecdysteroid on maturation of silkworm, *Bombyx mori* L. *Indian J. Seric.* **2003**, *42*, 75–77.

4. Mellon, D.F.; Greer, E. Induction of precocious molting and claw transformation in *Alpheid shrimps* by exogenous 20-hydroxyecdysone. *Biol. Bull.* **1987**, *172*, 350–356. [CrossRef]

5. Dinan, L.; Lafont, R. Effects and applications of arthropod steroid hormones (ecdysteroids) in mammals. *J. Endocrinol.* **2006**, *191*, 1–8. [CrossRef] [PubMed]

6. Thiem, B.; Kikowska, M.; Maliński, M.P.; Kruszka, D.; Napierała, M.; Florek, E. Ecdysteroids: Production in plant in vitro cultures. *Phytochem. Rev.* **2017**, *16*, 603–622. [CrossRef] [PubMed]

7. Mu, L.; Yang, S.C.; Guan, D.J.; Yang, T.; Wen, G.S.; Zhang, W.M. A study on accumulation of β-ecdyson and optimal harvest time for *Cyanotis arachnoidea* C. B. Clarke. *J. Yunnan Agric. Univ.* **2011**, *26*, 194–198. (In Chinese)

8. Tarkowská, D.; Strnad, M. Plant ecdysteroids: Plant sterols with intriguing distributions, biological effects and relations to plant hormones. *Planta* **2016**, *244*, 545–555. [CrossRef] [PubMed]

9. Rohmer, M. The discovery of a mevalonate-independent pathway for isoprenoid biosynthesis in bacteria, algae and higher plants. *Nat. Prod. Rep.* **1999**, *16*, 565–574. [CrossRef] [PubMed]

10. Devarenne, T.P.; Sen-Michael, B.; Adler, J.H. Biosynthesis of ecdysteroids in *Zea mays*. *Phytochemistry* **1995**, *40*, 1125–1131. [CrossRef]

11. Nakagawa, T.; Hara, N.; Fujimoto, Y. Biosynthesis of 20-hydroxyecdysone in *Ajuga* hairy roots: Stereochemistry of C-25 hydroxylation. *Tetrahedron Lett.* **1997**, *38*, 2701–2704. [CrossRef]

12. Adler, J.H.; Grebenok, R.J. Biosynthesis and distribution of insect-molting hormones in plants—A review. *Lipids* **1995**, *30*, 257–262. [CrossRef] [PubMed]

13. Tsukagoshi, Y.; Ohyama, K.; Seki, H.; Akashi, T.; Muranaka, T.; Suzuki, H.; Fujimoto, Y. Functional characterization of CYP71D443, a cytochrome P450 catalyzing C-22 hydroxylation in the 20-hydroxyecdysone biosynthesis of *Ajuga* hairy roots. *Phytochemistry* **2016**, *127*, 23–28. [CrossRef] [PubMed]

14. Yang, C.Q.; Fang, X.; Wu, X.M.; Mao, Y.B.; Wang, L.J.; Chen, X.Y. Transcriptional regulation of plant secondary metabolism. *J. Integr. Plant Biol.* **2012**, *54*, 703–712. [CrossRef] [PubMed]

15. Tomás, J.; Camps, F.; Claveria, E.; Coll, J.; Melé, E.; Messeguer, J. Composition and location of phytoecdysteroids in *Ajuga reptans* in vivo and in vitro cultures. *Phytochemistry* **1992**, *31*, 1585–1591. [CrossRef]

16. Tomás, J.; Camps, F.; Coll, J.; Melé, E.; Messeguer, J. Phytoecdysteroid production by *Ajuga reptans* tissue cultures. *Phytochemistry* **1993**, *32*, 317–324. [CrossRef]

17. Yang, L.; Ding, G.H.; Lin, H.Y.; Cheng, H.N.; Kong, Y.; Wei, Y.K.; Fang, X.; Liu, R.Y.; Wang, L.; Chen, X.Y.; et al. Transcriptome analysis of medicinal plant *Salvia miltiorrhiza* and identification of genes related to tanshinone biosynthesis. *PLoS ONE* **2013**, *8*, e80464. [CrossRef]

18. Gupta, P.; Goel, R.; Pathak, S.; Srivastava, A.; Singh, S.P.; Sangwan, R.S.; Asif, M.H.; Trivedi, P.K. De novo assembly, functional annotation and comparative analysis of *Withania somnifera* leaf and root transcriptomes to identify putative genes involved in the withanolides biosynthesis. *PLoS ONE* **2013**, *8*, e62714. [CrossRef] [PubMed]

19. Wang, Q.J.; Zheng, L.P.; Sima, Y.H.; Yuan, H.Y.; Wang, J.W. Methyl jasmonate stimulates 20-hydroxyecdysone production in cell suspension cultures of *Achyranthes bidentata*. *Plant Omics* **2013**, *6*, 116–120.

20. Wang, Q.J.; Zheng, L.P.; Zhao, P.F.; Zhao, Y.L.; Wang, J.W. Cloning and characterization of an elicitor-responsive gene encoding 3-hydroxy-3-methylglutaryl coenzyme A reductase involved in 20-hydroxyecdysone production in cell cultures of *Cyanotis arachnoidea*. *Plant Physiol. Biochem.* **2014**, *84*, 1–9. [CrossRef] [PubMed]

21. Zhu, G.D.; Zhang, Y.X.; Ye, J.Y.; Wu, C.L.; Yu, L.; Yang, X. Effect of different size of ground *Cyanotis arachnoidea* particles on the release of β-ecdysone. *Agric. Sci. Technol. Hunan* **2011**, *12*, 1318–1326.

22. Niwa, R.; Niwa, R.S. Enzymes for ecdysteroid biosynthesis: Their biological functions in insects and beyond. *Biosci. Biotechnol. Biochem.* **2014**, *78*, 1283–1292. [CrossRef] [PubMed]

23. Rewitz, K.F.; O'Connor, M.B.; Gilbert, L.I. Molecular evolution of the insect Halloween family of cytochrome P450s: Phylogeny, gene organization and functional conservation. *Insect Biochem. Mol. Biol.* **2007**, *37*, 741–753. [CrossRef] [PubMed]

24. Thagun, C.; Imanishi, S.; Kudo, T.; Nakabayashi, R.; Ohyama, K.; Mori, T.; Kawamoto, K.; Nakamura, Y.; Katayama, M.; Nonaka, S.; et al. Jasmonate-responsive ERF transcription factors regulate steroidal glycoalkaloid biosynthesis in tomato. *Plant Cell Physiol.* **2016**, *57*, 961–975. [CrossRef] [PubMed]

25. Kalia, R.K.; Rai, M.K.; Kalia, S.; Singh, R.; Dhawan, A.K. Microsatellite markers: An overview of the recent progress in plants. *Euphytica* **2011**, *177*, 309–334. [CrossRef]

26. Sharopova, N. Plant simple sequence repeats: Distribution, variation, and effects on gene expression. *Genome* **2008**, *51*, 79–90. [CrossRef] [PubMed]

27. Bao, S.; Corke, H.; Sun, M. Microsatellites in starch-synthesizing genes in relation to starch physicochemical properties in waxy rice (*Oryza sativa* L.). *Theor. Appl. Genet.* **2002**, *105*, 898–905. [PubMed]

MDPI

St. Alban-Anlage 66

4052 Basel

Switzerland

Tel. +41 61 683 77 34

Fax +41 61 302 89 18

www.mdpi.com

International Journal of Molecular Sciences Editorial Office

E-mail: ijms@mdpi.com

www.mdpi.com/journal/ijms

www.ingramcontent.com/pod-product-compliance
Lightning Source LLC
Chambersburg PA
CBHW051840210326
41597CB00033B/5726